T0210663

Lecture Notes in Computer Science 9675

Commenced Publication in 1973
Founding and Former Series Editors:
Gerhard Goos, Juris Hartmanis, and Jan van Leeuwen

More information about this series at http://www.springer.com/series/7407

Michael Butler · Klaus-Dieter Schewe
Atif Mashkoor · Miklos Biro (Eds.)

Abstract State Machines, Alloy, B, TLA, VDM, and Z

5th International Conference, ABZ 2016
Linz, Austria, May 23–27, 2016
Proceedings

 Springer

Editors
Michael Butler
University of Southampton
Southampton
UK

Atif Mashkoor
Software Competence Center
Hagenberg
Austria

Klaus-Dieter Schewe
Software Competence Center
Hagenberg
Austria

Miklos Biro
Software Competence Center
Hagenberg
Austria

ISSN 0302-9743　　　　　　　ISSN 1611-3349　(electronic)
Lecture Notes in Computer Science
ISBN 978-3-319-33599-5　　　ISBN 978-3-319-33600-8　(eBook)
DOI 10.1007/978-3-319-33600-8

Library of Congress Control Number: 2016936994

LNCS Sublibrary: SL1 – Theoretical Computer Science and General Issues

Printed on acid-free paper

This Springer imprint is published by Springer Nature
The registered company is Springer International Publishing AG Switzerland

Preface

The fifth edition of the international conference ABZ took place in Linz, Austria, during May 23–27, 2016. This conference records the latest research developments in state-based formal methods, abstract state machines, Alloy, B, Circus, Event-B, TLS$^+$, VDM and Z. It followed the success of the previous ABZ conferences in London, UK (2008), Orford, Canada (2010), Pisa, Italy (2012), and Toulouse, France (2014).

This ABZ conference celebrated two anniversaries: Egon Börger, one of the inventors of the ASM method (with Yuri Gurevich) and co-founder of the ABZ conference (with Jean-Raymond Abrial) turned 70, and was invited as keynote speaker; a mini symposium on "Abstract State Machines" was held in his honor as part of the conference. The second anniversary concerns Atelier-B, one of the toolsets supporting the Event-B method, which turned 20 this year. Thierry Lecomte, one of the master-minds of the Atelier-B toolset, was invited as keynote speaker, and a mini symposium on the use of B and Event-B in industry was held as part of the conference.

At ABZ 2016 four keynotes were presented. Egon Börger from the University of Pisa presented his research on modelling distributed algorithms using the ASM method in comparison with Petri nets. Richard Banach from the University of Manchester shed light on rigorous specification and refinement of hybrid and cyber-physical systems. Thierry Lecomte from Clearsy presented recent developments in Atelier-B. Klaus Reichl from Thales Austria addressed the rigorous modelling of safety-critical railway applications. In addition to these keynotes, the conference offered two tutorials on hybrid systems development and business process modelling with rigorous methods, which were given by Richard Banach and Bernhard Thalheim together with Felix Kossak, respectively. We are grateful to the four invited speakers and the tutorial presenters for contributing to the success of ABZ 2016.

After the successful installment of a case study track at ABZ 2014 addressing a "landing gear system" to be modelled with state-based rigorous methods, ABZ 2016 offered another challenging case study addressing a hemodialysis machine.

ABZ 2016 received 61 submissions covering the whole spectrum of rigorous methods within the scope of the conference. These papers ranged from fundamental contributions, applications in industrial contexts, tool developments and improvements, and contributions to the case study. Each paper was reviewed by at least four reviewers and the Program Committee accepted 12 regular research papers, 15 short papers presenting work in progress, and five papers on the case study.

ABZ 2016 would not have succeeded without the deep investment and involvement of the Program Committee members and the external reviewers who carefully reviewed all submissions and selected the best contributions. This event would not exist if authors and contributors did not submit their proposals. We extend our thanks to every person, reviewer, author, Program Committee members, and the local Organizing Committee members involved in the success of ABZ 2016. Many thanks for their support.

May 2016

Michael Butler
Klaus-Dieter Schewe
Atif Mashkoor
Miklos Biro

Organization

Program Committee

Jean-Raymond Abrial	Independent, France
Yamine Ait Ameur	IRIT/INPT-ENSEEIHT, France
Paolo Arcaini	Charles University in Prague, Czech Republic
Richard Banach	University of Manchester, UK
Miklos Biro	Software Competence Center Hagenberg, Austria
Egon Boerger	Università di Pisa, Italy
Eerke Boiten	University of Kent, UK
Frédéric Boniol	ONERA, France
Michael Butler	University of Southampton, UK
Ana Cavalcanti	University of York, UK
David Chemouil	ONERA, France
David Deharbe	Universidade Federal do Rio Grande do Norte, Brazil
John Derrick	University of Sheffield, UK
Juergen Dingel	Queen's University, Canada
Kerstin Eder	University of Bristol, UK
Roozbeh Farahbod	SAP Research, Germany
Flavio Ferrarotti	Software Competence Centre Hagenberg, Austria
Mamoun Filali-Amine	IRIT, France
John Fitzgerald	Newcastle University, UK
Marc Frappier	University of Sherbrooke, Canada
Leo Freitas	Newcastle University, UK
Vincenzo Gervasi	University of Pisa, Italy
Uwe Glässer	Simon Fraser University, Canada
Stefania Gnesi	ISTI-CNR, Italy
Gudmund Grov	Heriot-Watt University, UK
Lindsay Groves	Victoria University of Wellington, New Zealand
Stefan Hallerstede	Aarhus University, Denmark
Klaus Havelund	Jet Propulsion Laboratory, California Institute of Technology, USA
Ian J. Hayes	University of Queensland, Australia
Rob Hierons	Brunel University, UK
Thai Son Hoang	Yokohama Research Laboratory, Hitachi Ltd., Japan
Regine Laleau	Paris Est Creteil University, France
Peter Gorm Larsen	Aarhus University, Denmark
Thierry Lecomte	ClearSy, France
Michael Leuschel	University of Düsseldorf, Germany
Yuan-Fang Li	Monash University, Australia

Tiziana Margaria   Lero, Ireland
Atif Mashkoor   Software Competence Center Hagenberg, Austria
Dominique Mery   Université de Lorraine, LORIA, France
Stephan Merz   Inria Nancy, France
Mohamed Mosbah   LaBRI - University of Bordeaux, France
Cesar Munoz   NASA, USA
Uwe Nestmann   Technische Universität Berlin, Germany
Jose Oliveira   Universidade do Minho, Portugal
Luigia Petre   Åbo Akademi University, Finland
Andreas Prinz   University of Agder, Norway
Philippe Queinnec   IRIT/INPT, France
Alexander Raschke   University of Ulm, Germany
Elvinia Riccobene   University of Milan, Italy
Ken Robinson   The University of New South Wales, Australia
Tom Rodeheffer   Google Inc., USA
Alexander Romanovsky   Newcastle University, UK
Thomas Santen   Microsoft Research Advanced Technology Labs
                          Europe, Germany
Patrizia Scandurra   University of Bergamo, Italy
Gerhard Schellhorn   Universität Augsburg, Germany
Klaus-Dieter Schewe   Software Competence Center Hagenberg, Austria
Steve Schneider   University of Surrey, UK
Colin Snook   University of Southampton, UK
Jing Sun   The University of Auckland, New Zealand
Mana Taghdiri   KIT, Germany
Loredana Tec   Software Competence Center Hagenberg, Austria
Marcel Verhoef   European Space Agency, The Netherlands
Friedrich Vogt   University of Technology Hamburg-Harburg, Germany
Laurent Voisin   Systerel, France
Qing Wang   The Australian National University, Australia
Virginie Wiels   ONERA/DTIM, France
Kirsten Winter   University of Queensland, Australia

Additional Reviewers

Bonfanti, Silvia   Iliasov, Alexei
Boussabbeh, Maha   Karcher, David S.
Brodmann, Paul-David   Kossak, Felix
Brunel, Julien   Kromodimoeljo, Sentot
Cegielski, Patrick   Ladenberger, Lukas
Ciancia, Vincenzo   Leupolz, Johannes
Ernst, Gidon   Moscato, Mariano
Ferrari, Alessio   Pfähler, Jörg
Gervais, Frederic   Raggi, Daniel

Schneider, David
Smith, Graeme
Sousa Pinto, Jorge
Stankaitis, Paulius

Tounsi, Mohamed
Yaghoubi Shahir, Amir
Yaghoubi Shahir, Hamed
Zohrevand, Zahra

Local Organization

Atif Mashkoor	Co-chair
Klaus-Dieter Schewe	Co-chair
Loredana Tec	Communication with venue
Martina Höller	Social program
Doris Haghofer	Finances
Mircea Boris Vleju	ABZ website
Andreea Buga	ABZ website
Isabel Tober-Kastner	Registration
Flavio Ferrarotti	Local arrangements
Felix Kossak	Local arrangements

Abstracts of Keynote Talks

How to Brew Your Own Hybrid/Cyberphysical Formalism

Richard Banach

School of Computer Science, University of Manchester,
Oxford Road, Manchester, M13 9PL, UK
banach@cs.man.ac.uk

The enthusiasm for hybrid and cyberphysical systems is exploding in today's world of cheap processors, sensors and controllers, and leads to the cost-effectiveness of a smart-everywhere approach to new services and systems. The presence of control as first class citizen in these systems leads to the impingement of discrete techniques from the computing sphere with continuous techniques from the physical systems sphere. It is often claimed that completely new formalisms will be needed to reason about these systems, a view that is a little puzzling considering that every component of such systems comes with a well understood mathematical model that captures the predictability of its behaviour in engineering contexts.

These days, most design and development of cyberphysical systems is very focused on the integration of and cooperation between existing tools and techniques from different areas of computer science and different branches of engineering and technology. Overwhelmingly, such tools and techniques are focused on discrete descriptions of system behaviour, and usually pay scant regard to the continuous aspects of physical behaviour. Unsurprisingly, such approaches are fraught with problems of compatibility and unpredictable interworking, arising from the lack of precision with which they view issues which are fundamentally continuous, and about which they consequently either cannot speak at all, or can say very little. All of which is evidently undesirable.

The way to master the incompatibility of the various existing techniques is not to oversee a war between their incompatible features, but to design a framework in which all of these features can be faithfully embedded, allowing their interrelationships to be properly expressed. Such a job has to be done bottom-up. In addressing it, equal weight should be given to input from computing formalisms, physical modelling approaches, and the various branches of mathematics that underpin these disciplines. When this is done conscientiously, starting from a basis that treats each of these aspects with an equal level of rigour, the remaining room for manoeuvre is surprisingly limited.

We survey this 'requirements-led' approach to designing a foundational framework for hybrid and cyberphysical systems, and the consequences that ensue. Recurring guides in this process are the way that discrete event formalisms relate to real world behaviour, and the deep analogies that exist between discrete event transitions on the one hand, and descriptions of continuous behaviour on the other. Further considerations affect notions of refinement for such formalisms. Existing formalisms for modelling features of cyberphysical systems emerge as partial projections of the fuller picture we develop. We discuss the prospects for automated and interactive verification within our framework, especially when supported by calculational oracles like Mathematica.

Modeling Distributed Algorithms by Abstract State Machines Compared to Petri Nets

Egon Börger

Università di Pisa, Dipartimento di Informatica, 56125 Pisa, Italy
boerger@di.unipi.it

We show how to model distributed algorithms by Abstract State Machines (ASMs). Comparing these models with Petri nets (PNs) reveals a certain number of idiosyncrasies of PNs which complicate both model design and analysis. The ASMs we define illustrate how one can avoid such framework related technicalities.

Atelier B Has Turned Twenty

Thierry Lecomte

CLEARSY, 320 avenue Archimède, Les pléiades 3, Aix en Provence, France
thierry.lecomte@clearsy.com

Atelier B trajectory is tightly linked with industry. It was born at Gec Alsthom in the early 90s, developed by signaling engineers. With a strong support from RATP, SNCF and INRETS (now IFSTTAR) it was made ready for industry applications in 4 years with extensive debugging, improvement, documentation, test and validation performed by Digilog engineers and Jean-Raymond Abrial. Matra installed an Automatic Train Protection software developed in B aboard the Paris line 14 metro in 1998. In 2001, with the creation of ClearSy which managed to gather full Atelier B ownership and with the birth of Event-B, the tool was experimented in a number of studies covering a wide range of domains: air traffic control, automation, automotive, bank, military vehicles, nuclear plant, railways, smartcard and space. Useful feedback was collected from these studies: if the Event-B and Atelier B were sufficient to com-plete meaningful modelling in such various domains, with diverse functional and non-functional specification, and properties, technology acceptance and deployment were more diffi-cult to ensure because of severe issues. These issues have triggered several focused experiments with applicable results in the railways (formal data validation) and in the smartcard (certification). The development of Atelier B was also backed by several publically funded projects at regional, national and European levels. Several significant improvements were obtained: automatic refinement, traceable proof obligation generator, external provers. In this article, we present the twenty last years of Atelier B and its maturation as seen from different points of view: modelling, proof and code generation.

Modeling Safety Critical Railway Applications – An Industrial Experience

Klaus Reichl

Thales Ground Transport Division, Austria
klaus.reichl@thalesgroup.com

The presentation gives an overview on the vision of modeling for railway applications and a feedback to the current state of the art in the railway domain. We will show requirements to methodologies and tools for modeling and reasoning together with objectives and goals. These stem from industrial needs towards sustainable product developement beyond use cases discussed in the formal methods area. Traditionally, essential parts of railway funtionality are related to safety critical control and thus need to undergo rigorous reasoning (hard facts about the product). This process is guided by the CENELEC normatives, which highly recomment the use of formal methods. Moreover, we discuss aspects like modularity, large-scale reuse, composability and variability (hard facts about the developement of the product and the relation to the products properties). Addressing the problems above, we introduce a common domain model to ensure the consistency of viewpoints for different aspects on the same model.

Contents

Articles Contributing to the Hemodialysis Machine Case Study

Keynote Article

Modeling Distributed Algorithms by Abstract State Machines Compared to Petri Nets

Egon Börger[✉]

Dipartimento di Informatica, Università di Pisa, 56125 Pisa, Italy
boerger@di.unipi.it

Abstract. We show how to model distributed algorithms by Abstract State Machines (ASMs). Comparing these models with Petri nets (PNs) reveals a certain number of idiosyncrasies of PNs which complicate both model design and analysis. The ASMs we define illustrate how one can avoid such framework related technicalities.

1 Introduction

This paper is about modeling of distributed algorithms ands property verification using ASMs [9]. Since among theoretical computer scientists PNs seem to be considered suitable for distributed algorithms we compare the ASMs with PNs.

There is a variety of PNs with different semantics, expressiveness and complexity, depending on the incorporated features, e.g. priorities, time, colours, stochastic, continuous or hybrid discrete-continuous features, etc., see [25]. To make a concrete comparison feasable we stick to the PNs defined in [28] to "provide the expressive power necessary to model elementary distributed algorithms adequately, retaining intuitive clarity and formal simplicity" (ibid. p. VII).

To avoid any bias in selecting the comparison examples we follow the author of [28], who is considered an authority in the field, for the "choice of small and medium size distributed algorithms" proposed as representative "for a wide class of distributed algorithms" which "can help the practitioner to design distributed algorithms" [28, p. V]. We focus on the 'Advanced System Models' in [28, Parts B,D]; that they range among the simpler ones in the standard text book [22] should not diminish the relevance of a comparative analysis.

We discovered that the proposed PNs, compared to the ASMs shown below, are neither 'intuitively clear' nor 'formally simple' but hide the underlying intuition under various technicalities of the low-level token-based modeling approach and as a consequence make the mathematical analysis of model properties more complicated than necessary. We hope that the concrete comparison between PNs and ASMs in this paper will help the practitioner to see how ASMs allow one to efficiently design and analyse distributed algorithms without being detracted by extraneous technicalities of the underlying modeling framework.

This work was partially supported by the European Commission funded project BIOMICS, Grant no. 318202.

© Springer International Publishing Switzerland 2016
M. Butler et al. (Eds.): ABZ 2016, LNCS 9675, pp. 3–34, 2016.
DOI: 10.1007/978-3-319-33600-8_1

Caveat. I learnt PNs 50 years ago when Dieter Rödding at the Institute for Mathematical Logic and Foundational Studies in Münster started their systematic investigation and elaboration in his seminars on the theory of automata and networks, seminars which became a regular event at the universities of Münster, Dortmund and Paderborn until Rödding's premature death in 1984. I also heard Petri explain his ideas in person. My (at that time a logician's) interest in PNs was biased by their challenging mathematical theory, in particular by in the 60'ies/70'ies open expressiveness and decidability questions, but also by their use to study the semantics of concurrency (see [27] for a good textbook). Only in 1990 when I started systematic experiments to model computational systems with ASMs I looked at PNs through a practitioner's eyes, as I do here, namely to figure out how practical the language is for modeling distributed systems in combination with appropriate property verification methods.

2   Network Algorithms

In this section we investigate some distributed network algorithms for which one finds in [28] (carefully layed out!) PN formalizations we behaviorally compare with ASMs. For an unbiased statement of the requirements, unless we take them directly from [28] we resort to the problem descriptions in the standard textbook [22]. This allows the reader to evaluate to which degree the two frameworks support capturing requirements accurately, in a way that can be 'justified' and 'checked' (for epistemological reasons not mathematically verified!) to be 'correct' (*ground model problem* [3]), and to document design decisions in a transparent, easily accessible way [23]. To be able to also include visualization aspects into the behavioral comparison we define the ASM models using the traditional flowchart representation for control-state ASMs, an extension of FSMs with well-known meaning (precisely defined in [9, Fig. 2.5]). The phase (also called mode) structure of FSMs offers componentwise[1] definitions and to separate visualized control-flow (phase structure) elements from (better textually described) communication/data/resource-related predicates and actions. *mode* can also be interpreted as a flag so that for ASM rules a guard with $mode = wait$ (as used below) does not necessarily imply busy waiting.

Many of the algorithms below have as background structure finite directed graphs (*Process, Edge*) of agents each executing some program using (an abstract form of unless otherwise stated reliable) communication among *Neighb*ors where $Neighb_p = \{q \mid (p,q) \in Edge\}$ (outgoing neighbors). We use the following abstract operations on each agent's (initially empty) *mailbox*:

SEND(msg, p) with effect to (eventually) INSERT($msg, mailbox_p$)
Received(msg) iff $msg \in mailbox$
CONSUME(msg) = DELETE($msg, mailbox$)

We state explicitly if the graph is undirected or *mailbox* is considered to be a queue instead of a set (with corresponding refinement of its operations).

---

[1] See the use of modes in [23] as a means to structure the set of states.

2.1 Leader Election in Connected Graphs

For a directed connected graph of linearly ordered *Processes* p design and verify a distributed algorithm which uses communication only among *Neighbor* processes. Each p starts considering itself as a leader *candidate* ($cand_p = p$), PROPOSES *cand* to its *Neighbors* and checks it against its *Neighbors'* proposals s.t.:

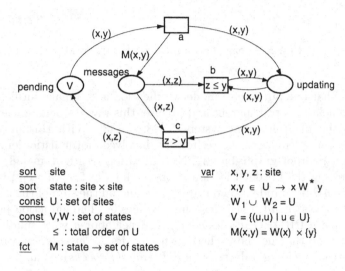

			var	
sort	site			x, y, z : site
sort	state : site × site			$x,y \in U \rightarrow x\,W^*\,y$
const	U : set of sites			$W_1 \cup W_2 = U$
const	V,W : set of states			$V = \{(u,u) \mid u \in U\}$
	≤ : total order on U			$M(x,y) = W(x) \times \{y\}$
fct	M : state → set of states			

Fig. 1. Basic leader election Petri net ([28, Fig. 32.1])

Leader Election Property ([22, 15.2]): In every asynchronous run of processes each with program FLOODINGLEADELECT, if every enabled process will eventually make a move, eventually for every $p \in Process$ holds:

- $cand = max(Process)$ (everybody 'knows' the leader wrt their order $<$)
- mailbox *Proposals* $= \emptyset$ (there is no more communication)

Before proceeding to a detailed comparison of Figs. 1 and 2 we invite the reader to grasp an understanding of the two diagrams (noticing the effort and time needed to comprehend each model) and to compare the definitions in Fig. 1 with the textual definitions of the actions and predicates occuring in Fig. 2: they mark the transition from a description in natural language to a mathematically rigorous model (a *ground model* in the sense described in [3,5]) and for this reason must be understandable and checkable by domain experts to 'correctly' capture the intended intuitive meaning.

PROPOSE = **forall** $q \in Neighb$ SEND($cand, q$)
$ProposalsImprove = (max(Proposals) > cand)$
IMPROVEBYPROPOSALS = ($cand := max(Proposals)$)

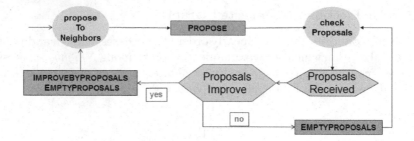

Fig. 2. 2-phase FLOODINGLEADELECT ASM

We agree that one purpose of model verification is "to make intuitive statements and conclusions transparent and precise, this way deepening the reader's insight into the functioning of systems" [28, p. 143]. With this in mind we invite the reader to compare the technically involved (formalistic, lengthy and hard to follow) 'proof graph'-based PN verification of Fig. 1 in [28, pp. 258–260] with proving the Leader Election Property for Fig. 2 by a step-properties-based, intuition-guided induction on ASM runs and on the sum of the differences $diff(max(Process), cand)$ until this sum becomes 0.[2] The induction progresses each time some p PROPOSEs its $cand_p$ to a neighbor q with smaller $cand_q$, so that next time that neighbor checks whether its $ProposalsImprove$, $cand_q$ increases (wrt $<$) yielding a decrease of $diff(max(Process), cand_q)$.

Idiosyncrasy 1. *Low-level token-based encoding* turns objects and executing agents indistinctly into (though abstract) tokens and actions into token manipulations, enforcing to carry agents around (often together with some of their attributes) to wherever they must perform an action. This complicates to grasp (via decoding) the intuitive behavioral meaning of actions of single agents.

For example to SEND the current leader idea *cand* to all *Neighb*ors (the PROPOSE action in Fig. 2) is implemented in Fig. 1 by a transition which

- deletes a token (x, y)—encoding a process x with leader candidate y—from place *pending* and adds it to place *updating*
  - This encodes the *mode* update for x from *send* (*proposeToNeighbors*) to *receive* (*checkProposals*): changing place x must drag along also y!
- adds the set $M(x, y) = W(x) \times \{y\}$ of tokens to place *messages*
  - $W(x)$ encodes the logical expression **forall** $q \in Neighb$: instead of the communication medium forwarding the *cand* msg y into each neighbors' *mailbox* it is x which moves around all its neighbors (coupled with y) for further processing at place *messages* (a global mailbox of all processes)!

---

[2] $diff(p, q) = orderNumber(p) - orderNumber(q)$ for processes numbered $1, 2, \ldots,$ in increasing order.

Correspondingly the initialization condition that *cand* = **self** for each process is encoded by the token set V in place *pending*.

See Sect. 2.5 for a structural diagram change due to the token encoding of adding an agent attribute to satisfy a small requirements change.

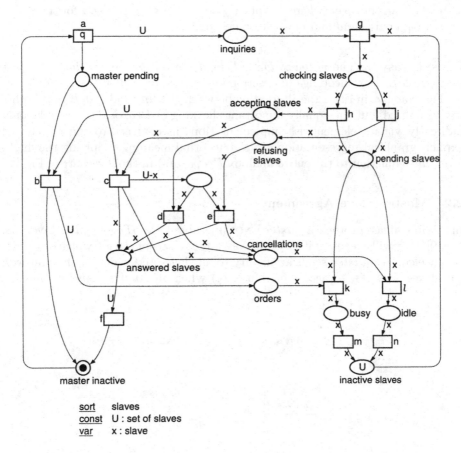

sort   slaves
const   U : set of slaves
var   x : slave

Figure 30.1. Distributed master/slave agreement

Idiosyncrasy 2. The *global overall process view*[3] obstructs the separation of concerns, obfuscates the architectural system view (here the structure of communicating agents) and burdens the net layout with background elements one better deals with in the background.

For example in Fig. 1 the *Neighbor*hood background structure is coded into the diagram together with the dynamic control structure instead of dealing with (frequently static) background elements separately using standard mathematical

---

[3] Not to be confused with the issue of a global vs local state view.

means. In *one* PN the initialization, actions and communications of each process are defined[4] instead of describing them locally as executed by agents with precise component interfaces, in component-based and stepwise refinable fashion. For ex. the components PROPOSE, CHECKPROPOSALS, IMPROVEBYPROPOSALS in Fig. 2 are defined with implicit (or where needed explicit **self** or p) parameter instead of carrying everywhere x explicitly around as done in Fig. 1 for the $\leq$ / $<$ test though only *cand* and message values y, z are needed.

Idiosyncrasy 3. The *visualization is helpful mostly for control flow* but must resort to less clear encodings to describe underlying data flow.[5]

For example, in Fig. 1 the alternation between *pending* and *updating* is explicitly visualized, but the equally important checking of *Proposals* is 'visible' only indirectly via its elementwise implementation, in contrast to the direct and explicit graphical representation of its conceptual meaning (applying the 'high-level' *max* function to the entire mailbox *Proposals*) in the flowchart of Fig. 2.

2.2   Master/Slave Agreement

In this four-phase protocol a *master* ENQUIRES about a *JobToAssign* to slaves [28, p. 119]; they will execute it if they all ANSWERed to accept the job which otherwise is canceled: (a) *master* sends an inquiry to all slaves and waits for their answer; (b) slaves ANSWER to accept or refuse; (c) when all *AnswersArrived master*

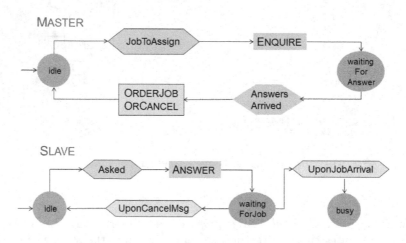

Fig. 3. 2/3-phase MASTER/SLAVE ASM programs

---

[4] The problem to separate business from coordination logic triggered a similar observation in [18, p. 133] that in a PN "every component must be modeled explicitly".

[5] One would expect that using colored tokens [16] may help, but the tokens in [28] are already of a most general nature, namely elements of abstract domains (represented by first-order logical terms) which comprise colored tokens.

will ORDERJOB execution (in case all accepted) ORCANCEL his request (otherwise); (d) slaves go busy (*UponJobArrival*) or return idle (*UponCancelMsg*). Thus eventually *master* returns *idle*, with each slave *idle* or each slave *busy*.

Compare Fig. 30.1 (copied from [28]) with Fig. 3 and its predicate/action definitions. *JobToAssign* describes the event that triggers a protocol round.

ENQUIRE = **forall** $s \in Slave$ SEND($enquire, s$)
AnswersArrived = **forall** $s \in Slave$
  $Received((accept, s))$ **or** $Received((refuse, s))$
ORDERJOBORCANCEL =
  if *SomeSlaveRefused* **then** **forall** $s \in Slave$ SEND($cancel, s$)
  else **forall** $s \in Slave$ SEND(job, s)
  CLEANUP // clean up work for next round
SomeSlaveRefused = **forsome** $s \in Slave$ $Received((refuse, s))$
CLEANUP = $\begin{cases} mailbox := \emptyset \\ JobToAssign := false \text{ // consume input event trigger} \end{cases}$

$Asked = Received(enquire)$
ANSWER = $\begin{cases} \textbf{choose } answer \in \{accept, refuse\} \text{ SEND}((answer, s), master) \\ \text{CONSUME}(enquire) \text{ // consume input msg} \end{cases}$
$UponCancelMsg = Received(cancel), \quad UponJobArrival = Received(job)$

75.2 State properties

Proof of (1) is based on the following place invariants of $\Sigma_{75.1}$:

 inv1: $E + L + F + G - D - U * |B| = 0$
 inv2: $F + G + H + J + N + P + K + L = U$
 inv3: $U * A + U * B + C + D = U$
 inv4: $F + G + J + H - M = 0$
 inv5: $H + J + N + P + K - E - U * A - C = 0$
 inv6: $L + M + N + P + K = U$

inv4 and inv6 imply $F + G + J + H \leq U$.

The surprisingly involved master/slave PN in Fig. 30.1 compared to the simple MASTERSLAVE ASMs in Fig. 3 illustrates once more the PN idiosyncrasies explained in Sect. 2.1 and adds a consequence of Idiosyncrasy 1.

Idiosyncrasy 3b. The *complexity of the graphical layout to define PNs, in particular where unrelated to the algorithmic problem structure*, complicates both the understanding of the model[6] and its verification.

---

[6] We disregard here the peculiar (global overall process view triggered?) design decision in Fig. 30.1 where the slaves organize the refusal case among themselves, triggering (via c) the master to eventually return to idle (via f), without further master involvement: no slave 'reports ... refusal to the master' and never 'the master sends a cancellation to each slave', contrary to the formulated requirements [28, p. 30].

$$\alpha = B \wedge (E + L + F + G \geq U). \tag{3}$$

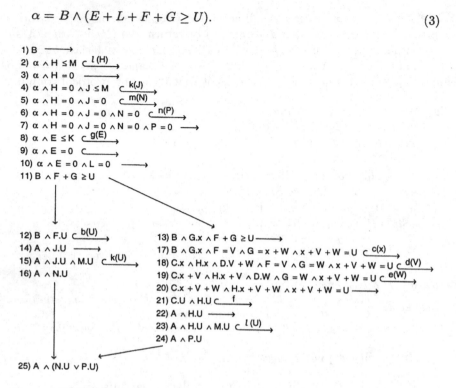

Figure 75.2. Proof graph for (1), with shorthand (3)

The verification of Fig. 30.1 in [28, pp. 255–257] is to be compared with an induction on concurrent ASM runs of MASTERSLAVE showing that if the master starts to ENQUIRE, then (assuming that every enabled agent will eventually make a move) eventually the master becomes idle and either all slaves become idle too or all slaves become busy. In [28, p. 255] the correctness property appears as:

$$(1) \quad \Sigma_{75.1} \models B \hookrightarrow A \wedge (N.U \vee P.U)$$

and is proved using the proof graph of Fig. 75.2 which in turn is based upon 6 invariants (Sect. 75.2 State properties, both copied from [28]) and is justified on pp. 256–258 using some net-representation-driven (syntactical) 'proof patterns'. This illustrates one of the author's goal, namely to reach "a maximally tight combination of modeling and analysis techniques" [28, p. 2]. It is the reader to judge whether such net-structure-driven verifications of properties which too are derived from the static net structure come up to the claim "to make intuitive statements and conclusions transparent and precise, this way deepening the reader's insight into the functioning of systems" [28, p. 143].

2.3   Acknowledged Broadcast (Echo Algorithm)

Here we illustrate the lack of support by PNs, mentioned in Sect. 2.2, to directly capture via their graphical layout the underlying algorithmic intuitions, which

makes it difficult to build satisfactory *ground models* domain experts can understand, justify and check to correctly reflect the requirements ([5]). We use
an algorithm which guarantees an initiator's message being broadcast (building a spanning tree) and acknowledged (echoed) through a connected, for bidirectional communication undirected graph of processes, using only communication between *Neighb*ours. The algorithmic idea ([22, 4.2.2]) is that

- the distinguished *initiator* upon *BroadcastTrigger* will BROADCAST an info
  msg to its *Neighb*ors and then *waitForAck* messages from them
- if a not yet informed non-initiator node *ReceivedInfoFromSomeNeighb*or it
  will PROPAGATEINFOTONONPARENTNEIGHBors and *waitForAck*
- once a non-initiator *ReceivedAckFromAllChildren* it in turn sends an
  ACKTOPARENTNEIGHB node by which it had been informed
- initiator TERMINATEs once it *ReceivedAckFromAllChildren*

The encoding scheme for the PN solution in Fig. 33.3 (copied from [28]) is
explained there on two pages [pp. 127–129]. In contrast, the ASM programs in
Fig. 4 directly reflect the algorithmic idea: the upper program lines (from left to
right) describe building a spanning tree, the lower lines (from right to left) the
notification of completion from leafs back to the *initiator*.

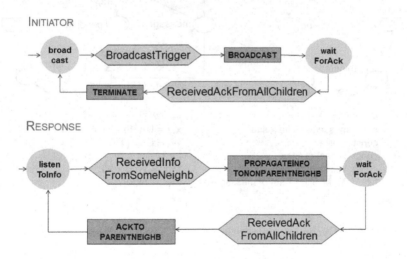

Fig. 4. 2-phase ECHO ASMs

In Fig. 4 event *BroadcastTrigger* triggers the *initiator* to start.

BROADCAST = **forall** $n \in Neighb$ SEND($infoFrom(initiator), n$)
TERMINATE = EMPTY($mailbox$) // clear for next round

Response actions capture spanning tree construction and navigation:

$ReceivedInfoFromSomeNeighb =$
  forsome $p \in Neighb\ Received(infoFrom(p))$
PROPAGATEINFOTONONPARENTNEIGHB = // *tree building step*
  choose $p \in Neighb$ **with** $Received(infoFrom(p))$
    forall $n \in Neighb \setminus \{p\}$ SEND($infoFrom(\mathbf{self}), n$)
    $parent := p$ // define receiver of later ackFrom msg
    INFORMABOUTCHILDRELATION(p)
$ReceivedAckFromAllChildren =$ // true at leafs
  $ChildKnowlIsComplete$ **and** **forall** $m \in Children\ Received(ackFrom(m))$
$Children_n = \{m \in Neighb_n \mid parent(m) = n\}$
ACKTOPARENTNEIGHB = // *pass notification along spanning tree*
  SEND($ackFrom(\mathbf{self}), parent$)
  $parent :=$ **undef**   EMPTY($mailbox$) // clear for next round
INFORMABOUTCHILDRELATION(p) =
  SEND($IamYourChild(\mathbf{self}), p$)
  forall $q \in Neigh \setminus \{p\}$ SEND($IamNotYourChild(\mathbf{self}), p$)
$ChildKnowlIsComplete$ **iff** **forall** $n \in Neighb$
  $Received(IamYourChild(n))$ **or** $Received(IamNotYourChild(n))$

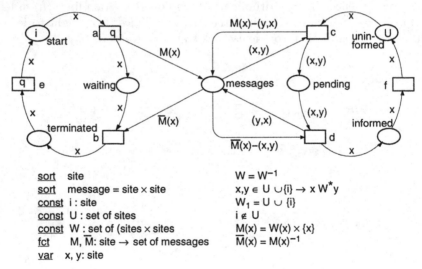

sort   site
sort   message = site × site
const   i : site
const   U : set of sites
const   W : set of (sites × sites
fct   M, $\overline{\text{M}}$: site → set of messages
var   x, y: site

$W = W^{-1}$
$x, y \in U \cup \{i\} \to x\,W^*y$
$W_1 = U \cup \{i\}$
$i \notin U$
$M(x) = W(x) \times \{x\}$
$\overline{M}(x) = M(x)^{-1}$

Figure 33.3. Cyclic echo algorithm

Comparing the verification of Figs. 33.3 and 4 illustrates once more the heavy burden the graphical PN complexity can put on the verification effort. The correctness proof for Fig. 33.3 in [28, pp. 260–266] is 6 pages long and hides the intuition underlying the fundamental spanning tree method to a large extent in the PN 'proof graphs'. In contrast one can show by induction on concurrent ECHO runs (where each enabled agent will eventually make a move):

- **Lemma 1.** Each PROPAGATEINFOTONONPARENTNEIGHBors increases in the tree of agents *waitingForAck* the distance to the *initiator* until leafs are reached. (Proof by downward induction.)
- **Lemma 2.** When an agent executes AckToParentNeighbor, in the tree the distance to the *initiator* of nodes with a subtree of informed agents shrinks, until the *initiator* is reached. (Proof by upward induction.)

The two lemmas imply that each time the *initiator* performs a BROADCAST of an *infoFrom* message it will eventually TERMINATE (termination), but only after all other agents have *Received* that *infoFrom* msg and have acknowledged this to their *parent* neighbor by an *ackFrom* msg (correctness). (See [22, 15.3])

2.4   Load Balancing in Rings

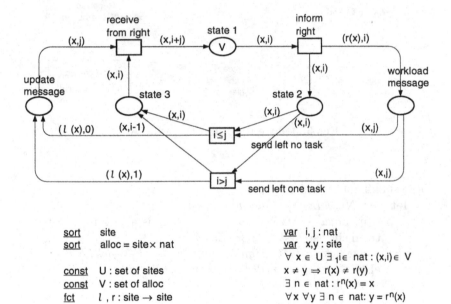

Figure 37.1. Distributed load balancing

Here the goal is to balance the *workLoad* (number of *Tasks* to be executed by a process) among a fixed set of (say at least 3) processes in a given ring using communication only with *leftNeighbor* and *rightNeighbor*, assuming a fixed total workload. The algorithmic idea (see [28, p. 140]) is that every process sends

- a *LeftNeighbLoad* message (i.e. its *workLoad*) to its *rightNeighbor*,
- a task *Transfer* message to its *leftNeighbor* to balance their workloads,

and when $ReceivedTransfer$ message, i.e. some $t \in Task \cup \{nothingToTrans-fer\}$ from its $rightNeighb$, accepts the task to balance their workloads (unless $transfer = nothingToTransfer$).

This 3-phase protocol is directly expressed by Fig. 5 with its (straightforward) action/predicate definitions. In Fig. 37.1 (copied from [28]) the control flow is encoded in the subgraph of places $state\ i$ ($i = 1, 2, 3$); the other two places are two global (!) mailboxes (each one used by all processes).

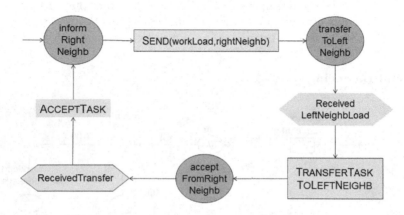

Fig. 5. 3-phase RingLoadBalance ASM

TransferTaskToLeftNeighb =
    let $\{leftNeigbLoad\} = mailbox \cap Nat$
      if $workLoad > leftNeighbLoad$ // there is a task to transfer
        then choose $task \in WorkLoad$
          Send$(task, leftNeighb)$ Delete$(task, WorkLoad)$
        else Send$(nothingToTransfer, leftNeighb)$
      Consume$(leftNeighbLoad)$
AcceptTask =
    let $\{transfer\} = mailbox \cap (Task \cup \{nothingToTransfer\})$
    if $transfer \in Task$ **then** Add$(transfer, WorkLoad)$
    Consume$(transfer)$ // msg removal from mailbox

The correctness property—that eventually the workload difference between two neighboring nodes becomes and remains at most 1, with constant total workload—follows by induction on the $workLoad$ count differences, to be compared with the lengthy verification of Fig. 37.1 in [28, pp. 291–297].

In [28] Fig. 37.2 is proposed for an adaptation to dynamic workload change triggered by the environment. Comparison with the ASM refinement in Fig. 6 illustrates the following problem with PN modeling of multi-agent systems.

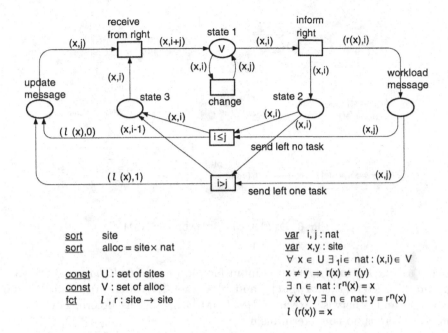

sort   site
sort   alloc = site × nat

const   U : set of sites
const   V : set of alloc
fct l , r : site → site

var   i, j : nat
var   x,y : site
$\forall\ x \in U\ \exists_1 i \in$ nat : (x,i) $\in$ V
$x \neq y \Rightarrow r(x) \neq r(y)$
$\exists\ n \in$ nat : $r^n(x) = x$
$\forall x\ \forall y\ \exists\ n \in$ nat: $y = r^n(x)$
$l\ (r(x)) = x$

Figure 37.2. Distributed load balancing in a floating environment

Idiosyncrasy 4. The lack of component structure[7] and more generally of the structure of communicating agents and/or their environment leads to *model environment actions by nondeterministic internal transitions* (abuse of nondeterminism, see also Sect. 2.9).

In fact Fig. 37.2 models the interaction of one (global?!) environment with any local process as a nondeterministic internal transition *change*:

> From the perspective of the local balance algorithm, this interference shines up as nondeterministic change of the cardinality of the site's workload. ([28, pp. 141–142]

Similarly in the PNs in Figs. 27.7, 28.1, 28.2 the message loss action of the communication medium is modelled as a nondeterministic internal action of the file transfer message protocol. Such a theoretically sound use of nondeterminism fits the semantical PN and verification framework, but is inappropriate from the modeling point of view. In asynchronous systems "input actions are assumed not to be under the automaton's control—they just arrive from the outside" [22, p. 200]. So we model the environment/process interaction in Fig. 6 by letting

---

[7] Compare this with the "illustration of the power of decomposition methods in enabling simple descriptions (and proofs) of complicated distributed algorithms" in [22], quote from p. 532.

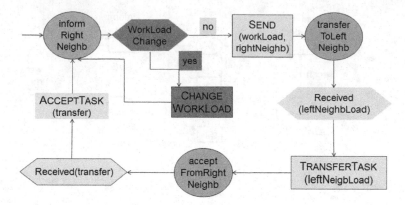

Fig. 6. DynRingLoadBalance ASM

each p a) watch a local event—an input location $workLoadChange_p$ with values in $\{add(T), delete(T), noInput\}$—and b) ChangeWorkLoad when triggered to do so and otherwise execute RingLoadBalance as before. For the local process there is no nondeterminism here.

> $WorkLoadChange$ iff $workLoadChange \in \{add(T), delete(T)\}$
> ChangeWorkLoad =
>   **if** $workLoadChange = add(T)$ **then** Add$(T, WorkLoad)$
>   **if** $workLoadChange = delete(T)$ **then** Delete$(T, WorkLoad)$
>   Consume$(workLoadChange)$ // input consumption

For the same reason we model in Figs. 3, 4 the trigger events which start the process (but do not appear in the PNs in Fig. 33.3, 30.1) and in Sects. 2.6, 2.7 the *timeout* events which trigger resending (but do not appear in the PN models in Fig. 27.7, 28.1, 28.2) as guards for local process actions. Treating timeout as nondeterminism makes interfacing scheduling mechanisms difficult the *timeout* location in Figs. 8, 9 prepares the ground for.

2.5   Consensus in Graphs

The algorithmic idea to "organize consensus about some contract or agreement among the sites of a network", using only communication between neighbors, without considering "neither the contents of messages nor the criteria for a site to accept or refuse a proposed contract" [28, p. 134] is that every agent (site, node) may

- spontaneously go to *agreed* (when without new requests),
- LaunchNewRequest to its *Neighb*ors and *waitForOk* from them,

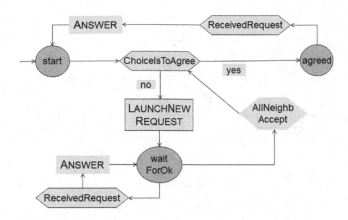

Fig. 7. 3-phase CONSENSUS ASM

- receive and ANSWER requests,
- if *AllNeighbAccept* its last launched request either go to *agreed* or once more
  LAUNCHNEWREQUEST

such that IF the algorithm terminates (maybe never), then all agents *agreed*
and there are no requests left.

The algorithmic idea can be directly traced in the control structure of Fig. 7,
to be compared with Fig. 35.1 (copied from [28]). In particular the predominant
role of nondeterminism in *start* mode is brought to the surface in Fig. 7 by the
explicit *ChoiceIsToAgree* option (whose definition below uses a *choice* function
as interface for "the criteria for a site to accept or refuse a proposed contract"
[28, p. 134]), whereas from the PN in Fig. 35.1 this feature has to be extracted
from the edge structure at pending sites. Also the encoding of requests/answers
by tokens $(y, x)/(x, y)$ in 'initiated/completed' places yields an artificial initialization (instead of an initially empty mailbox), described by (p. 135):

'Initially ... each msg is *completed* (i.e. in the hands of its sender)'

For the sake of comparison completeness we list here the (straightforward)
definitions for CONSENSUS predicates and actions.

$ChoiceIsToAgree$ iff $(choice(\{agree, propose\}) = agree)$
LAUNCHNEWREQUEST $=$ // broadcast new request to neighbors
  forall $n \in Neighb$ SEND($requestFrom(\textbf{self}), n$)
  REINITIALIZEREPLIES
REINITIALIZEREPLIES $=$
  forall $n \in Neighb$ CONSUME($replyFrom(n)$)
$AllNeighbAccept =$**forall** $n \in Neighb$ $Received(replyFrom(n))$
$ReceivedRequest =$
  forsome $n \in Neighb$ $Received(requestFrom(n))$
ANSWER $=$**choose** $n \in Neighb$ **with** $Received(requestFrom(n))$
  SEND($replyFrom(\textbf{self}), n$)   CONSUME($requestFrom(n)$)

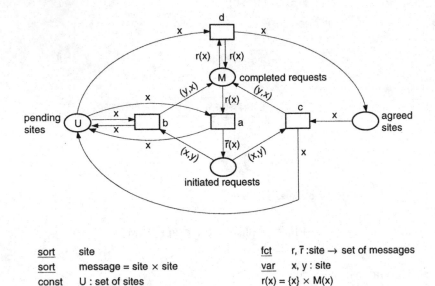

Figure 35.1. Basic algorithm for distributed consensus

An 'Advanced Consensus' PN with quiet/demanded sites (Fig. 35.2 copied from [28]) illustrates how to adapt Fig. 35.1 to the following requirements change request [28, p. 136]:

> ... two further states, *demanded sites* and *quiet sites*. All sites are initially *quiet*. Each newly sent message ... may cause its receiver ... to swap from *demanded* to *quiet* and vice versa... A demanded site *u* is not *quiet*. If *demanded* and *pending*, the immediate step to *agreed* is ruled out.

To reflect this in the ASM model it suffices to (textually!) add to the structurally unchanged Fig. 7 the needed new attribute, action and guard:

- $Quiet \in \{true, false\}$, $Demanded =$ **not** $Quiet$: attribute added to signature
- SWAP $= (Quiet :=$**not** $Quiet)$: action added to ANSWER[8]
- $Quiet = true$: constraint added to *ChoiceIsToAgree* guard

2.6   Alternating Bit Protocol

The Alternating Bit protocol [22, 22.3] transfers any sequence $F(1), \ldots, F(n)$ of files from a *sender* to a *receiver* s.t. eventually the *receiver* has a copy $G = F$, *assuming* that the communication medium may lose (but not change) finitely

---

[8] We interpret 'may cause' as 'causes'; otherwise add **choose** $s \in \{swap, noSwap\}$ **in   if** $s = swap$ **then** SWAP to ANSWER.

many consecutive messages.[9] The algorithmic idea is that in rounds (one per file with initially file number $currRound = 0$ at sender, $currRound = 1$ at receiver; the receiver remains round-ahead of the *sender*):

- the *sender* SENDs the current file and continues to RESENDFILE upon *timeout* until an acknowledgement of receipt arrives from the *receiver*, whereafter in the next $currRound + 1$ the *sender* will STARTNXTFILETRANSFER,
- when sending file $F(round)$ a synchronization bit $round$ mod 2 is
  - attached to file messages $(F(round), round$ mod 2),
  - extracted and resent by the *receiver* as acknowledgment message,
  - checked upon $ReceivedMsg$ by *sender/receiver* for $Matching$ its own synchronization bit and in case of matching is flipped for the next $round + 1$.

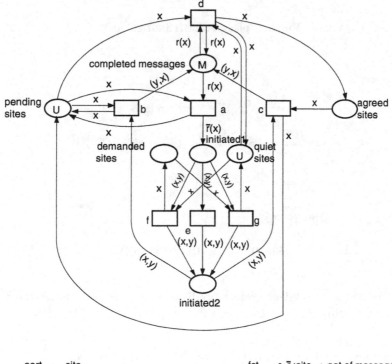

Figure 35.2. Distributed consensus with demanded negotiators

The protocol specification does not include a specification of the communication medium. It seems to be another example for PN Idiosyncrasy 2 that Fig. 27.7, 28.1, 28.2 also define the possible message loss by the transmission lines as an internal nondeterministic action of the PN. Message loss is not an action of the protocol agents but of the communication medium.

The algorithmic idea is easily traced in Figs. 8 and 9; compare with Fig. 27.7 (copied from [28, pp. 107–111] where its details are explained on 5 pages).

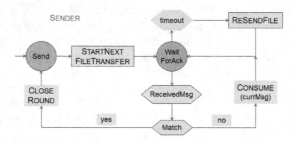

Fig. 8. 2-phase ALTBITSENDER ASM

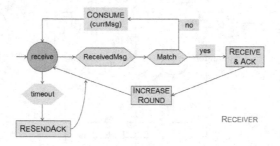

Fig. 9. ALTBITRECEIVER ASM

In the ASM the sender/receiver predicates/actions work under the no-msg-overtaking assumption, reflected by a FIFO-queue mailbox $MsgQueue$.

STARTNEXTFILETRANSFER =
    SEND(($nextFile, nextSyncBit$), $receiver$)   INCREASEROUND
        where $nextFile = F(currRound + 1)$
            $nextSyncBit = currRound + 1$ mod 2 // flipped sync bit
RESENDFILE = SEND(($F(currRound), currRound$ mod 2), $receiver$)
$ReceivedMsg =$ iff $MsgQueue \neq [\ ]$ // mailbox not empty
$Match$ iff $syncBit(currMsg) = currRound$ mod 2
CLOSEROUND = CONSUME($currMsg$) **where** $currMsg = head(MsgQueue)$
INCREASEROUND = ($currRound := currRound + 1$)

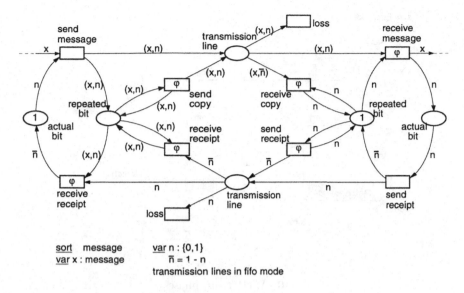

Figure 27.7. The alternating bit protocol

RECEIVE&ACK = STOREFILE SENDACK CONSUME($currMsg$)
SENDACK = SEND($syncBit(currMsg)$)
RESENDACK = // NB. receiver is round-ahead of sender
    SEND($flip(currRound$ mod 2), $sender$) // previous sync bit
STOREFILE = $(G(currRound) := file(currMsg))$

We found no PN verification for Fig. 27.7 in [28] although it is easy to prove the correctness by induction on ALTBIT run phases, see Fig. 10 and [22, 22.3].

Idiosyncrasy 1b. *The token-based transition view introduces algorithmically extraneous technicalities* (language specific details about checking the presence of tokens, token insertion/deletion) which are unrelated to the subject matter, obfuscate the intuitive understanding of the algorithm's behavior and complicate both its verification and further refinement to code (implementation).

For example the token-based transition view is responsible for:

- *moving around unchanged data* between places or worse deleting and simultaneously adding them from/to one and the same place (Fig. 27.7 shows many examples), a technicality that may produce an overwhelming effect on net size and readability and is analogous to the *frame problem* (avoided by ASMs!) of logical descriptions of the no-change part of actions,
- *doubling of locations for same data* involved in different transitions, possibly with different values; e.g. in Fig. 27.7 places *actualbit* and *repeatedbit* double the *syncBit* location at sender and receiver part of the net,
- *simulation of shared locations by token manipulation* which multiplies places and transitions. This point is illustrated in detail in Sect. 2.8.

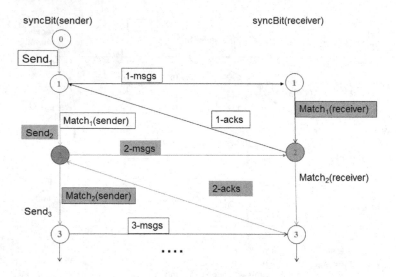

Fig. 10. AltBit run phases

2.7   Adapting AltBit to Sliding Window Protocol

The no-msg-overtaking assumption can be dismissed by replacing single file transfer rounds by (re-) sending in any order multiple files $F(i)$ (and corresponding acks) distinguished by their index i, to be used as (still called) syncBit instead of $i \mod 2$, in a window $[low, high]$ between low and $high$ s.t.

- initially $low = 1, high = 0$ at *sender* and *receiver*,
- *sender* can perform StartNextFileTransfer and IncreaseWindow (by the next syncBit $high := high + 1$) as long as **not** $FullWindow$ (where $FullWindow = high - low + 1 = maxWinSize$),

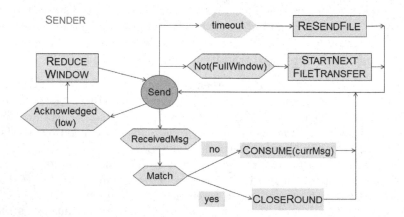

Fig. 11. SlidingWindowSender

- if *Acknowledged(low)* (i.e. that *F(low)* has been received) the *sender* will REDUCEWINDOW at its left end (*low := low + 1*),
- since $high_{receiver} \leq high_{sender}$, each time a file is received for the first time, its index *i* is larger than the *receiver*'s *high* window end, triggering to SLIDEWINDOW at the right end by setting *high := i* and updating *low*.

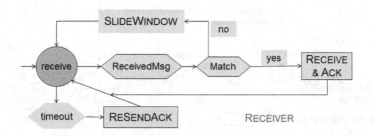

Fig. 12. SLIDINGWINDOWRECEIVER

To turn this description into an ASM model one can preserve the component structure of ALTBIT, except for adding the needed new REDUCEWINDOW component and collapsing the abandoned sequential send/waitForAck phases; the rest is data refining the send/receive predicates and actions as indicated below. This yields the definition in Figs. 11 and 12 we invite the reader to compare with the largely *structural PN redesign* of Fig. 27.7 to Fig. 28.1 (copied from [28]).

STARTNEXTFILETRANSFER
- is refined by

$$nextFile = F(high + 1), \quad nextSyncBit = high + 1$$
$$\text{INCREASEROUND} = (high := high + 1) \text{ // INCREASEWINDOW}$$

- *ReSendFile* = SEND((*F(low),low*), *receiver*)
- *Match* = (*low ≤ syncBit(currMsg) ≤ high*) // syncBit in window
- CLOSEROUND refines CONSUME(*currMsg*) by additionally recording that receipt of *currMsg* has been acknowledged, i.e.

$$Acknowledged(syncBit(currMsg)) := true$$

(initially *Acknowledged(i) = false* for each *i*).

In the SLIDINGWINDOWRECEIVER RECEIVE&ACK is not followed any more by INCREASEROUND and in case of no *Match* one cannot CONSUME the *currMsg* with *syncBit(currMmsg) > high* but instead the receiver must SLIDEWINDOW to let *currMsg Match*. Thus RESENDACK = SEND(*low, sender*) and

SLIDEWINDOW =
    let *s = syncBit(currMsg)*
        high := s and *low := max{1, s − maxWinSize + 1}*

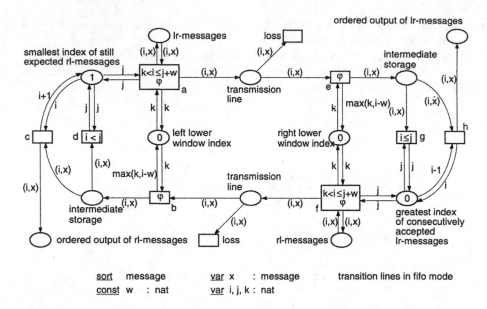

Figure 28.1. Balanced sliding window protocol with unbounded indices

Remark. In case of no message overtaking finitely many indices suffice using $+1 \bmod r$ for updating *low, high* for a sufficiently large r depending on *maxWinSize*: a *pure data refinement* in the ASM framework. Compare this with the sophisticated extension of Fig. 28.1 to Fig. 28.2 (copied from [28]).

2.8   Mutual Exclusion Problem

This is about allocation of one or more nonshareable resources to $n \geq 2$ processes. Peterson's Mutex algorithm ([22, 10.5]) works for one resource and n processes, satisfying also the **Lockout-Freedom Requirement** ([22, 10.4]):

- If each process always returns the resource, every process that reaches the *trying* region (where it competes for the resource) *eventually* will enter the *critical* region (where it uses the resource).
- Every process that reaches its *exit* region (where it stops using the resource) eventually will enter a *remainder* region (without interest in the resource).

In $\mathrm{MUTEXPETERSON}_n$ processes share some locations used to compete at successive *level* $= 1, \ldots, n-1$ (a local process variable, initially *level* $= 1$):

- for every *level* value a global location $stickAt_{level} \in \{1, \ldots, n\}$ (with arbitrary initial value) all processes can read and write. It must have been FETCHed by p (i.e. updated to p) and—in case some other process is interested—later be released (i.e. FETCHed by another interested process) before p becomes the *level Winner*,

- a local location $flag \in \{0, \ldots, n-1\}$ at every p (initially $flag = 0$) which is writable by p and readable by each other process. $flag_p = l > 0$ indicates that p started the competition ('is interested') at level l to get the resource.

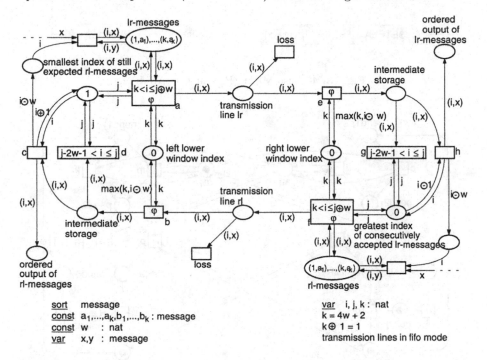

Figure 28.2. Balanced sliding window protocol with bounded indices

We first explain the case $n = 2$ treated in [28], with fixed competition $level = 1$. Compare Fig. 13.7 (copied from [28]) with Fig. 13 and its local actions (formulated with $level$ parameter to prepare their generalization to multiple levels):

RE/SETFLAG = $(flag := 0/level)$ // for $n = 2$ holds $level = 1$
FETCHSTICK = // for $n = 2$, one more guard below for $n > 2$
    if (**not** $HasStick_{level}$) **then** $stickAt_{level} :=$ **self** // **skip** if $HasStick_{level}$
        where $HasStick_{level}$ iff $stickAt_{level} =$ **self**
$Winner = NobodyElseInterested$ **or** $MeantimeSomebodyElseFetchedStick$
$NobodyElseInterested(level) =$ **forall** $p \neq$ **self** $flag_p < level$
    // for case $n = 2$ this means $flag_{theOtherProcess} = 0$
    where $theOtherProcess = \begin{cases} 1 \text{ if } \textbf{self} = 2 \\ 2 \textbf{ else} \end{cases}$ // only for case $n = 2$
$MeantimeSomebodyElseFetchedStick(level)$ iff $stickAt_{level} \neq$ **self**
    // meaning in case $n = 2$ that $stickAt_1 = theOtherProcess$

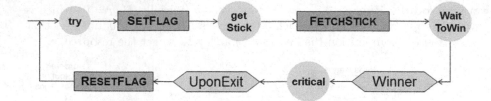

Fig. 13. 4-phase MUTEXPETERSON₂ ASM

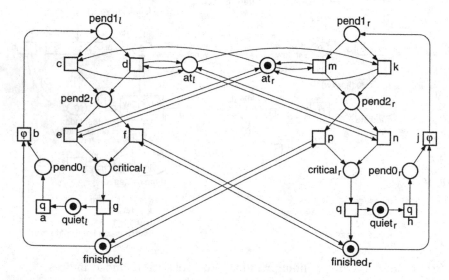

Figure 13.7. Peterson's *mutex* algorithm

The reader may also compare the detailed, easy-to-follow correctness proof for MUTEXPETERSON₂ in [22, pp. 281–282] with the technically involved PN verification in [28, pp. 180–182] which uses the 'evolution proof graph' in Fig. 49.6 (copied from [28]) whose 16 'nodes ... are justified' one by one with help of 4 invariants. These nets and proof graphs risk to explode when generalizing Fig. 13.7 from 2 to $n > 2$ processes (see [22, 10.5.2] and the ASM refinemen step below).

Figure 13.7 illustrates how the *simulation of shared locations by token manipulation* multiplies places and transitions (see Idiosyncrasy 1b, Sect. 2.6). Here it produces 8 places and 12 transitions to simulate updates of 3 locations:

- *stickAt* read/write is simulated by 6 places $pendi_l, pendi_r, at_l, at_r$ $(i = 1, 2)$ with
  - 2 token swapping transitions which simulate the 2 possible writes
  - 4 token checking read ('simultaneous delete/add token') transitions
- multi-reader single-writer *flags* are encoded by places $finished_l, finished_r$ with
  - for a reader a transition to simulate reading a value of the writer's *flag*

- for the writer for each possible update value one transition
  * here two transitions encoded as delete resp. add token transitions at
    place *finished* of each writer process

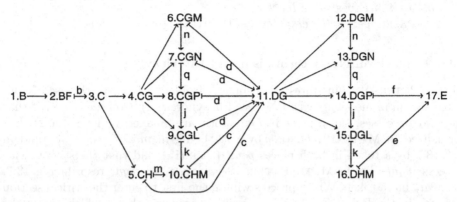

Figure 49.6. Evolution of Peterson's algorithm

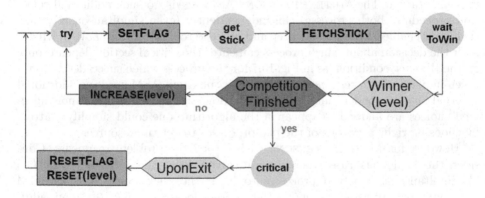

Fig. 14. MUTEXPETERSON$_n$ (competition thru levels) ASM

Multiplication of places/transitions *risks to result in spaghetti PNs* where
shared locations involve more than two values and/or processes. In comparison the ASM refinement MUTEXPETERSON$_n$ for $n > 2$ in Fig. 14 iterates the
MUTEXPETERSON$_2$ competition through *levels* 1 to $n - 1$, guaranteeing that at
each *level* there is a least one *loser*, i.e. a process that has to *waitToWin* until
NobodyElseInterested(level) any more. Thus at level k at most $n - k$ processes
can win so that at most one can win at level $n - 1$.

Refining Fig. 14 means to add an iterator component and a further guard to
FETCHSTICK guaranteeing that in each step at most one process p can write
stickAt$_{level}$. To permit an implementation by any component which computes
the *select*ion we formulate the additional guard using a choice function:

$chosenWriterFor(stickAt_{level}) =$ **self**
  where $chosenWriterFor(stickAt_l) =$
    $select(\{p \mid flag_p = l$ **and** $mode_p = getStick$ **and** $stickAt_l \neq p\})$
$CompetitionFinished$ iff $level = n - 1$
INCREASE$(level) = (level := level + 1)$ RESET$(level) = (level := 1)$

2.9   Remark on Nondeterminism and Interleaving

MUTEXPETERSON$_2$ illustrates an adequacy problem related to the use of non-determinism in modeling with PNs (see Idiosyncrasy 4). Consider the case that the two processes, say $left$ and $right$, both are enabled to fetch the stick—formalized in MUTEXPETERSON$_2$ by both being in $mode = getStick$ and in Fig. 13.7 by a token in both places $pend1_l, pend1_r$—and that the stick is with process $right$—i.e. in MUTEXPETERSON$_2$ $stickAt_1 = right$ resp. in Fig. 13.7. place at_r has a token. Which process will be the first to enter the critical section depends in the PN of Fig. 13.7 on a nondeterministic choice which one of the two enabled but conflicting transitions m and c is fired first; one will have to define this (centralized?) selection separately when refining the algorithm to an implementation. The MUTEXPETERSON$_2$ ASM needs no such additional selection procedure: both processes can independently (even simultaneously) make their next enabled FETCHSTICK step and proceed to $mode = waitToWin$ where-after the decision about which process can enter the critical section depends only on the $Winner$ condition, as in Fig. 13.7 for the process which succeeded to put a token in place $pend2_l$ resp. $pend2_r$.[10] So the concurrent ASM needs no additional external help by nondeterministic or other choices: if such not furthermore specified choices are allowed to appear in the algorithm one could simplify matters by choosing right away one of the two processes to get the resource.

However for MUTEXPETERSON$_n$ with $n > 2$ the problem reappears. Consider the analogous case where at any given $level$ there are $n - (level - 1)$ simultaneously enabled processes to fetch the stick. Since it is assumed that only one process can write the shared location $stickAt_{level}$ an additional selection procedure is needed[11] to decide upon who becomes the $chosenWriterFor(stickAt_{level})$ (which in the considered case will make it the loser at this $level$). By introducing the function $chosenWriterFor(stickAt_{level})$ into the ASM model in Fig. 14 this selection is at least made explicit. In [22, p. 286] the selection is hidden in the interleaving assumption which decides in each step upon which process among the enabled ones to write $stickAt_{level}$ (called there $turn(level)$) will be chosen to write that location.

---

[10] [22, p. 280] has a similar nondeterministic choice for the order in which the processes write the shared $stickAt_1$ (there called $turn$) location, a nondeterminism resulting from the interleaving assumption of the underlying asynchronous shared memory I/O automata execution model. But this extraneous nondeterminism could easily be avoided the same way as in the ASM model.

[11] This is already a form of mutual exclusion. For a genuine solution of the mutual exclusion problem for multiple writers see [19].

As refinement to compute such a choice function often the *head* function of a FIFO *queue* structure is used; this means that the queue is assumed to decide upon the insertion order if two or more processes request simultaneously to get INSERTed.[12] The question remains why instead of running a complicated algorithm such a queue structure is not used right away to organize the decision about which one among $n > 2$ simultaneously interested processes is selected to (be the first to) write a shared location which guards entering the critical section, choosing once a winner instead of choosing $n - 1$ times a loser (one per level).[13]

3   Lifting Petri Nets to Concurrent ASMs

The analysis of PNs in Sect. 2 reveals three major sources of the inadequacy of the PNs proposed in [28] as high-level models for distributed algorithms:

- *insufficient abstraction* and *introduction of algorithmically irrelevant technicalities* resulting from the low-level token-based view of objects (including executing agents), predicates and transitions—despite of the abstract data-interpretation of tokens in [28, Sect. 16] as elements of some abstract domain so that "local states of nets are ... propositions and predicates, thus providing the elements for logic-based analysis techniques" (ibid., p. VII),
- *lack of component structure (architectural system view)*[14] *and of separation of concerns*, mainly due to the global overall process view, thus failing to support appropriate modeling of interaction among agents and/or the environment[15] and complicating the implementation process,
- *complexity of graphical layout* even for small algorithms, indicating a too great esteem of the graphical 'nature' of PNs as a help to understand (or even to define)[16] them and the *lack of an appropriate combination of visual and textual description elements*.

ASMs are well-known not to suffer from any of these problems. In the contrary they provide a justifiably most general abstraction concept—see the various

---

[12] For a prominent example see [22, p. 346]. [20] uses the assumption only for concurrent atomic read/write operations of single digits.

[13] See [21, p. 561] for a distributed mutual exclusion algorithm which deals with the case where the order matters in which requests for the resource are made.

[14] See the observation in [11, p. 3]: "The core issue of Petri nets is that they model behavioral aspects of *distributed systems*, i.e., systems with components that are locally separated and communicate with each other. Surprisingly, neither components nor any notion of locality appears with the usual definition of a Petri net".

[15] Note that at the time of Petri's doctoral thesis [24] computers were monolithic mainframes, there were no agents interacting via pools of networks, servers, services, etc.

[16] In [28, p. V] it is even suggested that "The hurried reader may just study the pictures"! See also the analysis in [11, p. 2] where the characteristics that "Petri nets are a *graphical notion* and at the same time a *precise mathematical notion*" are taken as "the most important properties".

forms of ASM thesis ([8] contains over a dozen of references)—which is coupled to a correspondingly general refinement concept [4] that supports a strong separation of concerns and componentwise design discipline—see the numerous successful practical applications of the ASM refinement method (e.g. [29], for other programming language applications see the recent survey [6]).

The concept of multi-agent ASMs carries these abstraction and refinement capabilities to component-based multi-process modeling and concurrency [8].

Using control-state ASMs as we did in Sect. 2 allows one to explicitly separate and appropriately combine visual (control flow, run-time relevant) and textual (data structure related) description elements to express the underlying (whether static or dynamically changing) data, predicates on data, the effect of actions on data, communication and resources. Such a conceptually well-founded practical combination of these two fundamental elements of a complete system state is crucial for a practical modeling method.

How can PNs share these properties of concurrent ASMs? As has been analyzed already in [9, p. 297] (and elaborated in one more bit of detail in [8, Appendix C]) PNs represent a specific class of multi-agent ASMs, characterized by a particular notion of state and various concepts of (interleaved, lock-step, concurrent, also called distributed) run. In fact each PN P can be defined as a multi-agent ASM where each agent has exactly one transition t of P as its rule of form

if *Enabled*(t) **then** FIRE(t)

This view—which holds mutatis mutandis also for the version of PNs used in [28] and more generally for colored and so-called high-level PNs—allows one to enhance the modeling capabilities of PNs by abstraction, refinement, separation of concerns, component- and agent-based modeling features and combinability of visual and textual description elements the ASM method comes with. We wouldn't even be surprised to see (in particular PN-based) formulations of well-known concepts of concurrency theory [30] get simplified within an ASM-based mathematical framework [8],[17] in analogy to the simplifications one can obtain for classical computation theory [1] and the theory of algorithms by using ASMs instead of the multitude of specific (historically important) machine concepts and complexity measures [9, Sect. 7.1.1].

4   Related Criticism

Other researchers have observed the inadequacy of PNs to model complex computational systems in practice, in particular in the field of Business Process Modeling (BPM). One main issue there is the lack of sufficient support for a seamless inclusion of data, resources and communication among processes into control-flow-centric (e.g. PN inspired BPMN) models, for references see the survey in

---

[17] Right before sending this paper for the Proceedings Klaus-Dieter Schewe and myself discovered that distributed PN runs are distributed ASM runs in the sense of Gurevich [13]. A further investigation of the consequences of this surprising fact remains to be done.

[17, Chap. 2]. On the positive side the control-flow character of traditional PNs comes out clearly from the beautiful foundational result that computationally they are equivalent to asynchronous automata nets built up from a few rather elementary components [26]. But modeling BPs needs more expressive means. For further discussion of this issue see [10], the work at IBM on the data-centric Guard-State-Milestone approach [15],[18] Fleischmann's Subject-Oriented-BPM approach [12], the observation in [18, p. 132] that with (even colored) PNs "the modeling is quite low-level and ... does not provide any higher-level abstractions" resulting in unreadable diagrams (a fascinatingly useless (PN-tool-generated!) example is shown in op.cit. Fig. 9), the recent comparative BPM case study by YAWL experts reporting that "YAWL diagrams only depict the control flow and not the resource perspective" [14, Sect. 4], etc. Also in the domain of stochastic simulation of biological systems it has been recognized that the PN approach "makes it difficult to exploit compositionality of nets to build models incrementally" [31, p. 354].

5   Conclusion

The author of [28] warns the reader that the "book's scope is modest, as it sticks to a choice of small and medium size algorithms", furthermore the proposed design and analysis method is disclaimed to work for large systems where "systematic refinement of specifications and compositional proof techniques are inevitable"[28, p. 13]. However it is also stated that "nevertheless this book's claims are ambitious: Just as PASCAL-like programming structures and Hoare-style proof techniques appear optimal for a wide class of sequential algorithms, this book's formalism is suggested to be optimal for a wide classe of distributed algorithms" and "can help the practitioner to design distributed algorithms"(ibid., p. V). Neither our experience in teaching nor the cooperation with industrial partners nor the analysis above confirm these two claims. Furthermore if "large algorithms require adequate techniques for *small* algorithms" (ibid., p. 13), by contraposition an unsatisfactory framework for small algorithms will not be helpful for large systems. We hope that the above definition of ASMs shows that some improvement can be obtained for PNs, even for design and analysis of small distributed algorithms, by freeing oneself from the low-level token-based PN view and switching to a language where distributed control, data, resource and communication structures can be combined in a technically simpler way, exploiting for a better visualization also the old fashioned flowchart representation means. This holds even more because the ASM method is not exhausted by using control-state ASMs for small distributed algorithms but by its numerous (including industrial) applications in a variety of fields has contributed to reduce the huge gap between much of academic theory and the prevailing software and hardware practice (see [2,6], [9, Chap. 9] for a survey).

---

[18] In [7] ASM nets are defined through which IBM's Guard-State-Milestone approach [15] to BPM can be equipped with the systematic ASM refinement method.

6   Appendix: Reaction to Criticism

Some critic argued that using only one book for the comparison could relativize the conclusions because the idiosyncrasies may belong less to PNs than to the way the book's author uses them. This is possible, however this paper targets not a person but a proposed scientific method. To all practical purposes, in front of thousands of PN publications, what better one could have done—without being an active PN researcher but with the desire to see how PNs behave compared to ASMs when modeling distributed systems—than carefully study a book about a core use of PNs, written by an author who is considered an authority in the field and who promises nothing less than a new proof technique at the level of the Hoare-style proof method, furthermore a very carefully written book (especially concerning the graphical layout of the PNs) whose examples are well-known distributed algorithms every computer science student learns so that the comparison can be explained and followed without need to assume additional knowledge.

Some critics wondered whether comparing the graphical PN notation and the graphical flowchart-like notation used for (the non-textual part of) control-state ASMs is fair. This paper makes no such comparison at all:[19] for given PNs and ASMs we compare the intellectual effort (and the time needed) to define them, explain them, understand them, justify them as adequate (ground) models of the requirements, prove properties of interest for them; in making all these comparisons we *use the behavioral semantics of the graphical notations* in which the models are expressed and which are rigorously defined for PNs as they are for ASMs (see [9, Sect. 2.2.6]). For the behavioral comparison of models we point to figures which define the models, but we do NOT compare the figures as graphical objects. For the behavioral comparison of the PN/ASM *ground models* which are expected to satisfy given requirements and to "help the practitioner to design distributed algorithms" [28, p. V] it does not matter whether the graphical layout of the behavior defining diagrams is or can be furthermore formalized, e.g. to make the diagrams executable or machine checkable or subject to transformations, etc. That a rather formalized version of PN diagrams is needed for the proof-graph method is clear, which brings us to the next critical objection we heard.

Some critics argued that the reader finds no ASM-based proofs in this paper. Clearly not; the only thing one could reasonably do in this paper was to point the reader to proofs in [22]—the pseudo-code there is similar to our ASMs, so the proofs apply mutatis mutandis to concurrent ASMs—to be compared to the corresponding ones in [28]; in addition in a few places, where the proofs are so simple to be built directly from scratch we gave some indication which property or proof scheme to use for this. But also in these cases the proofs can be found in [22] and be compared with those in [28]. We are not talking here about the use of machine support for proof activities. The comparison in this paper was about how to (a) model distributed algorithms to "help the practitioner to design distributed algorithms" [28, p. V] and (b) how to verify their properties of interest

---

[19] Idiosyncrasies 3 and 3b which do speak about graphical PN features however discuss only possible deficiencies of PN diagrams; they are not related to flowcharts.

"to make intuitive statements and conclusions transparent and precise, this way deepening the reader's insight into the functioning of systems" [28, p. 143]. The doubt we epxress is whether using PN proof graphs helps to reach this goal.

Acknowledgement. I thank over a dozen of colleagues who read and commented upon the many previous versions of this paper; I do not mention names so that the entire responsibility of what is stated in this paper remains mine.

References

1. Börger, E.: Computability, Complexity, Logic (English translation of "Berechenbarkeit, Komplexität, Logik", Vieweg-Verlag 1985). Studies in Logic and the Foundations of Mathematics, vol. 128, North-Holland (1989)
2. Börger, E.: The origins and the development of the ASM method for high-level system design and analysis. J. Univ. Comput. Sci. **8**(1), 2–74 (2002)
3. Börger, E.: The ASM ground model method as a foundation of requirements engineering. In: Dershowitz, N. (ed.) Verification: Theory and Practice. LNCS, vol. 2772, pp. 145–160. Springer, Heidelberg (2004)
4. Börger, E.: The ASM refinement method. Formal Aspects Comput. **15**, 237–257 (2003)
5. Börger, E.: Construction and analysis of ground models and their refinements as a foundation for validating computer based systems. Formal Aspects Comput. **19**, 225–241 (2007)
6. Börger, E.: The abstract state machines method for modular design and analysis of programming languages. J. Logic Comput (2014). Special Issue Concepts and Meaning (Leitsch Festschrift). First published online 18 December 2014. doi:10.1093/logcom/exu077
7. Börger, E., Fleischmann, A.: Abstract state machine nets. closing the gap between business process models and their implementation. In: Proceeding S-BPM ONE 2015, ACM Digital Library. ACM, April 2015. ISBN 978-1-4503-3312-2
8. Börger, E., Schewe, K.-D.: Concurrent Abstract State Machines. Acta Informatica, pp. 1–24 (2015). http://link.springer.com/article/10.1007/s00236-015-0249-7, doi:10.1007/s00236-015-0249-7
9. Börger, E., Stärk, R.F.: Abstract State Machines. A Method for High-LevelSystem Design and Analysis. Springer, Heidelberg (2003)
10. Cohn, D., Hull, R.: Business artifacts: a data-centric approach to modeling business operations and processes. IEEE Data Eng. Bull. **32**, 3–9 (2009)
11. Desel, J., Juhás, G.: What is a Petri Net? In: Ehrig, H., Juhás, G., Padberg, J., Rozenberg, G. (eds.) APN 2001. LNCS, vol. 2128, pp. 1–25. Springer, Heidelberg (2001)
12. Fleischmann, A., Schmidt, W., Stary, C., Obermeier, S., Börger, E.: Subject-Oriented Business Process Management. Springer, Heidelberg (2012). http://www.springer.com/978-3-642-32391-1
13. Gurevich, Y.: Evolving algebras 1993: lipari guide. In: Börger, E. (ed.) Specification and Validation Methods, pp. 9–36. Oxford University Press, Oxford, (1995)
14. Hense, A.V., Malz, R.: Comparison of the subject-oriented and the Petri net based approach for business process automation. In: S-BPM ONE 2015. ACM (2015)

15. Hull, R., et al.: Introducing the guard-stage-milestone approach for specifying business entity lifecycles (invited talk). In: Bravetti, M. (ed.) WS-FM 2010. LNCS, vol. 6551, pp. 1–24. Springer, Heidelberg (2011)

16. Jensen, K., Kristensen, L.M.: Coloured Petri Nets. Springer, Heidelberg (2009)

17. Kossak, F., et al.: A Rigorous Semantics for BPMN 2.0 Process Diagrams. Springer, Switzerland (2014)

18. Kühn, E., Craß, S., Joskowicz, G., Marek, A., Scheller, T.: Peer-based programming model for coordination patterns. In: De Nicola, R., Julien, C. (eds.) COORDINA-TION 2013. LNCS, vol. 7890, pp. 121–135. Springer, Heidelberg (2013)

19. Lamport, L.: A new solution of Dijkstra's concurrent programming problem. Commun. ACM **17**(8), 453–455 (1974)

20. Lamport, L.: Concurrent reading and writing. Commun. ACM **20**(11), 806–811 (1977)

21. Lamport, L.: Time, clocks, and the ordering of events in a distributed system. Commun. ACM **21**(7), 558–565 (1978)

22. Lynch, N.A.: Distributed Algorithms. Morgan Kaufmann, Burlington (1996)

23. Parnas, D.L.: The use of precise documentation in software development. Tutorial at FM 2006, August 2006. http://fm06.mcmaster.ca/t8.htm

24. Petri, C.A.: Kommunikation mit Automaten. PhD thesis, Institut für Instrumentelle Mathematik, Universität Bonn, Schriften des IMM Nr.2 (1962)

25. http://www.informatik.uni-hamburg.de/TGI/PetriNets/

26. Priese, L.: Automata and concurrency. TCS **25**, 221–265 (1983)

27. Priese, L., Wimmel, H.: Theoretische Informatik. Petri-Netze. Springer, Heidelberg (2003)

28. Reisig, W.: Elements of Distributed Algorithms. Springer, Heidelberg (1998)

29. Stärk, R.F., Schmid, J., Börger, E.: Java and the Java Virtual Machine: Definition, Verification, Validation. Springer, Heidelberg (2001)

30. Winskel, G., Nielsen, M.: Models for concurrency. Handbook of Logic and the Foundations of Computer Science. Semantic Modelling, pp. 1–148. Oxford University Press, Oxford (1995)

31. Zunino, R., Nikolic, Ð., Priami, C., Kahramanogullari, O., Schiavinotto, T.: ℓ: an imperative DSL to stochastically simulate biological systems. In: Bodei, C., Ferrari, G.-L., Priami, C. (eds.) Degano Festschrift. LNCS, vol. 9465, pp. 354–374. Springer, Heidelberg (2015). doi:10.1007/978-3-319-25527-9_23

Regular Research Articles

A Universal Control Construct for Abstract State Machines

Michael Stegmaier, Marcel Dausend, Alexander Raschke[(✉)],
and Matthias Tichy

Institute of Software Engineering and Compiler Construction, Ulm University,
89069 Ulm, Germany
{michael-1.stegmaier,marcel.dausend,alexander.raschke,
matthias.tichy}@uni-ulm.de

Abstract. Abstract State Machines can be used to specify arbitrary system behaviour. However, when writing executable specifications one often has to write additional statements which organise how, e.g., in which order, the rules are executed. This reduces the readability and comprehensibility of specifications and can introduce additional defects to them. We propose a new syntax construct for the specification of control flow for the ASM language which improves the compactness and readability of specifications by providing syntactic elements for often manually realised behaviour. This construct enables to parametrise which rules shall be selected for execution and how the selected rules are executed. We illustrate how the control construct can improve the code's readability on some examples. The proposed control construct is also released as a plugin for CoreASM.

Keywords: Abstract State Machines · Control construct · Control flow

1 Introduction

Abstract State Machines (ASMs) (see [5]) allow a formal description of the functional requirements in the analysis and design phase. They are a state-based specification language, as they allow to model a software system or hardware system by states and possible state transitions.

Unlike in finite state machines, states in ASMs don't have names. They are general mathematical structures instead. These mathematical structures are universes (non-empty sets) together with functions operating on the sets. This underlying mathematical approach leads to an improvement of verifiability and reusability [3]. They offer a conceptually simple, yet flexible approach for specifying state transition systems.

In ASMs, control flow is realised through a combination of multiple basic control constructs. The specification of a semantically complex control flow is often hard to realise using the basic control constructs and, hence, results in high nesting depths. High nesting depths increase complexity and therefore deteriorate readability [15].

© Springer International Publishing Switzerland 2016
M. Butler et al. (Eds.): ABZ 2016, LNCS 9675, pp. 37–53, 2016.
DOI: 10.1007/978-3-319-33600-8_2

In this paper, we propose a universal control construct (UCC) that unites different step semantics (**parallel**, **sequence**, **rulebyrule**, **stepwise**) and conditional blocks such as **if**, **while** and **iterate**. Furthermore, it provides the possibility to limit the execution of a block to a given number of repetitions which can be useful for situations like initialisation. Last but not least, it provides a way to select and execute only a subset of rules which can be useful, for example, when choosing a strategy or a heuristic for an algorithm or for the simulation of errors.

In the next section, we review the current support for the specification of control flow in ASM and identify several shortcomings using concrete examples. In Sect. 3, we present the proposed control construct, its syntax and its semantics and compare it to related work in Sect. 4. We conclude the paper in Sect. 5 and give an outlook on future work.

2   Shortcomings of Current ASM Control Constructs

This section shows some shortcomings of current ASM specifications and motivates the introduction of a more powerful universal control construct by means of examples. The meaning of the presented control construct is quite intuitive for the reader and should be understandable without a precise definition of the semantics as given in Sect. 3. For better readability we use (parallel) nesting by indentation.

In complex specifications of real systems, the notion of basic ASMs as a list of guarded updates fired in parallel often does not fit. Usually, after an initialisation phase, several steps have to be performed in sequence. Introducing modes is a common pattern for specifying this behaviour. This class of ASM specifications is named "control state ASMs" [5]. Mode variables have to be defined and each rule is guarded by a mode condition such that only one rule is executed per ASM machine step.

An example for this applied pattern is given in Listing 1. It is the main rule of the specification of the operational semantics of the control construct proposed in this paper. The initial value of the mode is assumed to be INIT.

```
1   rule MAIN =
2       if mode = INIT then
3           INITIALISE
4       if mode = SELECT then
5           SELECTION
6       if mode = PREPARE_EXECUTION then
7           PREPAREEXECUTION
8       if mode = EXECUTION then
9           EXECUTION
10      if mode = RESET then
11          RESET
```

Listing 1. MAIN rule of UCC specification

One problem of the control state ASM pattern is that it is not easy to extract the order of modes from the specification. This is because the subsequent modes are set inside nested guards in separate rules.

In our example, the reader needs the whole specification to gain the insight that (in this particular case) the INIT mode is executed only once and after that, the remaining rules are executed rule by rule in an infinite loop.

Our proposed construct aims at overcoming this weakness as shown in Listing 2. UCC allows for defining that rules are executed only once (**at most 1 times**) and that the other rules are executed rule by rule per machine step (**stepwise**). In this example, the order of modes is non-linear which is also a common case. Depending on whether a new selection should be made, the mode following RESET either is SELECT or PREPARE_EXECUTION. This non-linearity cannot directly be specified using **stepwise**, therefore the guard in line 6 is needed. If this guard evaluates to **false** the conditional rule is treated as a **skip**. Thus, the UCC forces the user to make the condition under which rules are executed more explicit and more visible. Obviously, this circumstance is not always an advantage. In specifications realising very complex automata it might be better to not linearise the sequence of the modes.

```
1   rule MAIN =
2     perform always stepwise
3       perform at most 1 times
4         INITIALISE
5       end
6     if shouldSelect then
7       SELECTION
8     PREPAREEXECUTION
9     EXECUTION
10    RESET
```

Listing 2. Improved MAIN rule using UCC

Another example for better readability of specifications by hiding technical (yet necessary for execution) details is given in Listing 3. For testing and demonstration purposes, the specified system behaves normally or it simulates a subset of three different errors. The current behaviour shall be chosen non-deterministically for each step. Listing 3 specifies an environment for a safety-critical system that should be able to cope with different kinds of errors. An arbitrary subset of the error simulating rules is chosen and executed in parallel. If there was no error the environment should behave normally.

```
1   choose errorsToSimulate ⊆ {SIMULATESENSORERROR,
        SIMULATETEMPERATUREERROR, SIMULATECOMMUNICATIONERROR} do
2     if |errorsToSimulate| = 0 then
3       NORMALBEHAVIOUR
4     else
5       forall r ∈ errorsToSimulate do
6         r
```

Listing 3. A subset of all errors can occur simultaneously

Using the UCC, the complex rule of Listing 3 can be condensed into the succinct rule of Listing 4.

```
1   perform always single variable selection
2       NormalBehaviour
3   perform any nonempty variable selection
4       SimulateSensorError
5       SimulateTemperatureError
6       SimulateCommunicationError
```

Listing 4. Equivalent specification as in Listing 3 using UCC

The following example (Listing 5) shows excerpts of a specification of the A* algorithm [14] using ASMs. The A*-algorithm is a heuristic method to determine the shortest path between two nodes in a directed graph with only positive edge weights. To illustrate the algorithm, the sliding puzzle has been chosen, which is also known as 15-puzzle (see [12]). In this puzzle, there are fifteen numbered tiles and one free place. The goal of the game is to repeatedly move tiles into the free place until the desired state is reached. In this specification, multiple heuristics have been realised.

Listing 5 shows the initialisation rule of this specification. Besides the already mentioned typical initialisation mode, the heuristic to use for the algorithm is chosen in this rule. Since all functions in ASMs are globally accessible there is no simple way to ensure that the heuristic will not change throughout a complete run of the algorithm (see Listing 5, line 9).

```
1   rule InitialiseAstar =
2       if not initialised then
3           seq
4               MakePuzzle
5               FindEmpty (InitialState)
6               root ← CreateNode (InitialState, undef, undef)
7               OpenListEnqueue (root)
8           endseq
9           choose h ∈ HEURISTIC do heuristic := h
10          initialised := true
```

Listing 5. The rule InitialiseAstar of *Specification of A**

Using UCC, the rule InitialiseAstar becomes significantly shorter and now only consists of the actual initialisation of the algorithm. Using **perform at most 1 times in sequence** the contained ruleblock will be executed in sequence and at most once. This way we make sure that the algorithm is initialised only once. We can omit the function initialised now because UCC takes care of making sure that the initialisation is never re-executed.

```
1   rule InitialiseAstar =
2       perform at most 1 times in sequence
3           MakePuzzle
4           FindEmpty (InitialState)
5           root ← CreateNode (InitialState, undef, undef)
6           OpenListEnqueue (root)
```

Listing 6. The rule InitialiseAstar of *Specification of A** using the proposed construct

Furthermore, the heuristic does not have to be decided during the initialisation anymore (Listing 6). Instead, the heuristic can be chosen permanently at the point where it is needed (see Listing 7).

```
1   derived  GetHeuristicalValue(state)  =  return value in
2       perform single fixed selection
3           value := CalcGoalHeuristic(state)
4           value := CalcMisplacedTiles(state)
5           value := CalcManhattan(state)
```

Listing 7. The rule GetHeuristicalValue of *Specification of A** using the proposed construct

By using the **single fixed selection**, we ensure that a single rule (a single heuristic in this case) is selected and will always be selected (fixed) for the complete run of the specification. This use of the UCC has the following advantages:

- A permanent random decision does not need to take place during the initialisation anymore.
- The function storing the random permanent decision can be omitted. The UCC can remember its random decision and it ensures that its decision cannot be changed from outside.
- The decision is made at the point where it is needed. The reader won't need to search the specification in order to find out how the decision is being made.

In the next section, the syntax and semantics of the UCC is defined in details. It supports the control constructs already available in (Turbo-)ASM as well as additional description possibilities as shown in this section. This unification also reduces the nesting depth and, thus, improves readability [15].

3   A Universal Control Construct for ASM

The goal of the proposed control construct is to provide a succinct high-level description scheme, formulated in intuitive terms, one can use to specify complex control structures and reduce nesting depth. In this section, its syntax and semantics are defined.

3.1   Syntax

For our control construct, we propose the syntax defined by the following grammar shown in Listing 8. The nonterminals Term, ConstantTerm and Rule are defined as expected [7].

```
Selection = 'all' | SubSelection;
SubSelection = SubSelectionSize ('variable' | 'fixed') 'selection';
SubSelectionSize = ('any' ['nonempty']) | 'single';
Enabled = ('always' | EnabledAtMost | EnabledUntil) [EnabledReset];
EnabledAtMost = 'at most' ConstantTerm 'times';
```

```
EnabledUntil = 'until' ('no updates' | 'no change');
EnabledReset = 'resetting' 'on' ConstantTerm;
StepSemantics = 'in' ('parallel' | 'sequence') | 'rulebyrule' | 'stepwise';
Condition = (('if' | 'while') Term) | 'iterate';
RuleBlock = Rule {Rule};
Semantics = ?Any sequence of Selection, StepSemantics, Enabled?;
PerformRule = 'perform' [Semantics] [Condition] RuleBlock ['end'];
```

Listing 8. The resulting grammar for UCC

The syntax and the corresponding semantics are split up into four groups which cover different orthogonal aspects of the control construct: (1) *Selection* supports selecting a subset of rules from the rule block, (2) *Enabled* determines whether the construct should be enabled or not, (3) *StepSemantics* determines the step semantics to be used for the execution of the selected rules, and (4) *Condition* provides a condition that has to be **true** before the selected rules can be executed. The first group of keywords is the *Selection* group:

- **all** - The whole ruleblock is selected for execution.
- **any** - A random subset of rules from the ruleblock is selected for execution. It can be attributed by the keyword **nonempty** to avoid selecting an empty subset.
- **single** - A single random rule from the ruleblock is selected for execution.

An **any** or **single** selection can either be **variable**, i.e., a selection is made for every evaluation of the construct, or **fixed**, i.e., the selection is permanently made in the first evaluation of the construct and reused for consecutive evaluations. The second group of keywords is the *Enabled* group:

- **always** - The construct is always enabled.
- **at most** n **times** - The construct is enabled at most n times.
- **until no change** - The construct is only enabled until there's no update resulting from the evaluation of the construct or all the resulting updates are trivial [10], that is, no update (if any) resulting from the evaluation of the construct does change the value of any function.
- **until no updates** - The construct is enabled until there are no updates resulting from the evaluation of the construct.

If the construct is not enabled anymore, it will not do anything. The construct can only be re-enabled if a reset condition is specified in the **resetting on** part. The third group of keywords is the *StepSemantics* group:

- **in parallel** - The rules are executed in parallel.
- **in sequence** - The rules are executed in sequence.
- **rulebyrule** - The rules are executed rule by rule. That is, in the first evaluation of this construct the first rule is executed, in the second evaluation the second rule is executed and so on. After the last rule of the block the first rule is executed again and so on.

− **stepwise** - Similar to **rulebyrule** the rules are executed rule by rule. The difference is that with **stepwise**, the same rule is executed for every evaluation during the same machine step. In the next machine step, the next rule is executed and so on. This difference is explained in more detail on the next page.

The fourth group of keywords is the *Condition* group:

− **if** - The selected rules are only executed if the specified condition evaluates to **true**.
− **while** - The selected rules are executed in a loop as long as the specified condition evaluates to **true**. This corresponds to the semantics of the turbo rule **while**.
− **iterate** - The selected rules are executed in a loop as long as they produce updates. This corresponds to the semantics of the turbo rule **iterate**.

In general, the construct must not be confused with a loop. Most notably, the *Enabled* part does not specify a looping condition. For example, **at most** n **times** does not mean that the construct loops n times. Instead it means that the construct must not be evaluated more than n times at all. After the nth repetition, the construct is disabled and cannot be executed anymore. That is, it will behave like a **skip**. The construct is re-enabled if the condition provided for **resetting on** evaluates to **true**. But it is possible to make the construct loop by either using **while** or **iterate**. While there is a significant similarity in the descriptions of **iterate** and **until no updates**, there is a major difference. The keyword **iterate** causes the construct to loop while the keyword **until no updates** does not.

The difference between **rulebyrule** and **stepwise** occurs when using them in conjunction with loops. Loops allow a construct to be evaluated multiple times during the same machine step. With **rulebyrule**, one rule after another is executed within the loop. With **stepwise**, always the same rule is executed during the loop. Only if the UCC is reached again in another machine step, the next rule is executed.

```
1  forall i in {1, 2} do
2     perform rulebyrule r1, r2, r3
3     perform stepwise r1, r2, r3
```

Listing 9. Example for stepwise vs. rulebyrule

Figure 1 illustrates the difference between **stepwise** and **rulebyrule** using the example in Listing 9. It shows a time line on which each machine step is indicated by a vertical bar. Each row shows the rules that are to be executed by the perform rule in the respective iteration of the loop. The rule that is executed in the current iteration is marked. The rules to be executed with **rulebyrule** are above the time line and the rules to be executed with **stepwise** are beneath the arrow.

In our syntax, we allow every sequence of the groups of keywords (see Listing 8). This way we improve the learning curve of our construct by

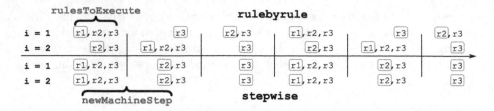

Fig. 1. Illustration of stepwise vs. rulebyrule

reducing the cognitive load of the specification author. He already has to remember all the keywords he wants to use, so at least he does not have to remember the sequence as well. This is not a threat to readability because the groups of keywords are semantically independent from each other.

An obvious criticism of the UCC would be the introduction of many different keywords. In terms of a programming language, this is a clear disadvantage. For one thing, all those keywords are reserved and cannot be used as identifiers anymore, for another, the author of a specification has to remember the keywords to effectively make use of the control construct. So initially, writing specifications with the UCC can be even more difficult. But at the same time, the control construct approaches a natural language even closer by using many different keywords. Getting closer to a natural language usually improves readability for non-experts. Since ASM specifications are often used for communication with customers, certain emphasis should be placed on readability.

3.2   Semantics

In the following specification of the operational semantics, we use the following auxiliaries to work with lists:

- *head* returns the head of the list, that is, the first element of the list.
- *tail* returns the tail of the list, that is, a list of all elements following the first element.

In general, there are different ways to describe the semantics of ASM constructs. One possibility is to define rules of inference for the update set for a construct [5]. Another possibility is the operational approach used in the design specification of CoreASM (see [6]). In this formalisation, the interpretation of each expression is described using ASMs producing a complex value containing the calculated update set and the result of the evaluation. In this approach, all partial update sets are collected and at a machine step applied to the current state.

In this paper, we also describe the semantics using ASMs. Instead of defining the partial interpretation of each part of each construct, we describe the interpreter of the UCC as a whole. Although it is not easily possible to integrate our definition into an execution engine like CoreASM, the translation process is straightforward. The advantage of the presented approach is the linearity of the description. We split the semantics into four phases that are described in detail in the following sections:

1. SELECTION A subset of rules is selected from the ruleblock.
2. PREPAREEXECUTION If the construct is still enabled, the selected rules are copied for execution.
3. EXECUTION The selected rules are executed and the copy of the selection is adapted according to the specified step semantics.
4. RESET The repetition state is reset if the given reset condition evaluates to **true**.

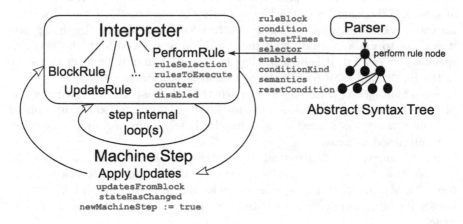

Fig. 2. Run Loop

For the description we assume that an occurrence of the UCC is already parsed and available in a proper model (see Fig. 2) that provides access to the following parts of the construct:

- The contained block of rules (accessed via the function ruleBlock).
- The condition of the *Condition* group (accessed via the function condition).
- The value of the term (n) provided to **at most** n **times** (accessed via the function atmostTimes).
- The selector ∈ {**ALL, SINGLE, ANY, ANY_NON_EMPTY**} (accessed via the function selector).
- The keyword from the *Enabled* group ∈ {**ATMOST_N_TIMES, UNTIL_NO_CHANGE, UNTIL_NO_UPDATES, ALWAYS**} (accessed via the function enabled).
- The keyword from the *Condition* group ∈ {**IF, WHILE, ITERATE**} (accessed via the function conditionKind).
- The step semantics ∈ {**IN_PARALLEL, IN_SEQUENCE, RULEBYRULE, STEPWISE**} (accessed via the function semantics).

– The reset condition of the *Enabled* group (accessed via the function resetCondition).

Furthermore, the following functions are used to maintain the state of the construct:

– ruleSelection holds the current selection and results from the SELECTION phase.
– rulesToExecute holds the rules that are to be executed in the EXECUTION phase. It results from the PREPAREEXECUTION phase. It is initialised as [].
– counter keeps track of the number of executions when **at most** n **times** is used. Its value is only relevant for the PREPAREEXECUTION phase. In that phase it gets increased for each execution. It is initialised as 0 and can only be reset to 0 in the RESET phase (if resetCondition evaluates to **true**).
– disabled is a flag that keeps track of whether the construct has been disabled. Its value is only relevant for the PREPAREEXECUTION phase. In that phase it gets set to **true** depending on the condition associated with the keyword specified for enabled. It is initialised as **false** and can only be reset to **false** in the RESET phase (if resetCondition evaluates to **true**).
– updatesFromBlock is a flag indicating whether the last RUN of the rules has produced updates. Its value is relevant for the PREPAREEXECUTION phase and the EXECUTION phase. Its value implicitly results from running the rules. It is initialised as **true**.
– stateHasChanged is a flag indicating whether the last RUN of the rules has changed the state, i.e., the updates resulting from running the rules change the value of a function. Its value is only relevant for the PREPAREEXECUTION phase. Its value implicitly results from running the rules. It is initialised as **true**.
– newMachineStep is a flag indicating whether the last interpretation of the construct was in another machine step than the current. Its value is only relevant for the PREPAREEXECUTION phase. Its value is set to **true** by the environment as soon as a new machine step is started. The construct sets its value to **false** in the PREPAREEXECUTION phase in order to remember that the construct has already been interpreted in the current machine step. This distinction is required for the difference between the semantics of **stepwise** and **rulebyrule**.

3.3 SELECTION Phase

The SELECTION phase is specified by the rule SELECTION (see Listing 10). It selects the rules to execute from the ruleBlock. With the keyword **all**, the selection corresponds to the whole ruleBlock (see Listing 10, line 4). With the keyword **single**, only an arbitrary single rule is selected (see Listing 10, line 6). With the keyword **any**, an arbitrary subset of rules is selected (see Listing 10, line 8). This selection can be empty. If the keyword **any** is attributed by the keyword **nonempty** an arbitrary subset of rules that is not empty is selected from the ruleBlock (see Listing 10, line 10).

```
1   rule SELECTION =
2     case selector of
3       ALL:
4         ruleSelection := ruleBlock
5       SINGLE:
6         choose r ∈ ruleBlock do ruleSelection := [r]
7       ANY:
8         choose s ⊆ ruleBlock do ruleSelection := s
9       ANY_NON_EMPTY:
10        choose s ⊆ ruleBlock with s ≠ ∅ do ruleSelection := s
11    endcase
12    mode := PREPARE_EXECUTION
```

Listing 10. Selection rule

3.4   PREPAREEXECUTION phase

The PREPAREEXECUTION phase is specified by the rule PREPAREEXECUTION (see Listing 11). It determines the rules to execute. With the keyword **always**, this always is the whole selection (see Listing 11, line 19). With the keyword **at most**, the rules to execute are the selection repeated n times with n being a constant natural number (see Listing 11, lines 5–7). With the keyword **until no updates**, the rules to execute only are the selection if there are updates resulting from the previous step (see Listing 11, line 14–17). Otherwise a flag is set to disable the construct (see Listing 11, lines 14, 15). With the keyword **until no change**, the rules to execute are the selection only if the state has changed in the previous step, i.e., there are updates from the previous step that actually change the value of a function in the state (see Listing 11, lines 9–12). Otherwise a flag is set to disable the construct (see Listing 11, lines 9, 10).

```
1   rule PREPAREEXECUTION =
2     if rulesToExecute = [] then
3       case enabled of
4         ATMOST_N_TIMES:
5           if counter < atmostTimes then
6             rulesToExecute := ruleSelection
7             counter := counter + 1
8         UNTIL_NO_CHANGE:
9           if not stateHasChanged then
10            disabled := true
11          else if not disabled then
12            rulesToExecute := ruleSelection
13        UNTIL_NO_UPDATES:
14          if not updatesFromBlock
15            disabled := true
16          else if not disabled then
17            rulesToExecute := ruleSelection
18        ALWAYS:
19          rulesToExecute := ruleSelection
```

```
20          endcase
21      else if semantics = STEPWISE and newMachineStep then
22          rulesToExecute := tail(rulesToExecute)
23      newMachineStep := false
24      mode := EXECUTION
```

Listing 11. PrepareExecution rule

3.5   EXECUTION Phase

The EXECUTION phase is specified by the rule EXECUTION (see Listing 12). With the keyword **if**, the rules to execute are executed if the specified condition evaluates to **true** (see Listing 12, line 4). With the keyword **while**, the rules to execute are executed as long as the specified condition evaluates to **true** (see Listing 12, line 5). With the keyword **iterate**, the rules to execute are executed as long as they produce at least one update (see Listing 12, line 6). At the same time the rules to execute are adjusted according to the specified step semantics. That is, in case of **rulebyrule** only the very first rule of the current rule selection is consumed (see Listing 12, lines 8, 9).

```
1    rule EXECUTION =
2        if rulesToExecute ≠ [] then
3            case conditionKind of
4                IF: if condition then RUN
5                WHILE: while condition do RUN
6                ITERATE: while updatesFromBlock do RUN
7            endcase
8            if semantics = RULEBYRULE then
9                rulesToExecute := tail(rulesToExecute)
10           else if semantics ≠ STEPWISE
11               rulesToExecute := []
12       mode := RESET
13
14   rule RUN =
15       case semantics of
16           IN_PARALLEL:
17               forall r ∈ rulesToExecute do
18                   r
19           IN_SEQUENCE:
20               foreach r ∈ rulesToExecute do
21                   r
22           RULEBYRULE, STEPWISE:
23               if rulesToExecute ≠ [] then
24                   let r = head(rulesToExecute) in
25                       r
26       endcase
```

Listing 12. Execution rule and Run rule

The rule RUN actually executes the rules. With the keyword **in parallel**, the rules are executed in parallel (see Listing 12, lines 17 – 19). With the keyword **in sequence**, the rules are executed in sequence (see Listing 12, lines 21, 22). The **foreach** rule is the sequential counterpart to the **forall** rule, i.e., each iteration of the loop is computed in sequence instead of in parallel. With the keyword **rulebyrule**, or the keyword **stepwise** the rules are executed one by one (see Listing 12, lines 24 – 26), i.e., with each evaluation only exactly one rule is executed.

3.6   RESET Phase

The RESET phase is specified by the rule RESET (see Listing 13). It resets the state of the *Enabled* part if the provided condition evaluates to **true** (see Listing 13, lines 2 – 4). If the keyword **variable** is used the next phase will be the SELECTION phase (see Listing 13, lines 5, 6). If the keyword **fixed** is used the next phase will be the PREPAREEXECUTION phase (see Listing 13, line 8). So in case of a **fixed** selection the selection stays untouched thus will be permanent.

```
1  rule RESET =
2     if resetCondition then
3        disabled := false
4        counter := 0
5     if selection = VARIABLE_SELECTION and rulesToExecute = []
       then
6        mode := SELECT
7     else
8        mode := PREPARE_EXECUTION
```

Listing 13. Reset rule

4   Related Work

In the following, we compare the presented UCC with the control structures found in other formal specification languages.

TurboASM [4] is an extension of basic ASMs that introduces control constructs with sequential step semantics. We compare our construct with this common extension in order to show that the UCC covers the possibilities of Turbo-ASMs completely.

The Vienna Development Method (VDM) [9] is a well established formal specification language which was originally developed by IBM. The specification language VDM-SL hides the theoretical background from less-experienced users. The state is defined by data structures built on abstract data types. VDM does not support the selection of a random subset of rules or based on priorities. The loop constructs do not limit the execution of rules for the complete run, but only for the current construct execution.

Henshin [1] is a formal language based on the graph transformation formalism [13]. Henshin uses graph transformation rules for the specification of state changes and provides different control constructs to specify which rules to execute and in which order. While Henshin supports all of the control construct features shown in Table 1, graph transformations are a different formalism compared to abstract state machines.

The Very High Speed Integrated Circuit Hardware Description Language (VHDL) [2] is a hardware description language for electronic design automation to describe systems such as integrated circuits. It's a high level specification language that can also be used as a general purpose parallel programming language.

Table 1 shows whether these languages explicitly support seven different control construct features. The first five control construct features cover how rules are executed, e.g., whether it can be specified to execute rules in parallel or in sequence. The last two aspects cover how rules are selected for execution, e.g., whether it is possible to specify priorities to select the rules to execute.

Note that all these specification languages support to manually realise the different control construct features, e.g., one could realise a random selection of rules by manually calling a random method to decide whether the rule should be executed for each rule. Hence, the table shows whether the UCC is *explicitly* realised by a syntax element. For example, Henshin specifically provides a so-called *IndependentUnit* to non-deterministically select rules for execution.

Table 1. Control constructs in formal specification languages

Supported Feature	UCC	TurboASM	VDM	Henshin	VHDL
Parallel execution	yes	yes	yes	yes	yes
Sequential execution	yes	yes	yes	yes	yes
Limit execution count	yes	no	no	yes	no
Sequential Loop (while)	yes	yes	yes	yes	yes
Conditional execution	yes	yes	yes	yes	yes
Random selection of rules	yes	yes	no	yes	no
Priority-based selection of rules	no	no	no	yes	no

The table shows that there is a common subset of features, i.e., parallel execution, sequential execution, conditional execution and sequential loops. But limiting the execution count or selecting a subset of rules to execute or even selecting a random subset of rules is only supported by some languages.

Control constructs can also be found in every imperative programming language. Simple loop constructs like `for` and `while` are covered in UCC by **while** and **at most** n **times**. Continuation with the next iteration (`continue`) and early exit of a loop (`break`) are not directly supported in our construct but could be simulated by appropriate guards. The programming languages Perl[1] and Ruby[2] support additional loop constructs. While `redo` restarts the current iteration, `retry` (in Ruby only) resets the entire loop. Restarting the current iteration is not possible in UCC, but the entire loop can be restarted by **resetting on**.

Random selection of rules has already been addressed by Gurevich in the context of bounded-choice nondeterminism [10]. He introduces the construct **choose among** to support non-deterministic choice algorithms like non-deterministic Turing machines. Applications of this construct are found in [11] where the actions of a thread are modeled as non-deterministic bounded choice between different rules (cf. Listing 14).

```
1   rule EXECUTE PROGRAM : choose among
2        WM—EE transfer
3        Create var
4        Create thread
```

Listing 14.)]Example usage of **choose among** (from [11])

A similar construct is used by Börger in [5, p. 294] to describe computations in Cold [8]. He chooses one or multiple rules from a given set of rules to realize non-deterministic rule execution (cf. Listing 15).

```
1   COLDUSE(Proc) = choose n ∈ ℕ, choose p₁, ..., pₙ ∈ Proc
2        p₁ seq ... seq pₙ
```

Listing 15.).]Random selection of rules that are executed in sequence (from [5]).

These kinds of non-deterministic selection from a set of rules are provided by our construct, too. In contrast to Gurevich and Brger, we allow a selection of rules that is permanent for the current run of a machine. For example, this extension can be used to specify heuristics that are modeled by different rules as in Listing 7.

In [5, p. 39] a rule CYCLETHRU is introduced, that cycles through a sequence of rules and executes them one by one. The "stepwise" execution of UCC can be simulated with that rule but resetting on a condition is not easily possible. A conditional update of the current position in parallel may result in an inconsistent update because this location is also set (and most likely to a different value) during the execution of CYCLETHRU.

5   Conclusion and Future Work

The goal of our approach is to improve the specification of *Abstract State Machines*. The proposed construct has been validated in several ways. On one hand, by defining transformation rules which transform any expression using the proposed control construct into a semantically equivalent block of standard ASM,

---

[1] https://www.perl.org.
[2] https://www.ruby-lang.org.

on the other hand, by implementing it as a CoreASM Plugin[3]. Additionally, a test suite has been developed which can be executed using the provided implementation. While this test suite primarily validates the implementation itself, it also demonstrates that the proposed semantics are actually applicable. Furthermore, the implementation allows to run any specification that uses UCC.

Using some examples, we have shown that in several situations UCC helps to simplify definitions. In future it should be determined and explored what other use cases for this control construct can be found and how existing specifications can be simplified by using it.

Furthermore, specifications using the UCC should be presented to non-software engineers to measure the level of understanding. These experimental studies must be conducted in order to obtain reliable results.

Acknowledgements. First ideas to introduce a UCC originate from a CoreASM workshop in Ulm some time ago. We thank Vincenzo Gervasi, Roozbeh Farahbod, and Simone Zenzaro for inspiring discussions and valuable comments on a preliminary version of this paper. We also thank the anonymous reviewers for their extensive and valuable reviews that helped us improve the paper significantly.

References

1. Arendt, T., Biermann, E., Jurack, S., Krause, C., Taentzer, G.: Henshin: advanced concepts and tools for in-place EMF model transformations. In: Petriu, D.C., Rouquette, N., Haugen, Ø. (eds.) MODELS 2010, Part I. LNCS, vol. 6394, pp. 121–135. Springer, Heidelberg (2010)
2. Ashenden, P.J.: The Designer's Guide to VHDL, 3rd edn. Morgan Kaufmann Publishers Inc., San Francisco (2008)
3. Börger, E.: The origins and the development of the ASM method for high level system design and analysis. J. UCS **8**(1), 2–74 (2002)
4. Börger, E., Bolognesi, T.: Remarks on Turbo ASMs for functional equations and recursion schemes. In: Börger, E., Gargantini, A., Riccobene, E. (eds.) ASM 2003. LNCS, vol. 2589, pp. 218–228. Springer, Heidelberg (2003)
5. Börger, E., Stärk, R.F.: Abstract State Machines. A Method for High-Level System Design and Analysis. Springer, Heidelberg (2003)
6. Farahbod, R.: CoreASM: an extensible modeling framework & Tool environment for high-level design and analysis of distributed systems. Ph.D. thesis, Simon Fraser University.(2009)
7. Farahbod, R., Gervasi, V., Glässer, U.: CoreASM: an extensible ASM execution engine. In: Proceedings of the 12th International Workshop on Abstract State Machines, ASM 2005. pp. 153–166 (2005)
8. Feijs, L.M.G., Jonkers, H.B.M.: Formal Specification and Design. Cambridge Tracts in Theoretical Computer Science. Cambridge University Press, Cambridge (1992)
9. Fitzgerald, J.S., Larsen, P.G., Verhoef, M.: Vienna Development Method. Wiley Encyclopedia of Computer Science and Engineering (2008)

---

[3] http://github.com/coreasm/coreasm.plugins/tree/master/org.coreasm.plugins. universalcontrol.

10. Gurevich, Y.: Sequential abstract-state machines capture sequential algorithms. ACM Trans. Comput. Logic (TOCL) **1**(1), 77–111 (2000)
11. Gurevich, Y., Schulte, W., Wallace, C.: Investigating Java concurrency using abstract state machines. In: Gurevich, Y., Kutter, P.W., Odersky, M., Thiele, L. (eds.) ASM 2000. LNCS, vol. 1912, pp. 151–176. Springer, Heidelberg (2000)
12. Johnson, W.W., Story, W.E.: Notes on the "15" puzzle. Am. J. Math. **2**(4), 397–404 (1879)
13. Rozenberg, G. (ed.): Handbook of Graph Grammars and Computing by Graph Transformation : Foundations, vol. 1. World Scientific Pub Co, Singapore (1997)
14. Russell, S., Norvig, P.: Artificial Intelligence: A Modern Approach, 3rd edn. Prentice Hall Press, Upper Saddle River (2009)
15. Schroeder, A.: Integrated program measurement and documentation tools. In: Proceedings of the 7th International Conference on Software Engineering, ICSE 1984, pp. 304–313. IEEE Press (1984)

Encoding TLA$^+$ into Many-Sorted First-Order Logic

Stephan Merz[1,2] and Hernán Vanzetto[3(✉)]

[1] Inria, Villers-lès-Nancy, France
[2] CNRS, Université de Lorraine, LORIA, UMR 7503, Vandoeuvre-lès-Nancy, France
stephan.merz@loria.fr
[3] Yale University, New Haven, CT, USA
hernan.vanzetto@yale.edu

Abstract. This paper presents an encoding of a non-temporal fragment of the TLA$^+$ language, which includes untyped set theory, functions, arithmetic expressions, and Hilbert's ε operator, into many-sorted first-order logic, the input language of state-of-the-art SMT solvers. This translation, based on encoding techniques such as boolification, injection of unsorted expressions into sorted languages, term rewriting, and abstraction, is the core component of a back-end prover based on SMT solvers for the TLA$^+$ Proof System.

1 Introduction

The specification language TLA$^+$ [10] combines variants of Zermelo-Fraenkel set theory with choice (ZFC) and of linear-time temporal logic for modeling, respectively, the data manipulated by an algorithm, and its behavior. The TLA$^+$ Proof System (TLAPS) provides support for mechanized reasoning about TLA$^+$ specifications, integrating back-end provers for making automatic reasoners available to users of TLAPS. The work reported here is motivated by the development of an SMT backend, through which users of TLAPS interact with off-the-shelf SMT (satisfiability modulo theories) solvers for non-temporal reasoning. More specifically, TLAPS is built around a so-called Proof Manager [4] that interprets the TLA$^+$ proof language, generates corresponding proof obligations, and passes them to external automated verifiers, which are the back-end provers of TLAPS.

Previous to this work, three back-end provers with different capabilities were available for non-temporal reasoning: Isabelle/TLA$^+$, a faithful encoding of TLA$^+$ set theory in the Isabelle proof assistant, which provides automated proof methods based on first-order reasoning and rewriting; Zenon, a tableau prover for first-order logic with equality that includes extensions for reasoning about sets and functions; and a decision procedure for Presburger arithmetic called SimpleArithmetic (now deprecated). The Isabelle and Zenon backends have very limited support for arithmetic reasoning, while SimpleArithmetic handles only pure arithmetic formulas, requiring the user to manually decompose the proofs until the corresponding proof obligations fall within the respective fragments.

Beyond its integration as a semi-automatic backend, Isabelle/TLA$^+$ serves as the most trusted back-end prover. Accordingly, it is also intended for certifying

© Springer International Publishing Switzerland 2016
M. Butler et al. (Eds.): ABZ 2016, LNCS 9675, pp. 54–69, 2016.
DOI: 10.1007/978-3-319-33600-8_3

proof scripts produced by other back-end provers. When possible, backends are expected to produce a detailed proof that can be checked by Isabelle/TLA$^+$. Currently, only the Zenon backend has an option for exporting proofs that can be certified in this way.

In this paper we describe the foundations of a back-end prover based on SMT solvers for non-temporal proof obligations arising in TLAPS.[1] When verifying distributed algorithms, proof obligations are usually "shallow", but they still require many details to be checked: interactive proofs can become quite large without powerful automated back-end provers that can cope with a significant fragment of the language. Sets and functions are at the core of modeling data in the TLA$^+$ language. Tuples and records, which occur very often in TLA$^+$ specifications, are defined as functions. Assertions mixing first-order logic (FOL) with sets, functions, and arithmetic expressions arise frequently in safety proofs of TLA$^+$ specifications. Accordingly, we do not aim at proofs of deep theorems of mathematical set theory but at good automation for obligations mixing elementary set expressions, functions, records, and (linear) integer arithmetic. Our main focus is on SMT solvers, although we have also used the techniques described here with FOL provers. The de-facto standard input language for SMT solvers is SMT-LIB [2], which is based on many-sorted FOL (MS-FOL [11]).[2]

In Sect. 3 we present the core of the SMT backend: a translation from TLA$^+$ to MS-FOL. Although some of our encoding techniques can be found in similar tools for other set-theoretic languages, the particularities of TLA$^+$ make the translation non-trivial:

- Since TLA$^+$ is untyped, "silly" expressions such as $3 \cup \text{TRUE}$ are legal; they denote some (unspecified) value. TLA$^+$ does not even distinguish between Boolean and non-Boolean expressions, hence Boolean values can be stored in data structures just like any other value.
- Functions, which are defined axiomatically, are total and have a domain. This means that a function applied to an element of its domain has the expected value but for any other argument, the value of the function application is unspecified. Similarly, the behavior of arithmetic operators is specified only for arguments that denote numbers.
- TLA$^+$ is equipped with a deterministic choice operator (Hilbert's ε operator), which has to be soundly encoded.

The first item is particularly challenging for our objectives: whereas an untyped language is very expressive and flexible for writing specifications, MS-FOL reasoners rely on types for good automation. In order to support TLA$^+$ expressions in a many-sorted environment, we introduce a "boolification" step for distinguishing between Boolean and non-Boolean expressions, and use a single sort for encoding non-Boolean TLA$^+$ expressions. We therefore call this translation the "untyped" encoding of TLA$^+$; it essentially delegates type inference of sorted expressions such as arithmetic to the solvers.

---

[1] Non-temporal reasoning is enough for proving safety properties and makes up the vast majority of proof steps in liveness proofs.

[2] In this paper we use the terms *type* and *sort* interchangeably.

The paper is structured as follows: Sect. 2 describes the underlying logic of TLA$^+$, Sect. 3 is the core of the paper and explains the encoding, Sect. 4 provides experimental results, Sect. 5 discusses related work, and Sect. 6 concludes and gives directions for future work.

2   A Non-temporal Fragment of TLA$^+$

In this section we describe a fragment of the language of proof obligations generated by the TLA$^+$ Proof System that is relevant for this paper. This language is a variant of FOL with equality, extended in particular by syntax for set, function and arithmetic expressions, and a construct for a deterministic choice operator. For a complete presentation of the TLA$^+$ language see [10, Sect. 16].

We assume given two non-empty, infinite, and disjoint collections $\mathcal{V}$ of *variable* symbols, and $\mathcal{O}$ of *operator* symbols,[3] each equipped with its arity. The only syntactical category in the language is the *expression*, but for presentational purposes we distinguish terms, formulas, set objects, *etc.* An expression e is inductively defined by the following grammar:

$$
\begin{aligned}
e ::=\ & v \mid w(e,\ldots,e) && \text{(terms)}\\
      & \mid \text{FALSE} \mid e \Rightarrow e \mid \forall v\colon e \mid e = e \mid e \in e && \text{(formulas)}\\
      & \mid \{\} \mid \{e,\ldots,e\} \mid \text{SUBSET}\ e \mid \text{UNION}\ e\\
      & \mid \{v \in e : e\} \mid \{e : v \in e\} && \text{(sets)}\\
      & \mid \text{CHOOSE}\ x\colon e && \text{(choice)}\\
      & \mid e[e] \mid \text{DOMAIN}\ e \mid [v \in e \mapsto e] \mid [e \to e] && \text{(functions)}\\
      & \mid 0 \mid 1 \mid 2 \mid \ldots \mid Int \mid -e \mid e + e \mid e < e \mid e\, ..\, e && \text{(arithmetic)}\\
      & \mid \text{IF}\ e\ \text{THEN}\ e\ \text{ELSE}\ e && \text{(conditional)}
\end{aligned}
$$

A *term* is a variable symbol v in $\mathcal{V}$ or an application of an operator symbol w in $\mathcal{O}$ to expressions. *Formulas* are built from FALSE, implication and universal quantification, and from the binary operators $=$ and $\in$. From these formulas, we can define the constant TRUE, the unary $\neg$ and the binary connectives $\wedge$, $\vee$, $\Leftrightarrow$, and the existential quantifier $\exists$. Also, $\forall x \in S\colon e$ is defined as $\forall x\colon x \in S \Rightarrow e$.

In contrast to standard set theory, TLA$^+$ has explicit syntax for *set objects* (empty set, enumeration, power set, generalized union, and two forms of set comprehension derived from the standard axiom schema of replacement), whose semantics are defined by the following axioms:

(extensionality)	$(\forall x\colon x \in S \Leftrightarrow x \in T) \Rightarrow S = T$	(2.1)
(empty set)	$x \in \{\} \Leftrightarrow \text{FALSE}$	(2.2)
(enumeration)	$x \in \{e_1,\ldots,e_n\} \Leftrightarrow x = e_1 \vee \ldots \vee x = e_n$	(2.3)
(power set)	$S \in \text{SUBSET}\ T \Leftrightarrow \forall x \in S\colon x \in T$	(2.4)
(union)	$x \in \text{UNION}\ S \Leftrightarrow \exists T \in S\colon x \in T$	(2.5)

---

[3] TLA$^+$ operator symbols correspond to the standard function and predicate symbols of first-order logic but we reserve the term "function" for TLA$^+$ functional values.

$$(\text{comprehension}_1) \qquad x \in \{y \in S: P(y)\} \Leftrightarrow x \in S \wedge P(x) \qquad (2.6)$$

$$(\text{comprehension}_2) \qquad x \in \{e(y): y \in S\} \Leftrightarrow \exists y \in S: x = e(y) \qquad (2.7)$$

We consider that the free variables in these formulas are universally closed, except for P and e in the comprehension axioms that are schematic variables, meaning that they can be instantiated by countably infinite expressions.[4]

Another primitive construct of TLA⁺ is Hilbert's choice operator ε, written CHOOSE $x: P(x)$, that denotes an arbitrary but fixed value x such that $P(x)$ is true, provided that such a value exists. Otherwise the value of CHOOSE $x: P(x)$ is some fixed, but unspecified value. The semantics of CHOOSE is expressed by the following axiom schemas. The first one gives an alternative way of defining quantifiers, and the second one expresses that CHOOSE is deterministic.

$$(\exists x: P(x)) \Leftrightarrow P(\text{CHOOSE } x: P(x)) \qquad (2.8)$$

$$(\forall x: P(x) \Leftrightarrow Q(x)) \Rightarrow (\text{CHOOSE } x: P(x)) = (\text{CHOOSE } x: Q(x)) \qquad (2.9)$$

From axiom (2.9) note that if there is no value satisfying some predicate P, then $(\text{CHOOSE } x: P(x)) = (\text{CHOOSE } x: \text{FALSE})$. Consequently, the expression CHOOSE $x:$ FALSE and all its equivalent forms represent a unique value.

Certain TLA⁺ values are *functions*. Unlike standard ZFC set theory, TLA⁺ functions are not identified with sets of pairs, but TLA⁺ provides primitive syntax associated with functions. The expression $f[e]$ denotes the result of applying function f to e, DOMAIN f denotes the domain of f, and $[x \in S \mapsto e]$ denotes the function g with domain S such that $g[x] = e$, for any $x \in S$. For $x \notin S$, the value of $g[x]$ is unspecified. A TLA⁺ value f is a function if and only if it satisfies the predicate $IsAFcn(f)$ defined as $f = [x \in \text{DOMAIN } f \mapsto f[x]]$. The fundamental law governing TLA⁺ functions is

$$f = [x \in S \mapsto e] \Leftrightarrow IsAFcn(f) \wedge \text{DOMAIN } f = S \wedge \forall x \in S: f[x] = e \quad (2.10)$$

Natural numbers $0, 1, 2, \ldots$ are primitive symbols of TLA⁺. Standard modules of TLA⁺ define Int to denote the set of integer numbers, the operators $+$ and $<$ are interpreted in the standard way when their arguments are integers, and the interval $a \mathbin{..} b$ is defined as $\{n \in Int : a \leq n \wedge n \leq b\}$.

As a set theoretic language, every TLA⁺ expression—including formulas, functions, and numbers—denotes a set.

---

[4] Both axioms (2.6) and (2.7) for set comprehension objects are instances of the standard axiom schema of replacement: taking the two single-valued predicates $\phi_1(x,y) \triangleq x = y \wedge P(y)$ and $\phi_2(x,y) \triangleq x = e(y)$, we can define $\{y \in S : P(y)\} \triangleq \mathcal{R}(S, \phi_1)$ and $\{e(y) : y \in S\} \triangleq \mathcal{R}(S, \phi_2)$. The replacement axiom says that, given an expression S and a binary predicate ϕ, such that ϕ is *single-valued* for any x in S, that is, $\forall x \in S: \forall y, z: \phi(x,y) \wedge \phi(x,z) \Rightarrow y = z$, then there exists a set object $\mathcal{R}(S, \phi)$, and that $x \in \mathcal{R}(S, \phi) \Leftrightarrow \exists y \in S : \phi(x, y)$.

3   Untyped Encoding into Many-Sorted First-Order Logic

The translation from TLA^+ to MS-FOL is as follows: given a TLA^+ proof obligation, we generate a collection of equi-satisfiable SMT-LIB formulas (restricted to
the AUFLIA logic) whose proof can be attempted by SMT solvers.

First, all expressions having a truth value are mapped to the sort Bool, and
we declare a new sort U (for TLA^+ universe) for all non-Boolean expressions,
including sets, functions, and numbers (Sect. refsec:boolify). Then we proceed in
two main steps. A preprocessing phase applies satisfiability-preserving transformations in order to remove expressions not supported by the target language
(Sect. 3.2). The result is an intermediate *basic* TLA^+ formula, *i.e.*, a TLA^+
expression that has an obvious counterpart in SMT-LIB. We define basic TLA^+
as a subset of TLA^+ consisting of terms, formulas, equality and set membership
relations, primitive arithmetic operators, and IF-THEN-ELSE expressions. The
second step is a shallow embedding of basic expressions into MS-FOL (Sect. 3.3).
Finally, we explain how the encoding of functions (Sect. 3.4) and CHOOSE expressions (Sect. 3.5) fit in the translation.

3.1   Boolification

Since TLA^+ has no syntactic distinction between Boolean and non-Boolean
expressions, we first need to determine which expressions are used as propositions. TLAPS adopts the so-called liberal interpretation of TLA^+ Boolean expressions [10, Sect. 16.1.3] where any expression with a top-level connective among
logical operators, $=$, and $\in$ has a Boolean value.[5] Moreover, the result of any
expression with a top-level logical connective agrees with the result of the expression obtained by replacing every argument e of that connective with $e = \text{TRUE}$.

For example, consider the expression $\forall x \colon (\neg\neg x) = x$, which is not a theorem.
Indeed, x need not be Boolean, whereas $\neg\neg x$ is necessarily Boolean, hence we
may not conclude that the expression is valid. However, $\forall x \colon (\neg\neg x) \Leftrightarrow x$ is valid
because it is interpreted as $\forall x \colon (\neg\neg(x = \text{TRUE})) \Leftrightarrow (x = \text{TRUE})$. Observe that
the value of $x = \text{TRUE}$ is a Boolean for any x, although the value is unspecified
if x is non-Boolean.

In order to identify the expressions used as propositions we use a simple
algorithm that recursively traverses an expression searching for sub-expressions
that should be treated as formulas. Expressions e that are used as Booleans,
i.e., that could equivalently be replaced by $e = \text{TRUE}$, are marked as e^b, whose
definition can be thought of as $e^b \triangleq e = \text{TRUE}$. This only applies if e is a

---

[5] The standard semantics of TLA^+ offers three alternatives to interpret expressions [10, Sect. 16.1.3]. In the liberal interpretation, an expression like $42 \Rightarrow \{\}$ always
has a truth value, but it is not specified if that value is true or false. In the conservative and moderate interpretations, the value of $42 \Rightarrow \{\}$ is completely unspecified.
Only in the moderate and liberal interpretation, the expression FALSE $\Rightarrow \{\}$ has a
Boolean value, and that value is true. In the liberal interpretation, all the ordinary
laws of logic, such as commutativity of $\wedge$, are valid, even for non-Boolean arguments.

term, a function application, or a CHOOSE expression. If an expression which is known to be non-Boolean by its syntax, such as a set or a function, is attempted to be boolified, meaning that a formula is expected in its place, the algorithm aborts with a "type" error. In SMT-LIB we encode x^b as boolify(x), with boolify : U $\rightarrow$ Bool. The above examples are translated as $\forall x^{\mathsf{U}}\colon (\neg\neg\mathsf{boolify}(x)) = x$ and $\forall x^{\mathsf{Bool}}\colon (\neg\neg x) \Leftrightarrow x$, revealing their (in)validity.

3.2 Preprocessing

Though a series of transformations to a boolified TLA$^+$ proof obligation, we obtain an equi-satisfiable formula that can be straightforwardly passed to the solvers using the direct encoding of basic expressions described below. The main motivation is to get rid of those TLA$^+$ expressions that cannot be expressed in first-order logic. Namely, they are $\{x \in S : P\}$, $\{e : x \in S\}$, CHOOSE $x\colon P$, and $[x \in S \mapsto e]$, where the predicate P and the expression e, both of which may have x as free variable, become second-order variables when quantified.

3.2.1 Normalization by Rewriting

We define a rewriting process that systematically expands definitions of non-basic operators. Instead of letting the solver find instances of the background axioms introduced in Sect. 2, it applies the "obvious" instances of those axioms during the translation. In most cases, we can eliminate all non-basic operators. For instance, the axioms (2.5) for the UNION operator and (2.6) for the first form of comprehension yield, respectively, the rewriting rules

$$x \in \text{UNION } S \longrightarrow \exists T \in S\colon x \in T, \text{ and}$$
$$x \in \{y \in S\colon P\} \longrightarrow x \in S \land P.$$

The other cases not covered by rewriting are left to the abstraction mechanism in the next subsection.

All rewriting rules defined in this paper apply equivalence-preserving transformations. To ensure soundness, we derive each rewriting rule from a theorem already proved in Isabelle/TLA$^+$. This is comparable to how rules are obtained in Isabelle's rewrite system, though in a manual way. More specifically, the theorem corresponding to a rule $a \longrightarrow b$ is $\forall \mathbf{x} : a \Leftrightarrow b$ when a and b are Boolean expressions, and $\forall \mathbf{x} : a = b$ otherwise, where $\mathbf{x}$ denotes all free variables in the rule. Most of these theorems exist already in Isabelle/TLA$^+$'s library.

The standard ZF extensionality axiom for sets (2.1) is unwieldy because it introduces an unbounded quantifier, which can be instantiated by any value of sort U. We therefore decided not to include it in the default background theory. Instead, we instantiate the extensionality property for expressions $x = y$ whenever x or y has a top-level operator that constructs a set. In these cases, we say that we *expand* equality. For each set expression T we derive rewriting rules for equations $x = T$ and $T = x$. For instance, the rules

$$x = \text{UNION } S \longrightarrow \forall z\colon z \in x \Leftrightarrow \exists T \in S\colon z \in T, \text{ and}$$
$$x = \{z \in S\colon P\} \longrightarrow \forall z\colon z \in x \Leftrightarrow z \in S \land P$$

are derived from set extensionality (2.1) and the axioms of UNION (2.5) and of bounded set comprehension (2.6).

By not including general extensionality, the translation becomes incomplete. Even if we assume that the automated theorem provers are semantically complete, it may happen that the translation of a semantically valid TLA^+ formula becomes invalid when encoded. In these cases, the user will need to explicitly add the extensionality axiom as a hypothesis to the TLA^+ proof.

We also need to include a rule for the *contraction* of set extensionality:

$$(\forall z \colon z \in x \Leftrightarrow z \in y) \longrightarrow x = y,$$

which we apply with higher priority than the expansion rules.

All rules of the form $a \longrightarrow b$, including those introduced below for functions and CHOOSE expressions, define a term rewriting system $(\text{TLA}^+, \longrightarrow)$, where $\longrightarrow$ is a binary relation over well-formed TLA^+ expressions.

Theorem 1. $(\text{TLA}^+, \longrightarrow)$ *terminates and is confluent.*

Proof (sketch). Termination is simply proved by embedding $(\text{TLA}^+, \longrightarrow)$ into another reduction system that is known to terminate, typically $(\mathbb{N}, >)$ [1]. The embedding is through an ad-hoc monotone mapping μ such that $\mu(a) > \mu(b)$ for every rule $a \longrightarrow b$. We define it in such a way that every rule instance strictly decreases the number of non-basic and complex expressions such as quantifiers. Confluence is proved by Newman's lemma [1], thus it suffices to prove that all critical pairs are joinable. By enumerating all combinations of rewriting rules, we can find all critical pairs $\langle e_1, e_2 \rangle$ between them. Then we just need to prove that e_1 and e_2 are joinable for each such pair. In particular, the contraction rule is necessary to obtain a strongly normalizing system.   □

3.2.2 Abstraction

Applying rewriting rules does not always suffice for obtaining formulas in basic normal form. As a toy example, consider the valid proof obligation $\forall x \colon P(\{x\} \cup \{x\}) \Leftrightarrow P(\{x\})$. The non-basic sub-expressions $\{x\} \cup \{x\}$ and $\{x\}$ do not occur in the form of a left-hand side of any rewriting rule, so they must first be transformed into a form suitable for rewriting.

We call the technique described here *abstraction* of non-basic expressions. After applying rewriting, some non-basic expression ψ may remain in the proof obligation. For all occurrences of ψ with free variables $x_1, \ldots, x_n$, we introduce in their place a fresh term $k(x_1, \ldots, x_n)$, and add the formula $k(x_1, \ldots, x_n) = \psi$ as an assumption in the appropriate context. The new term acts as an *abbreviation* for the non-basic expression, and the equality acts as its *definition*, paving the way for a transformation to a basic expression using normalization. Note that we replace non-basic expressions occurring more than once by the same symbol.

In our example the expressions $\{x\} \cup \{x\}$ and $\{x\}$ are replaced by fresh constant symbols $k_1(x)$ and $k_2(x)$. Then, the abstracted formula is

$$\wedge\ \forall x\colon k_1(x) = \{x\} \cup \{x\}$$
$$\wedge\ \forall x\colon k_2(x) = \{x\}$$
$$\Rightarrow \forall x\colon P(k_1(x)) \Leftrightarrow P(k_2(x)).$$

which is now in a form where it is possible to apply the instances of extensionality to the equalities in the newly introduced definitions. In order to preserve satisfiability of the proof obligation, we have to add as hypotheses instances of extensionality contraction for every pair of definitions where extensionality expansion was applied. The final equi-satisfiable formula in basic normal form is

$$\wedge\ \forall x, z\colon z \in k_1(x) \Leftrightarrow z = x \vee z = x$$
$$\wedge\ \forall x, z\colon z \in k_2(x) \Leftrightarrow z = x$$
$$\wedge\ \forall x, y\colon (\forall z\colon z \in k_1(x) \Leftrightarrow z \in k_2(y)) \Rightarrow k_1(x) = k_2(y)$$
$$\Rightarrow \forall x\colon P(k_1(x)) \Leftrightarrow P(k_2(x)).$$

3.2.3   Eliminating Definitions

To improve the encoding, we introduce a procedure that eliminates definitions, having the opposite effect of the abstraction method where definitions are introduced and afterwards expanded to basic expressions. This process collects all definitions of the form $x = \psi$, and then simply applies the rewriting rules $x \longrightarrow \psi$ to substitute every occurrence of the term x by the non-basic expression ψ in the rest of the context. The definitions we want to eliminate typically occur in the original proof obligation, that is, they do not result from the abstraction step.

This transformation produces expressions that can eventually be normalized to their basic form. To avoid rewriting loops and ensure termination, it can only be applied if x does not occur in ψ. For instance, the two equations $x = y$ and $y = x+1$ will be transformed into $y = y+1$, which cannot further be rewritten. After applying the substitution, we can safely discard from the resulting formula the definition $x = \psi$, when x is a variable. However, we must keep the definition if x is a complex expression. Suppose we discard an assumption DOMAIN $f = S$, where the conclusion is $f \in [S \rightarrow T]$. Only after applying the rewriting rules, the conclusion will be expanded to an expression containing DOMAIN f, but the discarded fact required to simplify it to S will be missing.

3.2.4   Preprocessing Algorithm

Now we can put together boolification and the encoding techniques described above in a single algorithm called *Preprocess*.

$$Preprocess(\phi) \stackrel{\Delta}{=} \phi \qquad\qquad Reduce(\phi) \stackrel{\Delta}{=} \phi$$
$$\qquad \triangleright Boolify \qquad\qquad\qquad\qquad \triangleright \text{FIX } (Eliminate \circ Rewrite)$$
$$\qquad \triangleright \text{FIX } Reduce \qquad\qquad\qquad \triangleright \text{FIX } (Abstract \circ Rewrite)$$

Here, FIX $\mathcal{A}$ means that step $\mathcal{A}$ is executed until reaching a fixed point, the combinator $\triangleright$, used to chain actions on a formula ϕ, is defined as $\phi \triangleright f \stackrel{\Delta}{=} f(\phi)$, and function composition $\circ$ is defined as $f \circ g \stackrel{\Delta}{=} \lambda\phi.\, g(f(\phi))$.

Given a TLA^+ formula ϕ, the algorithm boolifies it and then applies repeatedly the step called *Reduce* to obtain its basic normal form. Only then the resulting formula is ready to be translated to the target language using the embedding of Sect. 3.3. In turn, *Reduce* first eliminates the definitions in the given formula (Sect. 3.2.3), applies the rewriting rules (Sect. 3.2.1) repeatedly, and then applies abstraction (Sect. 3.2.2) followed by rewriting repeatedly.

The *Preprocess* algorithm is sound, because it is composed of sound sub-steps, and terminates, meaning that it will always compute a basic normal formula.

Theorem 2. *The Preprocess algorithm terminates.*

Proof (idea). Observe that the elimination step is in some sense opposite to the abstraction step: the first one eliminates every definition $x = \psi$ by using it as the rewriting rule $x \longrightarrow \psi$, while the latter introduces a new symbol x in the place of an expression ψ and asserts $x = \psi$, where ψ is non-basic in both cases. That is why we apply elimination before abstraction, and why each of those is followed by rewriting. We have to be careful that *Abstract* and *Eliminate* do not repeatedly act on the same expression. *Eliminate* does not produce non-basic expressions, but *Abstract* generates definitions that can be processed by *Eliminate*, reducing them again to the original non-basic expression. That is the reason for *Rewrite* to be applied after every application of *Abstract*: the new definitions are rewritten, usually by an extensionality expansion rule. In short, termination depends on the existence of extensionality rewriting rules for each kind of non-basic expression that *Abstract* may catch. Then, for any TLA^+ expression there exists an equi-satisfiable basic expression in normal form that the algorithm will compute. □

3.3 Direct Embedding

The preprocessing phase outputs a boolified basic TLA^+ expression that we will encode essentially using FOL and uninterpreted functions, without substantially changing its structure. In short, our encoding maps the given basic expression to corresponding formulas in the target language in an (almost) verbatim way.

For first-order TLA^+ expressions it suffices to apply a shallow embedding into first-order MS-FOL formulas. Non-logical TLA^+ operators are declared as function or predicate symbols with U-sorted arguments. For instance, the primitive relation $\in$ is encoded in SMT-LIB as the function in : $U \times U \rightarrow$ Bool. This is the only set theoretic operator that can appear in a basic formula. Expressions like IF c THEN t ELSE u can be conveniently mapped verbatim using SMT-LIB's conditional operator to ite(c, t, u), where c is of sort Bool (or boolified), and t and u have the same sort.

In order to reason about the theory of arithmetic, an automated prover requires type information, either generated internally, or provided explicitly in the input language. The operators and formulas that we have presented so far are expressed in FOL over uninterpreted function symbols over the sorts U and Bool. Because we want to benefit from the prover's native capabilities for arithmetic reasoning, we declare an injective function i2u: Int → U that embeds built-in integers into the sort U.[6] Integer literals k are simply encoded as i2u(k). For example, the formula $3 \in Int$ is translated as in(i2u(3), tla_Int), for which we have to declare tla_Int : U and add to the translation the axiom for Int

$$\forall x^{\mathsf{U}} : \text{in}(x, \text{tla\_Int}) \Leftrightarrow \exists n^{\mathsf{Int}} : x = \text{i2u}(n).$$

Observe that this axiom introduces two quantifiers to the translation. We can avoid the universal quantifier by encoding expressions of the form $x \in Int$ directly into $\exists n^{\mathsf{Int}} : x = \text{i2u}(n)$, but the existential quantifier remains. Arithmetic operators over TLA$^+$ values are defined homomorphically over the image of i2u by axioms such as

$$\forall m^{\mathsf{Int}}, n^{\mathsf{Int}} : \text{plus}(\text{i2u}(m), \text{i2u}(n)) = \text{i2u}(m + n),$$

where $+$ denotes the built-in addition over integers. For other arithmetic operators we define analogous axioms.

As a result, type inference in all these cases is, in some sense, delegated to the back-end prover. The link between built-in operations and their TLA$^+$ counterparts is effectively defined only for values in the range of the function i2u.

TLA$^+$ strings are encoded using the same technique: for every string literal that occurs in a proof obligation, we declare it as a constant of a newly declared sort Str, and assert that these constants are different from each other. Then, we use an injective function str2u : Str → U to lift string expressions.[7]

If we call *BasicEncode*(ϕ) to the embedding of a basic TLA$^+$ formula ϕ into MS-FOL, we can define the whole process of encoding TLA$^+$ into MS-FOL as:

$$
\begin{aligned}
Tla2MsFol(\phi) &\overset{\Delta}{=} \phi \\
&\quad \triangleright \textit{Preprocess} \\
&\quad \triangleright \textit{BasicEncode}
\end{aligned}
$$

3.4   Encoding Functions

A TLA$^+$ function $[x \in S \mapsto e(x)]$ is akin to a "bounded" λ-abstraction: the function application $[x \in S \mapsto e(x)][y]$ reduces to the expected value $e(y)$ if the argument y is an element of S, as stated by the axiom (2.10). As a consequence, e.g., the formula

$$f = [x \in \{1, 2, 3\} \mapsto x * x] \Rightarrow f[0] < f[0] + 1, \tag{3.1}$$

---

[6] The typical injectivity axiom $\forall m^{\mathsf{Int}}, n^{\mathsf{Int}} : \text{i2u}(m) = \text{i2u}(n) \Rightarrow m = n$ generates instantiation patterns for every pair of occurrences of i2u. Noting that i2u is injective iff it has a partial inverse u2i, we use instead the axiom $\forall n^{\mathsf{Int}} : \text{u2i}(\text{i2u}(n)) = n$, which generates a linear number of i2u(n) instances, where u2i : U → Int is unspecified.

[7] This encoding does not allow us to implement the standard TLA$^+$ interpretation of strings, which are considered as tuples of characters. Fortunately, characters are hardly used in practice.

although syntactically well-formed, should not be provable. Indeed, since 0 is not in the domain of f, we cannot even deduce that $f[0]$ is an integer.

We represent the application of an expression f to another expression x by two distinct first-order terms depending on whether the *domain condition* $x \in \text{DOMAIN } f$ holds or not: we introduce binary operators α and ω defined as

$$x \in \text{DOMAIN } f \Rightarrow \alpha(f, x) = f[x] \quad \text{and} \quad x \notin \text{DOMAIN } f \Rightarrow \omega(f, x) = f[x].$$

From these conditional definitions, we can derive the theorem

$$f[x] = \text{IF } x \in \text{DOMAIN } f \text{ THEN } \alpha(f, x) \text{ ELSE } \omega(f, x) \tag{3.2}$$

that gives a new defining equation for function application. In this way, functions are just expressions that are conditionally related to their argument by α and ω.

Using theorem (3.2), the expression $f[0]$ in the above example (3.1) is encoded as IF $0 \in \text{DOMAIN } f$ THEN $\alpha(f, 0)$ ELSE $\omega(f, 0)$. The solver would have to use the hypothesis to deduce that DOMAIN $f = \{1, 2, 3\}$, reducing the condition $0 \in \text{DOMAIN } f$ to false. The conclusion can then be simplified to the formula $\omega(f, 0) < \omega(f, 0) + 1$, which cannot be proved, as expected. Another example is $f[x] = f[y]$ in a context where $x = y$ holds: the formula is valid irrespective of whether the domain conditions hold or not.

Whenever possible, we try to avoid the encoding of function application as in the definition (3.2). From (2.10) and (3.2), we deduce the rewriting rule

$$[x \in S \mapsto e][a] \longrightarrow \text{IF } a \in S \text{ THEN } e[x \leftarrow a] \text{ ELSE } \omega([x \in S \mapsto e], a)$$

where $e[x \leftarrow a]$ denotes e with a substituted for x. This rule replaces two non-basic operators (function application and the function expression) in the left-hand side by only one non-basic operator in the right-hand side (the first argument of ω), which is required for termination of $(\text{TLA}^+, \longrightarrow)$ (Theorem 1).

In sorted languages like MS-FOL, functions have no notion of function domain other than the types of their arguments. Because explicit functions $[x \in S \mapsto e]$ cannot be mapped directly to first-order expressions, we treat them as any other non-basic expression. The following rewriting rule derived from axiom (2.10) replaces the function construct by a formula containing only basic operators:

$$f = [x \in S \mapsto e] \longrightarrow IsAFcn(f) \wedge \text{DOMAIN } f = S \wedge \forall x \in S \colon \alpha(f, x) = e$$

Observe that we have simplified $f[x]$ to $\alpha(f, x)$, because $x \in \text{DOMAIN } f$.

In order to prove that two functions are equal, we need to add a background axiom that expresses the extensionality property for functions:

$$\begin{aligned} \forall f, g \colon &\wedge IsAFcn(f) \wedge IsAFcn(g) \\ &\wedge \text{DOMAIN } f = \text{DOMAIN } g \\ &\wedge \forall x \in \text{DOMAIN } g \colon \alpha(f, x) = \alpha(g, x) \\ &\Rightarrow f = g \end{aligned}$$

Again, note that $f[x]$ and $g[x]$ were simplified using α. Unlike set extensionality, this formula is guarded by $IsAFcn$, avoiding the instantiation by expressions that are not considered functions. To prove that DOMAIN f = DOMAIN g, we still need to add to the translation the set extensionality axiom, which we abstain from. Instead, reasoning about the equality of domains can be solved by adding to the translation an instance of set extensionality for DOMAIN expressions only:

$$\forall f, g \colon \wedge\ IsAFcn(f) \wedge IsAFcn(g)$$
$$\wedge\ \forall x \colon x \in \text{DOMAIN}\ f \Leftrightarrow x \in \text{DOMAIN}\ g$$
$$\Rightarrow \text{DOMAIN}\ f = \text{DOMAIN}\ g$$

TLA$^+$ defines n-tuples as functions with domain $1 .. n$ and records as functions whose domain is a fixed finite set of strings. By treating them as non-basic expressions, we just need to add suitable rewriting rules to (TLA$^+$, $\longrightarrow$), in particular those for extensionality expansion. For instance, a tuple $\langle e_1, e_2, \ldots, e_n \rangle$ is defined as the function

$$[i \in 1 .. n \mapsto \text{IF}\ \ i = 1\ \text{THEN}\ e_1\ \text{ELSE} \left(\text{IF}\ \ i = 2\ \text{THEN}\ e_2\ \text{ELSE}\ (\ \cdots\ \text{ELSE}\ e_n))\right],$$

so that $\langle e_1, \ldots, e_n \rangle [i] = e_i$ when $i \in 1 .. n$. The following rule is derived from these definitions and from the axioms of extensionality (2.1) and functions (2.10):

$$t = \langle e_1, \ldots, e_n \rangle \longrightarrow \wedge\ IsAFcn(t)$$
$$\wedge\ \text{DOMAIN}\ t = 1 .. n$$
$$\wedge\ \bigwedge\nolimits_{e_i : \mathsf{U}} \alpha(t, i) = e_i$$
$$\wedge\ \bigwedge\nolimits_{e_i : \mathsf{Bool}} \alpha(t, i)^b \Leftrightarrow e_i$$

In order to preserve the satisfiability of expressions considered as terms from those considered as formulas, we treat differently the tuple elements e_i that are Booleans (noted $e_i : \mathsf{Bool}$) from those that are not (noted $e_i : \mathsf{U}$).

3.5   Encoding CHOOSE

The CHOOSE operator is notoriously difficult for automatic provers to reason about. Nevertheless, we can exploit CHOOSE expressions by using the axioms that define them. By introducing a definition for CHOOSE $x \colon P(x)$, we obtain the theorem

$$\left(y = \text{CHOOSE}\ \ x \colon P(x)\right) \Rightarrow \left((\exists x \colon P(x)) \Leftrightarrow P(y)\right),$$

where y is some fresh symbol. This theorem can be conveniently used as a rewriting rule after abstraction of CHOOSE expressions, and for CHOOSE expressions that occur negatively, in particular, as hypotheses of proof obligations.

For determinism of choice (axiom (2.9)), suppose an arbitrary pair of CHOOSE expressions $\phi_1 \triangleq$ CHOOSE $x \colon P(x)$ and $\phi_2 \triangleq$ CHOOSE $x \colon Q(x)$ where the free variables of ϕ_1 are $x_1, \ldots, x_n$ (noted $\mathbf{x}$) and those of ϕ_2 are $y_1, \ldots, y_m$ (noted $\mathbf{y}$). We need to check whether formulas P and Q are equivalent for every pair of expressions ϕ_1 and ϕ_2 occurring in a proof obligation. By abstraction of ϕ_1 and ϕ_2, we obtain the axiomatic definitions $\forall \mathbf{x} \colon f_1(\mathbf{x}) = $ CHOOSE $x \colon P(x)$ and $\forall \mathbf{y} \colon f_2(\mathbf{y}) = $ CHOOSE $x \colon Q(x)$, where f_1 and f_2 are fresh operator symbols of suitable arity. Then, we just need to state the extensionality property for the pair f_1 and f_2 as the axiom $\forall \mathbf{x}, \mathbf{y} \colon \left(\forall x \colon P(x) \Leftrightarrow Q(x)\right) \Rightarrow f_1(\mathbf{x}) = f_2(\mathbf{y})$.

4   Evaluation

In order to validate our approach we reproved several test cases that had been proved interactively using the previously available TLAPS backend provers Zenon, Isabelle/TLA$^+$ and the decision procedure for Presburger arithmetic. We will refer to the combination of those three backends as ZIP for short.

For each benchmark, we compare two dimensions of an interactive proof: size and time. We define the *size* of an interactive proof as the number of non-trivial proof obligations generated by the Proof Manager, which is proportional to the number of interactive steps and therefore represents the user effort for making TLAPS check the proof. The *time* is the number of seconds required by the Proof Manager to verify those proofs on a 2.2 GHz Intel Core i7 with 8 GB of memory.

Table 1. Evaluation benchmarks results. An entry with the symbol "-" means that the solver has reached the timeout without finding the proof for at least one of the proof obligations. The backends were executed with a timeout of 300 seconds.

	size	ZIP	CVC4	Z3
Peterson	3	-	0.41	0.34
Peterson	10	5.69	0.78	0.80
Bakery	19	-	36.86	15.20
Bakery	223	52.74		
Memoir-T	1	-	-	1.99
Memoir-T	12	-	3.11	3.21
Memoir-T	424	7.31		
Memoir-I	8	-	3.84	9.35
Memoir-I	61	8.20		
Memoir-A	27	-	11.31	11.46
Memoir-A	126	19.10		

Finite Sets	ZIP		Zenon+SMT	
	size	time	size	time
CardZero	11	5.42	5	0.48
CardPlusOne	39	5.35	3	0.49
CardOne	6	5.36	1	0.35
CardOneConv	9	0.63	2	0.35
FiniteSubset	62	7.16	21	5.94
PigeonHole	42	7.07	20	7.01
CardMinusOne	11	5.44	5	0.75

Table 1 presents the results for four case studies: type correctness and mutual exclusion of the Peterson and Bakery algorithms, type correctness (T) and refinement proofs (I, A) of the Memoir security architecture [7], and proofs of theorems about finite sets and cardinalities. We compare how proofs of different sizes are handled by the backends. Each line corresponds to an interactive proof of a given size. Columns correspond to the running times for a given SMT solver, where each prover is executed on all generated proof obligations. For our tests we have used off-the-shelf SMT solvers CVC4 v1.3 and Z3 v4.3.2.

In all cases, the use of the new SMT backend leads to significant reductions in proof sizes and running times compared to the original interactive proofs. In particular, the "shallow" proofs of the first three case studies required only minimal interaction. For instance, in the Peterson case, SMT solvers can cope with a proof that generates 3 obligations, while the ZIP backends time out in at least one of them. Instead, ZIP requires a more fine-grained proof of size 10.

In the Finite Sets benchmarks, some proof obligations generated from big structural high-level formulas can be proved only by Zenon. Beyond these benchmark problems, the SMT backend has become the default backend of TLAPS.

5   Related Work

In previous publications [13,14] we presented a primitive encoding of TLA$^+$ into SMT-LIB where boolification, normalization and abstraction were not made explicit in the translation, and CHOOSE expressions were not fully supported. This paper supersedes them. Some of our encoding techniques (Sect. 3) were already presented before (injection of unsorted expressions [14]) or are simply folklore (*e.g.*, abstraction), but to our knowledge they have not been combined and studied in this way. Moreover, the idiosyncrasies of TLA$^+$ render their applicability non-trivial. For instance, axiomatized TLA$^+$ functions with domains, including tuples and records, are deeply rooted in the language.

The B and Z languages are also based on ZF set theory, although in a somewhat weaker version, because terms and functions have (monomorphic) types in the style of MS-FOL, thus greatly simplifying the translations to SMT languages. Another difference is that functions are defined as binary relations, as is typical in set theory. There are two SMT plugins for the Rodin tool set for Event-B. The SMT solvers plugin [5] directly encodes simple sets (*i.e.*, excluding set of sets) as polymorphic λ-expressions, which are non-standard and are only handled by the parser of the veriT SMT solver. The ppTrans plugin [9] generates different SMT sorts for each combination of simple sets, power sets and cartesian products found in the proof obligation. Therefore, there is one membership operator for every declared set sort, with the advantage that it further partitions the proof search space, although this requires that the type of every term be known beforehand. (In TLA$^+$, this can only be achieved through *type synthesis*; see [15].) Additionally, when ppTrans detects the absence of set of sets, the translation is further simplified by encoding sets by their characteristic predicates.

Similarly, Atelier-B discharges proof obligations to different SMT solvers based on Why3 [12], with sets encoded using polymorphic types. ProB includes a translation between TLA$^+$ and B [8], allowing TLA$^+$ users to use ProB tools. It relies on Kodkod, the Alloy Analyzer's backend, to do constraint solving over the first-order fragment of the language, and on the ProB kernel for the rest [16].

More recently, Delahaye et al. [6] proposed a different approach to reason about set theory, instead of a direct encoding into FOL. The theory of deduction modulo is an extension of predicate calculus that includes rewriting of terms and propositions. It is well suited for proof search in axiomatic theories such as Peano arithmetic or Zermelo set thery, as it turns axioms into rewrite rules.

MPTP [18] translates Mizar to the unsorted first-order format TPTP/FOF [17]. The Mizar language, targeted at formalized mathematics, provides second-order predicate variables and abstract terms derived from replacement and comprehension, such as the set $\{n - m$ where m, n is *Integer* : $n < m\}$. During preprocessing, MPTP replaces them by fresh symbols, with their definitions at the top level. Similar to our abstraction technique, it resembles Skolemization.

6   Conclusions

We have presented a sound and effective way of discharging TLA$^+$ proof obligations using automated theorem provers based on many-sorted first-order logic. This encoding forms the core of a back-end prover that integrates external SMT solvers as oracles to the TLA$^+$ Proof System (TLAPS). The main component of the backend is a generic translation framework that makes available to TLAPS any SMT solver that supports the de facto standard format SMT-LIB/AUFLIA. Within the same framework, we have also integrated automated theorem provers based on unsorted FOL [17], such as those based on the superposition calculus.

Our translation enables the backend to successfully handle a useful fragment of the TLA$^+$ language. The untyped universe of TLA$^+$ is represented as a universal sort in MS-FOL. Purely set-theoretic expressions are mapped to formulas over uninterpreted symbols, together with relevant background axioms. The built-in integer sort and arithmetic operators are homomorphically embedded into the universal sort, and type inference is in essence delegated to the solver. Functions, tuples, records, and the CHOOSE operator (Hilbert's choice) are encoded using a preprocessing mechanism that combines term rewriting with abstraction. The soundness of the encoding is immediate: all rewriting rules and axioms about sets, functions, records, tuples, *etc.* are theorems in the background theory of TLA$^+$ that exist in the Isabelle encoding.

Encouraging results show that SMT solvers significantly reduce the effort of interactive reasoning for verifying "shallow" TLA$^+$ proof obligations, as well as some more involved formulas including linear arithmetic expressions. Both the time required to find automatic proofs and, more importantly, the size of the interactive proof, which reflects the number of user interactions, can be remarkably reduced with the new back-end prover.

The translation presented here forms the basis for further optimizations. In [15] we have explored the use of (incomplete) type synthesis for TLA$^+$ expressions, based on a type system with dependent and refinement types. Extensions for reasoning about real arithmetic and finite sequences would be useful. What is more important, we rely on the soundness of external provers, temporarily including them as part of TLAPS's trusted base. In future work we intend to reconstruct within Isabelle/TLA$^+$ (along the lines presented in [3]) the proof objects that many SMT solvers can produce. Such a reconstruction would have to take into account not only the proofs generated by the solvers, but also all the steps performed during the translation, including rewriting and abstraction.

References

1. Baader, F., Nipkow, T.: Term rewriting and all that. Cambridge University Press, Cambridge (1999)
2. C. Barrett, A. Stump, and C. Tinelli. The Satisfiability Modulo Theories Library (SMT-LIB)(2010). www.SMT-LIB.org
3. Blanchette, J.C., Böhme, S., Paulson, L.C.: Extending Sledgehammer with SMT solvers. J Autom. Reasoning **51**(1), 109–128 (2013)

4. Cousineau, D., Doligez, D., Lamport, L., Merz, S., Ricketts, D., Vanzetto, H.: TLA$^+$ proofs. In: Giannakopoulou, D., Méry, D. (eds.) FM 2012. LNCS, vol. 7436, pp. 147–154. Springer, Heidelberg (2012)
5. Déharbe, D., Fontaine, P., Guyot, Y., Voisin, L.: SMT solvers for Rodin. In: Derrick, J., Fitzgerald, J., Gnesi, S., Khurshid, S., Leuschel, M., Reeves, S., Riccobene, E. (eds.) ABZ 2012. LNCS, vol. 7316, pp. 194–207. Springer, Heidelberg (2012)
6. Delahaye, D., Doligez, D., Gilbert, F., Halmagrand, P., Hermant, O.: Zenon Modulo: when Achilles Outruns the tortoise using deduction modulo. In: McMillan, K., Middeldorp, A., Voronkov, A. (eds.) LPAR-19 2013. LNCS, vol. 8312, pp. 274–290. Springer, Heidelberg (2013)
7. Douceur, J.R., Lorch, J.R., Parno, B., Mickens, J., McCune, J.M.: Memoir-Formal Specs and Correctness Proofs. Technical report MSR-TR-19, Microsoft Research (2011)
8. Hansen, D., Leuschel, M.: Translating TLA$^+$ to B for validation with PROB. In: Derrick, J., Gnesi, S., Latella, D., Treharne, H. (eds.) IFM 2012. LNCS, vol. 7321, pp. 24–38. Springer, Heidelberg (2012)
9. Konrad, M., Voisin, L.: Translation from set-theory to predicate calculus. Technical report, ETH Zurich (2012)
10. Lamport, L.: Specifying Systems: The TLA$^+$ Language and Tools for Hardware and Software Engineers. Addison-Wesley, Boston (2002)
11. Manzano, M.: Extensions of First-Order Logic. Cambridge Tracts in Theoretical Computer Science, 2nd edn. Cambridge University Press, Cambridge (2005)
12. Mentré, D., Marché, C., Filliâtre, J.-C., Asuka, M.: Discharging proof obligations from Atelier B using multiple automated provers. In: Derrick, J., Fitzgerald, J., Gnesi, S., Khurshid, S., Leuschel, M., Reeves, S., Riccobene, E. (eds.) ABZ 2012. LNCS, vol. 7316, pp. 238–251. Springer, Heidelberg (2012)
13. Merz, S., Vanzetto, H.: Automatic verification of TLA$^+$ proof obligations with SMT solvers. In: Bjørner, N., Voronkov, A. (eds.) LPAR-18 2012. LNCS, vol. 7180, pp. 289–303. Springer, Heidelberg (2012)
14. Merz, S., Vanzetto, H.: Harnessing SMT Solvers for TLA$^+$ Proofs. Electron. Commun. Eur. Assoc. Softw. Sci. Tech., **53** (2012)
15. Merz, S., Vanzetto, H.: Refinement types for TLA$^+$. In: Badger, J.M., Rozier, K.Y. (eds.) NFM 2014. LNCS, vol. 8430, pp. 143–157. Springer, Heidelberg (2014)
16. Plagge, D., Leuschel, M.: Validating B,Z and TLA$^+$ Using PROB and Kodkod. In: Giannakopoulou, D., Méry, D. (eds.) FM 2012. LNCS, vol. 7436, pp. 372–386. Springer, Heidelberg (2012)
17. Sutcliffe, G.: The TPTP problem library and associated infrastructure. J. Autom. Reason. **43**(4), 337–362 (2009)
18. Urban, J.: Translating Mizar for first-order theorem. In: Asperti, A., Buchberger, B., Davenport, J.H. (eds.) MKM 2003. LNCS, vol. 2594, pp. 203–215. Springer, Heidelberg (2003)

Proving Determinacy of the PharOS Real-Time Operating System

Selma Azaiez[1], Damien Doligez[2], Matthieu Lemerre[1], Tomer Libal[3],
and Stephan Merz[4,5(✉)]

[1] CEA, Saclay, France
[2] Inria, Paris, France
[3] Inria, Saclay, France
[4] Inria, Villers-lès-Nancy, France
stephan.merz@loria.fr
[5] CNRS, Université de Lorraine, LORIA, UMR 7503,
Vandoeuvre-lès-Nancy, France

Abstract. Executions in the PharOS real-time system are deterministic in the sense that the sequence of local states for every process is
independent of the order in which processes are scheduled. The essential
ingredient for achieving this property is that a temporal window of execution is associated with every instruction. Messages become visible to
receiving processes only after the time window of the sending message
has elapsed. We present a high-level model of PharOS in TLA^+ and
formally state and prove determinacy using the TLA^+ Proof System.

1 Introduction

The outcome of an execution of a concurrent system depends not only on the
inputs provided from the system's environment, but also on the relative order in
which the system's processes are scheduled for execution. This order is largely
unpredictable, especially when the system executes on parallel hardware; it introduces an element of non-determinism even when every process behaves determin-
istically. Testing and debugging of concurrent systems is therefore challenging
and involves so-called "Heisenbugs" that are very difficult to reproduce.

For real-time systems, such as controllers of safety-critical components in
planes or cars, designers are very reluctant to admit systems that exhibit non-
deterministic behavior. Fortunately, it is possible to design concurrent real-time
systems such that their behavior does not depend on the order of scheduling,
as long as all components have access to a common time base. This hypothesis can be satisfied in local networks of embedded systems. For example, the
PharOS real-time system [9,10], commercialized[1] under the name Asterios®, has
been designed to ensure that system executions do not depend on the scheduling order of processes. The core idea is to associate every instruction that some

---

This work was supported by the French BGLE Project ADN4SE. It was also partly
funded by the Microsoft Research-Inria Joint Centre, France.
[1] http://www.krono-safe.com.

M. Butler et al. (Eds.): ABZ 2016, LNCS 9675, pp. 70–85, 2016.
DOI: 10.1007/978-3-319-33600-8_4

process wishes to execute with a temporal window of execution and to ensure that a message sent from one process to another can be received only if the execution window of the receiving instruction is strictly later than that of the sending instruction. Consider two executions that execute both the sending and the receiving instructions according to the given temporal constraints, then the message will either be received in both executions, or in neither of them. This argument is at the core of the determinacy proof for the PharOS model of execution [9].

In this work, we formally specify a high-level model of PharOS executions in the specification language TLA$^+$ [6] and use TLAPS, the TLA$^+$ Proof System [4], to formally prove determinacy of our model. Our proof is based on the paper-and-pencil proof of [9]. In contrast to that proof, TLA$^+$ proofs such as ours must be written in assertional style, i.e., based on inductive invariants, rather than making explicit references to different states of an execution. Moreover, the mere statement of determinacy is not entirely obvious in a linear-time framework such as TLA$^+$ because the property refers to the equivalence of different executions, whereas formulas of linear-time temporal logic are expressed in terms of a single, implicit execution.

Our work reinforces the confidence in the result that PharOS executions are indeed deterministic and makes explicit some hypotheses that were implicit in the original proof. It represents a significant case study for TLAPS and has also contributed several lemmas that are now included in the standard library of the TLAPS distribution.

Outline. Fundamental concepts of PharOS and of TLA$^+$ are presented in Sect. 2. Sections 3 and 4 are the core of the paper and describe the formal model of PharOS and the proof of the main theorem. We conclude in Sect. 5.

2   Background

2.1   PharOS

The PharOS model of execution [9,10] is based on the OASIS model [2,12] but relaxes the constraints on the precise instants of execution of individual processes. It was designed with the objectives of ensuring predictability of execution and of supporting formal reasoning. Avoiding race conditions between processes is crucial for achieving the first objective. It is assumed that the system consists of a fixed number of tasks (called *agents*), each of which executes instructions sequentially and atomically. All agents share a common time reference, and each instruction is associated with a fixed, non-empty time window of execution in the sense that it must be executed between an earliest and latest time instant (cf. Fig. 1). The only assumptions that are made about the scheduling policy for the agents are that the specified time windows are respected, and that deadlines are never missed. The latter property is in practice ensured by schedulability analysis of the implementation, different from the techniques discussed in the present paper [8].

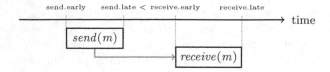

Fig. 1. Time constraints associated with instructions in PharOS.

Agents communicate with one another exclusively by asynchronous message passing. It can be seen (and follows from our proof) that determinacy of the execution model can therefore be reduced to determinacy of message reception. In order for a message m to be received by an agent, it is not enough that the corresponding send instruction was executed prior to reception: it must also be *visible* to the receiving instruction. As indicated in Fig. 1, the message can be visible only if the time window associated with the send instruction precedes the time window of the receiving instruction, i.e. the latest possible instant at which the message can be sent precedes the earliest instant at which it can be received. Hence, if there exists a schedule in which the message was not sent prior to the execution of the receiving instruction, the message cannot be received in any execution.

2.2   The TLA$^+$ Specification Language

TLA$^+$ [6] is a formal specification language that is mainly intended for modeling concurrent and distributed algorithms and systems, and that has successfully been used in academic and industrial environments [3,7,13]. It is based on untyped Zermelo-Fraenkel set theory for modeling the data manipulated by the system, and on the Temporal Logic of Actions, a variant of linear-time temporal logic, for describing executions. Formulas of temporal logic serve for specifying system behavior as well as properties of systems. Systems are modeled as state machines. In particular, the system state is represented by a tuple of variables. State predicates (i.e., first-order formulas containing state variables) represent sets of system states, such as the initial condition or system invariants. Transition predicates, also called *actions*, are first-order formulas that contain both ordinary (unprimed) and primed occurrences of state variables; they describe state transitions where unprimed variables denote the value in the first state and primed variables denote the value in the second state. For example, $x' = x - y'$ is true of any pair $\langle s, t \rangle$ of states such that the value of x in state t equals the difference between the values of x in state s and y in state t. The canonical form of the safety part of a system specification in TLA$^+$ is a temporal formula of the form

$$Init \wedge \Box[Next]_{vars}$$

where *Init* is a state predicate constraining the initial states, *Next* is an action that describes all possible system steps, and *vars* is a tuple of all variables used to represent the state of the system. Temporal formulas are evaluated over infinite sequences of states, and the above formula is true iff *Init* is true of the first state and all pairs of subsequent states either leave *vars* unchanged or satisfy *Next*. Fairness conditions can be added to the above formula in order to ensure liveness properties, but they play no role in this paper.

If P is a state predicate then P' denotes a copy of P in which all state variables x have been replaced by their primed counterparts x'. It is a transition formula that asserts that P is true of the second state of the pair of states. For example, the familiar proof obligation requiring that an invariant *Inv* be preserved by the system's next-state relation is written as the formula

$$Inv \wedge [Next]_{vars} \Rightarrow Inv'.$$

Sets and functions are central in TLA$^+$ for modeling systems and their data. Semantically, every TLA$^+$ value is a set. The notations for standard set-theoretic constructions are familiar. In particular, $\{\, e(x) : x \in S \,\}$ and $\{\, x \in S : p(x) \,\}$ are two forms of set comprehension: the first one denotes the set of values $e(x)$ for all elements x of set S, and the second one the subset of elements of S that satisfy predicate p. Also, SUBSET S and UNION S denote the powerset (set of all subsets) of S and the union of the elements of the family S of sets. A function f is a total mapping over its domain DOMAIN f, function application is written $f[e]$, and $[x \in S \mapsto e(x)]$ denotes the function with domain S such that $f[x] = e(x)$ for all $x \in S$. The set $[S \rightarrow T]$ denotes the set of functions f with domain S such that $f[x] \in T$ for all $x \in S$.

An n-tuple (or sequence of length n) $d = \langle d_1, \ldots, d_n \rangle$ is a function with domain $1 .. n$ such that $d[i] = d_i$ for all $i \in 1 .. n$. The set $Seq(S)$ denotes the set of all finite sequences whose elements are contained in the set S. Standard operations on sequences include *Append* (adding an element at the end of a sequence) and *Head* and *Tail* for accessing the first element and the remainder of a non-empty sequence. The predicate $IsPrefix(s, t)$ holds if s is a prefix of the finite sequence t.

Records are represented as functions whose domains are finite sets of strings. For example, $[id : String, bal : Int]$ denotes the set of records with two fields *id* and *bal* whose values are respectively a string and an integer. For such a record *accnt*, the fields are accessed as *accnt.id* and *accnt.bal* (short-hand forms for $accnt["id"]$ and $accnt["bal"]$). The expression $[id \mapsto "xyz", bal \mapsto 123]$ denotes a record in the above set.

For writing long formulas, TLA$^+$ adopts the convention of writing multi-line conjunctions and disjunctions as lists "bulleted" with $\wedge$ and $\vee$, and where indentation is used instead of parentheses to indicate precedence. For example,

$$
\begin{aligned}
&\wedge\ A \vee B \\
&\wedge\ \vee\ C \\
&\quad\ \vee\ D \\
&\wedge\ E \Rightarrow F
\end{aligned}
$$

is a conjunction of three formulas, the first and second of which are disjunctions.

2.3 TLA$^+$ Modules

TLA$^+$ specifications are structured as modules. A module declares parameters (for example, the set of processes of a multi-process algorithm), defines operators, and may state assumptions and theorems about the parameters and operators. In fact, standard integer and real arithmetic, as well as the operations on sequences mentioned above, are not part of the language itself, but are defined in modules of the standard library. Modules can be imported using the EXTENDS keyword, which corresponds to copying the contents of the imported module into the current module. A more elaborate form of import is provided through module instantiation, which allows substituting expressions for module parameters.

We define several modules that provide operations used in our specification. The module *Streams* defines an infinite sequence over a set S as a function from positive integers to S and provides several theorems about streams. In order to give a flavor of TLA$^+$, an excerpt of the module is shown here, omitting the proofs.[2]

─────────────── MODULE *Streams* ───────────────

EXTENDS *NaturalsInduction*, *Functions*, *SequenceTheorems*

$Natp \triangleq Nat \setminus \{0\}$

$Stream(S) \triangleq [Natp \to S]$

$take(w, n) \triangleq Restrict(w, 1 .. n)$

LEMMA *takeStream* $\triangleq$
    ASSUME NEW S, NEW $w \in Stream(S)$, NEW $n \in Nat$
    PROVE $take(w, n) \in Seq(S) \land Len(take(w, n)) = n$

LEMMA *takeStreamMonotonic* $\triangleq$
    ASSUME NEW S, NEW $w \in Stream(S)$,
           NEW $m \in Nat$, NEW $n \in Nat$, $m \leq n$
    PROVE $IsPrefix(take(w, m), take(w, n))$

Another module introduces the operation $filter(s, a)$. It takes as its first argument s a finite sequence of system states (cf. set *SystemState* introduced in Sect. 3 below); each system state is a record whose st field is an array containing the local states of all agents. The second argument of *filter* is an agent a. The operation projects s to the sequence of local states of agent a, and then removes finite repetitions of states: when s corresponds to a prefix of a system execution, repeated agent states typically correspond to steps where some other agent than a was executed. Its formal definition is thus given as the composition of a projection operator and another operator for removing finite stuttering:

$project(s, a) \triangleq [i \in 1 .. Len(s) \mapsto s[i].st[a]]$

$unstutter[s \in Seq(State)] \triangleq$

---

[2] The standard module *Functions* defines the domain restriction of a function as $Restrict(f, S) \triangleq [x \in S \mapsto f[x]]$.

$\qquad$ IF $Len(s) \leq 1$ THEN s
$\qquad$ ELSE IF $Last(s) = s[Len(s) - 1]$ THEN $unstutter[Front(s)]$
$\qquad$ ELSE $Append(unstutter[Front(s)], Last(s))$
$\quad filter(s, a) \triangleq unstutter[project(s, a)]$

The definition of *unstutter* illustrates the definition of recursive functions in TLA^+. We prove many facts about these operations that are used in our main proof, including those listed below.

$\quad$ LEMMA *filter_range* $\triangleq$
$\qquad$ ASSUME NEW $H \in Seq(SystemState)$, NEW $a \in Agent$
$\qquad$ PROVE $Range(filter(H, a)) = \{H[i].st[a] : i \in 1 .. Len(H)\}$
$\quad$ LEMMA *filter_IsPrefix* $\triangleq$
$\qquad$ ASSUME NEW $H1 \in Seq(SystemState)$, NEW $H2 \in Seq(SystemState)$,
$\qquad\qquad$ NEW $a \in Agent$, $IsPrefix(H1, H2)$
$\qquad$ PROVE $IsPrefix(filter(H1, a), filter(H2, a))$
$\quad$ LEMMA *IsPrefix_filter* $\triangleq$
$\qquad$ ASSUME NEW $H1 \in Seq(SystemState)$, NEW $H2 \in Seq(SystemState)$,
$\qquad\qquad$ NEW $a \in Agent$, $IsPrefix(filter(H1, a), filter(H2, a))$
$\qquad$ PROVE $\exists H \in Seq(SystemState) : \wedge\ IsPrefix(H, H2)$
$\qquad\qquad\qquad\qquad\qquad\qquad\qquad\quad \wedge\ filter(H, a) = filter(H1, a)$

2.4 Tool Support for TLA^+

The formal verification of TLA^+ specifications is supported by the TLC model checker and by TLAPS, the TLA^+ Proof System. TLC [16] is an explicit-state model checker that can verify properties of finite instances of TLA^+ specifications. Similar to the ProB model checker [11], TLC is notable for its capability to evaluate a highly expressive, set-based expression language.

For analysis with TLC, parameters of TLA^+ modules must be instantiated by fixed values (such as instantiating the set of processes to the set $\{1, 2, 3\}$). Additionally, it must be ensured that all values that variables take during any execution of the specified system belong to some finite set. This is not always possible: for example, a finite instance of a system may have an unbounded state space when communication channels are represented by unbounded sequences and messages may be resent. In such cases, analysis may be restricted to states satisfying a user-defined constraint. While this implies an under-approximation of the analyzed state space, any counter-example produced by TLC within the restricted search space is an actual system execution. Going further, the user may also override definitions that TLC either cannot evaluate or that lead to unbounded state spaces. In these cases, the semantics can be changed arbitrarily, and the significance of the results of verification must be ensured by the user, but skilled use of these features helps building confidence in the specification. In this project, we have mainly used TLC for validating definitions of complex operators such as the *filter* operation shown in Sect. 2.3, where it helped us to catch off-by-one and similar errors.

TLAPS   [4] is an interactive proof assistant for TLA$^+$. It allows users to develop proofs for lemmas and theorems asserted in a TLA$^+$ module, using a hierarchical proof language. These proofs are interpreted by the core of TLAPS, called the *proof manager*. Obligations corresponding to the steps in the proof are sent to automatic proof backends, including first-order provers, SMT solvers, and a decision procedure for propositional temporal logic. If no backend is able to prove the step, the user can write a more detailed, lower-level proof of the step. Proofs for the different steps can be developed in any order, which lets users concentrate on the most difficult or interesting part of a proof first and fill in details later. All proof obligations whose proof was already attempted during the current project are stored in a data base, and TLAPS allows users to quickly check the status of the proof and assess the impact of changes in definitions or assertions.

Both TLC and TLAPS are accessed from the TLA$^+$ Toolbox, an Eclipse-based GUI for editing TLA$^+$ specifications. The ability to use the same specifications for model checking and for proof is very valuable for validation. In particular, TLC can be used to check if an assertion appearing in a proof can be invalidated in a finite instance, before making a futile proof attempt.

3   A High-Level Model of PharOS in TLA$^+$

Our objective in specifying PharOS in TLA$^+$ is to provide a high-level model that abstracts from choices made in particular implementations. In particular, we do not wish to commit to any scheduling policy, nor fix the time taken by individual instructions, or indeed the set of instructions that an instance of PharOS executes. We thus obtain a highly non-deterministic specification that is intended to encompass all possible system executions.

A TLA$^+$ module representing the static model of PharOS appears in Fig. 2. PharOS coordinates executions of processes, called agents, each of which has local states. Correspondingly, the constants *Agent* and *State* are declared as parameters in the module. We assume that every state s identifies the instruction *instrOf*(s) that the agent will execute next. There are three kinds of instructions: local instructions simply modify the local state of an agent by applying an update function. Send instructions similarly update the local state but are also tagged with a message identifier and inserted into the message pool. Receive instructions attempt to retrieve the message[3] with the given identifier from the message pool and apply an update to the local state whose effect depends on whether the message could be received.

PharOS is a real-time system, and time is discrete, represented by natural numbers. In particular, the instructions manipulated by PharOS are equipped with a temporal execution window that indicates the earliest and latest points in time when the instruction can be executed. The functions updating the local states are assumed to be monotonic in the sense that the execution window of the instruction associated with the updated state cannot precede the execution

---

[3] For simplicity, messages are identified with the sending instruction.

$\text{---------- MODULE } \textit{Types} \text{ ----------}$

EXTENDS $\textit{Sequences, Streams, TLAPS}$

CONSTANT $\textit{Agent, State, initState, instrOf(\_), MsgId, Update, visible(\_,\_)}$

$\textit{Time} \triangleq \textit{Nat}$

$\textit{Instruction} \triangleq \quad [\textit{type} : \{ \text{"local"} \}, \textit{upd} : \textit{Update}]$
$\qquad\qquad\qquad \cup\ [\textit{type} : \{ \text{"send"} \}, \textit{msg} : \textit{MsgId, upd} : \textit{Update}]$
$\qquad\qquad\qquad \cup\ [\textit{type} : \{ \text{"receive"} \}, \textit{msg} : \textit{MsgId, bupd} : [\text{BOOLEAN} \to \textit{Update}] \ ]$

$\textit{DatedInstruction} \triangleq \{\ x \in [\textit{ins} : \textit{Instruction, early} : \textit{Time, late} : \textit{Time}] :$
$\qquad\qquad\qquad\qquad\qquad x.\textit{early} < x.\textit{late}\ \}$

$\textit{Message} \triangleq \{\ \textit{di} \in \textit{DatedInstruction} : \textit{di.type} = \text{"send"}\ \}$

ASSUME $\textit{instrOf\_type} \triangleq \forall s \in \textit{State} : \textit{instrOf}(s) \in \textit{DatedInstruction}$

$\textit{SystemState} \triangleq [\textit{st} : [\textit{Agent} \to \textit{State}], \textit{msgs} : \text{SUBSET } \textit{Message, t} : \textit{Time}]$

$\textit{GoodNextTime}(s, a, t) \triangleq\ \wedge\ s.t \le t$
$\qquad\qquad\qquad\qquad\quad \wedge\ \textit{instrOf}(s.\textit{st}[a]).\textit{early} \le t$
$\qquad\qquad\qquad\qquad\quad \wedge\ \forall b \in \textit{Agent} : t \le \textit{instrOf}(s.\textit{st}[b]).\textit{late}$

$\textit{GoodSystemState} \triangleq \{\ s \in \textit{SystemState} :$
$\qquad\qquad\qquad\qquad\qquad \exists a \in \textit{Agent}, t \in \textit{Time} : \textit{GoodNextTime}(s, a, t)\ \}$

ASSUME $\textit{goodInitState} \triangleq$
$\quad [\textit{st} \mapsto \textit{initState, msgs} \mapsto \{\}, t \mapsto 1] \in \textit{GoodSystemState}$

ASSUME $\textit{Update\_type} \triangleq \textit{Update} \in \text{SUBSET } [\textit{State} \to \textit{State}]$

ASSUME $\textit{Update\_monotonic} \triangleq \forall \textit{state} \in \textit{State, upd} \in \textit{Update} :$
$\qquad\qquad \wedge\ \textit{instrOf}(\textit{state}).\textit{early} \le \textit{instrOf}(\textit{upd}[\textit{state}]).\textit{early}$
$\qquad\qquad \wedge\ \textit{instrOf}(\textit{state}).\textit{late} \le \textit{instrOf}(\textit{upd}[\textit{state}]).\textit{late}$

ASSUME $\textit{visible\_cond} \triangleq \forall \textit{di, dj} \in \textit{DatedInstruction} :$
$\qquad\qquad \wedge\ \textit{visible}(\textit{di, dj}) \in \text{BOOLEAN}$
$\qquad\qquad \wedge\ \textit{visible}(\textit{di, dj}) \Rightarrow \textit{di.late} < \textit{dj.early}$

$\textit{msgReceived}(\textit{msgs, di}) \triangleq \exists i \in \textit{msgs} : \textit{di.ins.msg} = i.\textit{ins.msg} \wedge \textit{visible}(i, \textit{di})$

$\textit{exec}(\textit{state, msgs}) \triangleq$
$\quad$ LET $i \triangleq \textit{instrOf}(\textit{state})$
$\quad$ IN $\quad$ IF $i.\textit{ins.type} = \text{"update"}$ THEN $\langle i.\textit{ins.upd}[\textit{state}], \textit{msgs}\rangle$
$\qquad\quad$ ELSE IF $i.\textit{ins.type} = \text{"send"}$ THEN $\langle i.\textit{ins.upd}[\textit{state}], \textit{msgs} \cup \{i\}\rangle$
$\qquad\quad$ ELSE $\langle i.\textit{ins.bupd}[\textit{msgReceived}(\textit{msgs}, i)][\textit{state}], \textit{msgs}\rangle$

$\textit{executes}(a, \textit{pre, post}) \triangleq$
$\quad$ LET $\textit{pstate} \triangleq \textit{pre.st}[a]$
$\qquad\ \textit{res} \triangleq \textit{exec}(\textit{pstate, pre.msgs})$
$\quad$ IN $\quad \wedge\ \textit{GoodNextTime}(\textit{pre}, a, \textit{post.t})$
$\qquad\qquad \wedge\ \textit{post.st} = [\textit{pre.st} \text{ EXCEPT } ![a] = \textit{res}[1]]$
$\qquad\qquad \wedge\ \textit{post.msgs} = \textit{res}[2]$

ASSUME $\textit{TimeProgress} \triangleq$
$\quad \forall H \in \textit{Stream}(\textit{SystemState}) :$
$\qquad (\forall n \in \textit{Natp} : \exists a \in \textit{Agent} : \textit{executes}(a, H[n], H[n+1]))$
$\qquad \Rightarrow \forall T \in \textit{Time} : \exists k \in \textit{Natp} : H[k].t > T$

Fig. 2. Static model of PharOS.

window of the instruction for the original state. As specified in the definition of operator *msgReceived*, a message can be received if it is present in the message pool and if it is *visible*, which implies that the execution window of the sending instruction precedes the execution window of the receiving instruction.

The operator *exec* corresponds to the result of executing the pending instruction of the local state in the context of the message pool passed as the arguments. It returns a pair consisting of the updated state and the new message pool. Note that in our model, nothing is assumed about the delivery order of messages, and that a message may be received multiple times.

A *system state* is represented as a record with three fields corresponding to the array of local states per agent, the messages that have been sent, and the current time. Given system state s and agent a, the predicate *GoodNextTime*(s, a, t) identifies time instants t at which a can take a step without any deadline being missed. More precisely, it holds if t is at least as big as both the time recorded in system state s and the earliest execution time for the instruction that a is about to execute, but does not exceed the time window of the pending instruction of any agent b (including a itself). A *good system state* is one for which some such t exists, for some agent a. The initial states of every agent are given by parameter *initState*, and we assume that the system state formed by this state assignment, an empty message pool, and initial time 1 is good.

The predicate *executes*$(a, pre, post)$ is true if the system state *post* can be obtained from the system state *pre* by an execution of agent a at a good next time. It is easy to prove that whenever *pre* is a good system state then *executes*$(a, pre, post)$ implies that *post* is also good. As expressed by predicate *TimeProgress*, we assume that in any infinite sequence of system states such that every transition corresponds to the execution of some agent, the recorded time progresses beyond any bound. This corresponds to the familiar non-Zenoness assumption in the analysis of real-time systems [1].

Figure 3 presents the system specification of PharOS. The state of the system is represented by the variables *state*, *messages*, and *time*.[4] Additionally, variable *history* is used for the specification of determinacy; it records the sequence of all previous states of the system. The overall system behavior is specified in standard form as formula *Spec*. The definition of the initial condition is obvious. The next-state relation requires that time advances, without missing any deadline, to some value within the execution window for the instruction of some agent a, and that the state and message pool are updated according to the execution of that instruction. Moreover, the new system state is appended to the sequence *history*.

We start by proving that the predicate *TypeOK* is indeed an invariant of specification *Spec*, i.e. that *Spec* $\Rightarrow \Box$*TypeOK* is valid. Moreover, we prove the following invariants.

---

[4] Alternatively, we could have used a single variable and represented the system state as a record in set *SystemState*.

```
┌──────────────────────── MODULE System_Spec ────────────────────────┐
│ EXTENDS Types, Filters                                               │
│ VARIABLES state, messages, time, history                            │
│ vars ≜ ⟨state, messages, time, history⟩                             │
│ TypeOK ≜ ∧ state ∈ [Agent → State]                                  │
│           ∧ messages ∈ SUBSET Message                               │
│           ∧ time ∈ Time                                             │
│           ∧ history ∈ Seq(SystemState) \ {⟨⟩}                       │
│ Init ≜ ∧ state = initState ∧ messages = {} ∧ time = 1               │
│         ∧ history = ⟨[st ↦ state, msgs ↦ messages, t ↦ time]⟩       │
│ Next ≜                                                               │
│    ∧ time' ∈ {t ∈ Time : t ≥ time}                                  │
│    ∧ ∀a ∈ Agent : time' ≤ instrOf(state[a]).late                    │
│    ∧ ∃a ∈ Agent :                                                   │
│        ∧ instrOf(state[a]).early ≤ time'                            │
│        ∧ LET res ≜ exec(state[a], messages)                         │
│          IN   ∧ state' = [state EXCEPT ![a] = res[1]]               │
│               ∧ messages' = res[2]                                  │
│    ∧ history' = Append(history, [st ↦ state', msgs ↦ messages', t ↦ time']) │
│ Spec ≜ Init ∧ □[Next]_vars                                          │
└─────────────────────────────────────────────────────────────────────┘
```

Fig. 3. Dynamic model of PharOS executions.

$$Inv \triangleq \wedge Last(history) = [st \mapsto state, msgs \mapsto messages, t \mapsto time]$$
$$\wedge \forall i \in 1 .. Len(history) : history[i].t \leq time$$
$$\wedge \forall a \in Agent : time \leq instrOf(state[a]).late$$

$$MsgInv \triangleq \text{LET } instr(a) \triangleq \{instrOf(history[i].st[a]) : i \in \text{DOMAIN } history\}$$
$$sends(a) \triangleq \{di \in instr(a) : di.ins.type = \text{"send"}\}$$
$$toSend \triangleq \text{UNION } \{sends(a) : a \in Agent\}$$
$$expired \triangleq \{m \in toSend : m.late < time\}$$
$$\text{IN} \quad \wedge messages \subseteq toSend$$
$$\wedge expired \subseteq messages$$

Predicate *Inv* asserts that the last element of the *history* sequence records the current system state, that the time recorded at any history entry cannot exceed the current time, and that no agent attempts to execute an instruction whose deadline has passed. Predicate *MsgInv* states bounds on the contents of the message pool. First, any message that was sent corresponds to some send instruction of some agent in the execution history. Second, send instructions whose deadline has expired must indeed have been executed, and hence be in the message pool. The proofs of these invariants are straightforward.

4 Stating and Proving Determinacy

Our main result about the formal model of PharOS is that its executions are deterministic in the sense that the sequence of local states of every agent is the same in any execution, independently of the order of scheduling.

4.1 Witness Executions

Since we cannot directly refer to different executions in a linear-time formalism such as TLA^+, we will state and prove determinacy of executions in PharOS by relating any execution as specified by formula *Spec* to a fixed, statically chosen "witness" execution. The witness execution is represented as a stream of system states. We will choose the witness such that every agent executes infinitely often. Formally, we introduce the predicate

$$IsWitness(H) \triangleq$$
$$\wedge\ H \in Stream(GoodSystemState)$$
$$\wedge\ H[1] = [\,st \mapsto initState, msgs \mapsto \{\}, t \mapsto 1\,]$$
$$\wedge\ \forall n \in Natp : \exists a \in Agent : executes(a, H[n], H[n+1])$$
$$\wedge\ \forall a \in Agent : \forall n \in Natp :$$
$$\exists m \in Natp : m \geq n \wedge executes(a, H[m], H[m+1])$$

In words, a witness is a stream of good system states that starts in the initial system state, where all transitions correspond to the execution of some agent, and where every agent executes infinitely often. Observe that the witness execution is represented as a TLA^+ value, independent of the actual system execution, and that the predicate *IsWitness* does not refer to any state variable. In particular, system states that appear in the witness are explicitly numbered.

In order to ensure that our subsequent results are not vacuous, we prove the existence of a witness execution.

THEOREM *witnessExistence* $\triangleq$ $\exists w \in Stream(SystemState) : IsWitness(w)$

The proof relies on the inductive definition of an execution according to a specific scheduling strategy, for which we then prove that the resulting execution is a witness. The time progress assumption stated in Fig. 2 ensures that every agent must be scheduled infinitely often. We also prove some properties of witness executions similar to the invariants stated in Sect. 3.

4.2 Determinacy of Executions

In order to express determinacy, we define a predicate $Det(w)$ that relates the current execution and the witness w. More precisely, the predicate holds if for every agent a, the sequence of local states recorded in the history of the current execution agrees with the local states predicted by w.

$$Det(w) \triangleq \exists U \in Natp : \forall a \in Agent :$$
$$IsPrefix(filter(history, a), filter(take(w, U), a))$$

Since w is a stream of system states, the formal definition requires the existence of a sufficiently long initial subsequence of w such that the sequence of local states in the history is a prefix of the local states in that prefix of w. The determinacy theorem states that given any witness execution w, the predicate $Det(w)$ holds throughout the execution of the actual system.

THEOREM $Determinacy \triangleq$
  ASSUME NEW $w \in Seq(SystemState)$, $IsWitness(w)$
  PROVE $Spec \Rightarrow \Box Det(w)$

The theorem is proved by induction, relying on the previously proved invariants of the specification and the witness. From the definition of the witness predicate and the initial condition $Init$, the two executions start in the same system state, hence the predicate Det holds trivially, choosing $U = 1$. Inductively, assume that the predicate holds for some prefix of w up to U, and that the system takes a non-stuttering step, with agent a executing its current instruction. Since $filter(history, a)$ is a prefix of $filter(take(w, U), a)$ and a takes infinitely many steps in w, there is some N such that

$$filter(history, a) = filter(take(w, N), a) \tag{1}$$

and a performs a step in w from system state $w[N]$ to $w[N+1]$. Defining $nU \triangleq Max(U, N + 1)$, it suffices to show that

$$IsPrefix(filter(history', b), filter(take(w, nU), b))$$

holds for all agents b. This is easy to see for $b \neq a$, since $nU \geq U$ and the local state of b does not change in the transition of the system. Now assume $b = a$. Because $N + 1 \leq nU$, it is enough to show that

$$filter(history', a) = filter(take(w, N + 1), a).$$

From (1) it follows that $state[a] = w[N].st[a]$, i.e., the local states of a in the current execution and in the N-th configuration of the witness execution are the same. In particular, the same instruction is performed in both executions. For local updates and send instructions, this is enough for proving that $state'[a] = w[N + 1].st[a]$, since the same update is applied to the same state, and this implies the conclusion. For receive instructions, we must moreover show

$$msgReceived(w[N].msgs, instrOf(w[N].st[a]))$$
$$\Leftrightarrow msgReceived(messages, instrOf(state[a]))$$

in order to ensure that the updates are the same. For "$\Rightarrow$", if the message is received in the witness execution it must be in the message pool $w[N].msgs$, and it is visible. In particular, it was sent by some agent c, and the execution window of the send instruction strictly precedes the execution window of the current (receive) instruction. Therefore, the state at which c performed the send instruction is contained in the prefix $1 .. U$ of w and therefore also in the history of the system execution. Since no deadline is missed, the message must have

Table 1. Sizes of the different modules.

Module	Definitions (lines)	Proofs (lines)	# Theorems
Streams	2	50	5
Filters	9	624	27
System model	78	226	10
Witness	22	832	11
Main proof	6	241	1
Total	117	1973	54

been sent in the system execution (recall invariant $MsgInv$), and will therefore be received.

Conversely, if the message is received in the system execution, it must have been sent by some agent c, and the deadline of the send instruction strictly precedes the execution window of the receive instruction. By invariant $MsgInv$ and theorem $filter\_Range$, the message appears in $sent(Range(filter(history, c)))$, hence by induction hypothesis also in $sent(Range(filter(take(w, U), c)))$. Because the deadline for sending the message has expired when the witness executes the receive instruction in the step from $w[N]$ to $w[N + 1]$, and using the analogue of $MsgInv$ for the witness execution, the message must be contained in $witness[N].msgs$, and is therefore received.

4.3   Evaluation

Table 1 summarizes the sizes of the definitions and proofs (measured as the numbers of lines of TLA$^+$ code) that make up the different modules that we developed for the proof of determinacy, as well as the number of theorems proved in each module. The overall development required almost 2000 lines of proof, of which the main effort was devoted to proving auxiliary results about filters and the witness execution. The system specification (static and dynamic model) contains the bulk of the definitions; some basic invariants are also proved there.

The case study presented here also fed back into parts of the standard library of TLAPS. In particular, many lemmas about finite sequences were informed by the development of the proof of determinacy; these lemmas are being reused in other developments and do not appear in Table 1.

Compared to the paper-and-pencil proof of [9], although the main arguments are the same, the styles of proof differ in several respects.

- The original proof compares two possible system executions; determinism then means that for every agent a, the projection of one execution to the states of a must be a prefix of the other. This gives rise to two symmetric situations and would likely have led to duplications of proofs, which we avoided by choosing a fixed, infinite witness execution, of which the actual execution is always a prefix. We also prove the existence of a witness execution by exhibiting an actual scheduling strategy.

- The original proof is behavioral and refers to past and future instants of an execution, whereas the TLA$^+$ proof is assertional, relying on a central inductive invariant whose statement is distributed among the lemmas of the paper proof. This style is imposed by the formalism, but also follows established good practice on formalizing proofs of safety properties.
- The need for writing formal definitions in TLA$^+$ led us to introduce certain abstractions that were not explicit in the paper proof. For example, the notion of "good" execution times or system states proved to be very helpful in the machine-checked proof but are implicit in [9], where the assumption that deadlines are not missed is inferred from the assumption that agents execute infinitely often. Observe that our model does not impose a liveness condition on the specification of the system execution, and only requires non-Zenoness for the witness execution.
- Many of our auxiliary lemmas on filters and witness do not have a counterpart in the original proof, and in fact they may appear obvious to a mathematician. However, several mistakes that we made in the initial formal statements of these lemmas convinced us of the added value of a fully formal proof.

Our TLA$^+$ specification is written at the same level of abstraction as the model of computation of PharOS that was considered in [9]. In future work, it would be interesting to extend the specification and proof in two directions. First, we assume that no deadlines is ever missed. In typical instances of real-time systems, schedulability analysis ensures that this assumption is met. However, as discussed in [9], the result can be generalized to executions in which deadlines may actually be missed, or abrupt termination occurs for another reason. Second, it would be challenging to show that an actual implementation, as in the Asterios$^®$ system, refines our high-level model, and thus formally establish determinacy for an implementation. While TLA$^+$ includes a natural notion of refinement as trace inclusion, such a project would constitute a significant effort given the complexity of an actual implementation.

5 Conclusion

We have formally proved determinacy of the model of execution underlying PharOS, a real-time system that is now being commercialized as Asterios$^®$. Based on an existing paper-and-pencil proof, our machine-checked proof reinforces the confidence in the result and clarifies some of the underlying assumptions. Moreover, our proof represents a sizable case study for the TLA$^+$ Proof System, which is still actively being developed. The overall proof effort appears to be reasonable. The final TLA$^+$ model and proof was obtained in several iterations, mainly focusing on the introduction of auxiliary abstractions. The hierarchical style of TLA$^+$ proofs helped us focus on the main argument and let us fill in proofs of auxiliary lemmas only when we knew that they were actually needed. Access to the TLC model checker was helpful for validating intermediate definitions and lemmas.

Several proposals exist in the literature for making computations of real-time systems deterministic, and we refer to [9] for a detailed discussion. The time-triggered architecture (TTA) [5] is probably the most widely known one. In comparison, PharOS imposes fewer static constraints on when tasks are scheduled, and it is designed to run on off-the-shelf hardware without a specific, deterministic communication substrate. TTA has also been the object of formal verification using the PVS proof assistant [14,15]. In contrast to our work, which focuses on a high-level property of the execution model, these proofs focus on algorithms that underly the implementations of mechanisms such as clock synchronization or group membership. Relating our high-level model of execution with actual implementations of PharOS, down to actual code written in PsyC (a real-time extension of the C language) is left as a challenge for future work.

Acknowledgements. Jael Kriener contributed to this work by writing initial specifications and proofs of PharOS executions.

References

1. Alur, R., Dill, D.: A theory of timed automata. Theoret. Comput. Sci. **126**, 183–235 (1994)
2. Aussaguès, C., David, V.: A method and a technique to model and ensure timeliness in safety critical real-time systems. In: 4th International Conference Engineering of Complex Computer Systems (ICECCS 1998), Monterey, CA, U.S.A., pp. 2–12. IEEE Computer Society (1998)
3. Azmy, N., Merz, S., Weidenbach, C.: A rigorous correctness proof for pastry. In: Butler, M., Schewe, K.-D., Mashkoor, A., Biro, M. (eds.) Abstract State Machines, Alloy, B, TLA, VDM, and Z (ABZ). LNCS, vol. 9675, pp. 86–101. Springer, Heidelberg (2016)
4. Cousineau, D., Doligez, D., Lamport, L., Merz, S., Ricketts, D., Vanzetto, H.: TLA$^+$ proofs. In: Giannakopoulou, D., Méry, D. (eds.) FM 2012. LNCS, vol. 7436, pp. 147–154. Springer, Heidelberg (2012)
5. Kopetz, H., Bauer, G.: The time-triggered architecture. Proc. IEEE **91**(1), 112–126 (2003)
6. Lamport, L.: Specifying Systems. Addison-Wesley, Boston (2002)
7. Lamport, L.: Byzantizing paxos by refinement. In: Peleg, D. (ed.) Distributed Computing. LNCS, vol. 6950, pp. 211–224. Springer, Heidelberg (2011)
8. Lemerre, M., David, V., Aussagus, C., Vidal-Naquet, G.: Equivalence between schedule representations: theory and applications. In: Real-Time and Embedded Technology and Applications Symposium, RTAS 2008, pp. 237–247. IEEE, April 2008
9. Lemerre, M., Ohayon, E.: A model of parallel deterministic real-time computation. In : Proceedings of 33rd IEEE Real-Time Systems Symposium (RTSS 2012), San Juan, PR, U.S.A., pp. 273–282. IEEE Computer Society (2012)
10. Lemerre, M., Ohayon, E., Chabrol, D., Jan, M., Jacques, M.-B.: Method and tools for mixed-criticality real-time applications within PharOS. In: 14th IEEE International Symposium Object/Component/Service-Oriented Real-Time Distributed Computing Workshops, Newport Beach, CA, U.S.A., pp. 41–48. IEEE Computer Society (2011)

11. Leuschel, M., Butler, M.J.: ProB: an automated analysis toolset for the B method. Softw. Tools Technol. Transfer **10**(2), 185–203 (2008)
12. Louise, S., Lemerre, M., Aussaguès, C., David, V.: The OASIS kernel: a framework for high dependability real-time systems. In: 13th IEEE International Symposium High-Assurance Systems Engineering (HASE 2011), Boca Raton, FL, U.S.A., pp. 95–103. IEEE Computer Society (2011)
13. Newcombe, C., Rath, T., Zhang, F., Munteanu, B., Brooker, M., Deardeuff, M.: How amazon web services uses formal methods. CACM **58**(4), 66–73 (2015)
14. Pfeifer, H., von Henke, F.W.: Modular formal analysis of the central guardian in the time-triggered architecture. Reliab. Eng. Syst. Saf. **92**(11), 1538–1550 (2007)
15. Rushby, J.: An overview of formal verification for the time-triggered architecture. In: Damm, W., Olderog, E.-R. (eds.) Formal Techniques in Real-Time and Fault-Tolerant Systems. LNCS, vol. 2469, pp. 83–105. Springer, Heidelberg (2002)
16. Yu, Y., Manolios, P., Lamport, L.: Model checking TLA+ specifications. In: Pierre, L., Kropf, T. (eds.) CHARME 1999. LNCS, vol. 1703, pp. 54–66. Springer, Heidelberg (1999)

A Rigorous Correctness Proof for Pastry

Noran Azmy[1,2,3]($\boxtimes$), Stephan Merz[2,3], and Christoph Weidenbach[1]

[1] Max Planck Institute for Informatics, Saarbrücken, Germany
azmy@mpi-inf.mpg.de
[2] Inria, Villers-lès-Nancy, France
[3] CNRS, Université de Lorraine, LORIA, UMR 7503, Vandoeuvre-lès-Nancy, France

Abstract. Peer-to-peer protocols for maintaining distributed hash tables, such as Pastry or Chord, have become popular for a class of Internet applications. While such protocols promise certain properties concerning correctness and performance, verification attempts using formal methods invariably discover border cases that violate some of those guarantees. Tianxiang Lu reported correctness problems in published versions of Pastry and also developed a model, which he called LuPastry, for which he provided a partial proof of correct delivery assuming no node departures, mechanized in the TLA⁺ Proof System. Lu's proof is based on certain assumptions that were left unproven. We found counter-examples to several of these assumptions. In this paper, we present a revised model and rigorous proof of correct delivery, which we call LuPastry⁺. Aside from being the first complete proof, LuPastry⁺ also improves upon Lu's work by reformulating parts of the specification in such a way that the reasoning complexity is confined to a small part of the proof.

1 Introduction

In a peer-to-peer network, individual nodes – called *peers* – communicate directly with each other and act as both suppliers and users of a given service. One such service that can be implemented by a peer-to-peer protocol is a *distributed hash table* (DHT): a decentralized distributed system that provides a lookup service similar to a hash table, but where the responsibility for storing key-value pairs is divided among the different nodes on the network. Nodes can efficiently retrieve the value associated with a given key by sending a lookup message for that key. Correct routing guarantees that lookup messages arrive at the node responsible for storing the key-value pair. Pastry [7] and Chord [8] are well-known examples of protocols implementing DHTs.

Using Alloy to formally model and verify Chord, Zave [10] showed that the join protocol is correct provided that no node leaves the network, but that the full version of the protocol may not maintain the claimed invariants. In [11], she presented a version of Chord with a partially-automated proof of correctness. Lu discovered similar correctness problems for Pastry using the TLA⁺ proof assistant [6]. Some other work has been done in this area that does not rely on mechanized verification, *i.e.*, model checking or theorem proving. Borgström et al. [3]

© Springer International Publishing Switzerland 2016
M. Butler et al. (Eds.): ABZ 2016, LNCS 9675, pp. 86–101, 2016.
DOI: 10.1007/978-3-319-33600-8_5

used CCS to verify correctness of the DKS look-up protocol, assuming that the network remains stable (i.e., nodes neither join nor leave). Bakhshi et al. [2] used process algebra to formally verify the stabilization process of Chord.

Like Lu, we are interested in the formal verification of Pastry using the TLA$^+$ proof system, w.r.t. the safety property *correct delivery* [6]: *At any point in time, there is at most one node that answers a lookup request for a key, and this node must be the closest live node to that key.*

TLA$^+$ [5] is a formal specification language that mainly targets concurrent and distributed systems. It is based on untyped Zermelo-Fraenkel set theory for specifying data structures, and the Temporal Logic of Actions, a variant of linear temporal logic, for describing system behavior. Systems are specified as state machines over a tuple of state variables by defining a state predicate *Init* and a transition predicate *Next* that constrain the possible initial states and the next-state relation. Transition predicates (also called *actions*) are first-order formulas that contain unprimed and primed state variables for denoting the values of the variables in the state before and after the transition. Validation of TLA$^+$ specifications is mechanized by TLC [9], an explicit-state model checker for finite instances of TLA$^+$ specifications, and formal verification by TLAPS, the TLA$^+$ Proof System [4]. TLAPS is based on a hierarchical proof language; the user writes a TLA$^+$ proof in the form of a hierarchy of *proof steps*, each of which is interpreted by the *proof manager*, which generates corresponding proof obligations and passes them to automatic back-end provers, including Zenon, Isabelle/TLA$^+$, and SMT solvers. Larger steps that cannot be proven directly by any of the back-end provers can be broken further into sub-steps. Because the language is untyped, part of the proof effort consists in proving a typing invariant that expresses the shapes of functions and operators.

Using model checking and theorem proving, Lu discovered several problems in the original Pastry protocol and presented a variant of the protocol, called LuPastry, for which he verified correct delivery under the strong assumption that nodes never fail (*i.e.*, leave the network). Notably, his Pastry variant enforces that a live node may only facilitate the joining of one new node at a time. Lu's proof reduces correct delivery to a set of around 50 claimed invariants, which are proven with the help of TLAPS. As such, LuPastry represents a major effort in the area of computer-aided formal verification of distributed algorithms. Still, the proof relies on many unproven assumptions relating to arithmetic and to protocol-specific data structures. Upon examining these assumptions, we discovered counter-examples to several of them. While we were able to prove stronger variants of many assumptions, this was not possible for others. In fact, we were able to find a counter-example to one of Lu's claimed main invariants, for which the TLA$^+$ proof was only possible because of incorrect assumptions. Fixing these problems led to a redesign of the overall proof.

Our contribution in this paper is LuPastry$^+$: a revised specification and complete proof of correct delivery for LuPastry. While the essentials of the actual protocol are not changed, we improve on the LuPastry specification by fixing some unhandled border cases and introducing abstractions in some operator definitions that make the specification more modular and confine the reasoning complexity to a small part of the proof. The new abstractions typically lead to a significant reduction of the size of higher-level proofs. Moreover, our proof does not rely on unproven assumptions, because all low-level lemmas have been proved using TLAPS.

The paper is organized as follows. Section 2 describes the main aspects of the original LuPastry model, which are also in LuPastry$^+$. We explain our contribution and the structure of the LuPastry$^+$ proof in Sect. 3. A sketch of the machine-checked proof is given in Sect. 4. Section 5 summarizes our results and our experience in using TLAPS.

2   The (Lu)Pastry Model

The Pastry network can be visualized as a ring of keys $I \triangleq 0 \ldots 2^M - 1$ for some positive integer M (see Fig. 1). Each live node is assigned a unique key $k \in I$ as an identifier and needs to determine its *coverage*: a contiguous range of keys, including the node's own ID, that the node is responsible for, or *covers*. If a node i covers key k, then i considers itself (1) the proper recipient of all look-up messages addressed to k, and (2) the node responsible for facilitating the joining of any new node with ID k. In the absence of a central server and shared memory, live nodes need to rely on message passing and local information to agree on a proper division of coverage.

Let *ready* nodes be the live nodes that are not in the process of joining the network (they are fully-joined). Ready nodes are of particular interest since only ready nodes may accept look-up messages or facilitate the joining of new nodes. Ideally, the coverage ranges computed by all ready nodes (1) do not overlap, (2) cover the whole range of keys, and (3) are computed based on the smallest absolute distance to the node: if a ready node i covers key k, then k is closer to i in terms of absolute ring distance than it is to any other ready node $j \neq i$, with a rule for breaking ties. These conditions all hold for the ring illustrated in Fig. 1. Condition (2) may be temporarily violated when a new node joins but is not yet ready, thus the safety property that we are interested in verifying requires only (1) and (3).

For two nodes x and y, we may be interested in the *clockwise* distance from x to y, or the *absolute* (shortest) distance between x and y.

$$ClockwiseDistance(x, y) \triangleq$$
$$\quad \text{IF } y \geq x \text{ THEN } y - x \text{ ELSE } RingSize - x + y$$
$$AbsoluteDistance(x, y) \triangleq$$
$$\quad \text{LET } d1 \triangleq ClockwiseDistance(x, y)$$
$$\qquad\quad d2 \triangleq ClockwiseDistance(y, x)$$
$$\quad \text{IN } \text{ IF } d1 \leq d2 \text{ THEN } d1 \text{ ELSE } d2$$

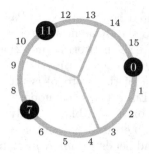

Fig. 1. A Pastry ring of size 16 with three live nodes 0, 7 and 11. The nodes should divide key coverage among them, as indicated by the separators.

A node i computes its coverage by maintaining a *leaf set*: a set containing what i believes to be its L live neighbor nodes on both the left and right sides, where the positive integer L is a parameter of the specification.[1]

$LeafSet \triangleq \{ ls \in [node : I, left : \text{SUBSET } I, right : \text{SUBSET } I] :$
$\wedge\ ls.node \notin ls.left \qquad \wedge\ Cardinality(ls.left) \leq L$
$\wedge\ ls.node \notin ls.right \qquad \wedge\ Cardinality(ls.right) \leq L \}$

The *neighbors* of i are the closest nodes to i in its leaf set ls.

$RightNeighbor(ls) \triangleq$
IF $ls.right = \{\}$ THEN $ls.node$
ELSE CHOOSE $n \in ls.right : \forall\, p \in ls.right :$
$ClockwiseDistance(ls.node, p) \geq ClockwiseDistance(ls.node, n)$

LeftNeighbor is defined analogously. CHOOSE denotes Hilbert's choice operator and will be discussed further in Sect. 3. Node i considers its coverage range as the interval $[LeftCoverage(ls), RightCoverage(ls)]$, where the key $LeftCoverage(ls)$ is the midpoint between $LeftNeighbor(ls)$ and i, and similarly, $RightCoverage(ls)$ is the midpoint between i and $RightNeighbor(ls)$.

$LeftCoverage(ls) \triangleq$
IF $LeftNeighbor(ls) = ls.node$ THEN $ls.node$
ELSE $(LeftNeighbor(ls) +$
$(ClockwiseDistance(LeftNeighbor(ls), ls.node) \div 2 + 1))\% RingSize$
$RightCoverage(ls) \triangleq$
IF $RightNeighbor(ls) = ls.node$
THEN $(RingSize + ls.node - 1)\% RingSize$
ELSE $(ls.node +$
$ClockwiseDistance(ls.node, RightNeighbor(ls)) \div 2)\% RingSize$

In Fig. 1, assuming up-to-date leaf sets, node 0's left and right neighbors are 11 and 7, respectively. Therefore, its coverage is the interval $[14, 3]$ (*i.e.*, the set of keys $\{14, 15, 0, 1, 2, 3\}$).

---

[1] Nodes also maintain routing tables for the purpose of efficient message routing, but these are irrelevant to our discussion.

The Pastry model presented here is called LuPastry [6], in which two main restrictions are introduced to the dynamic behavior of the protocol: (1) nodes are assumed to never fail (leave the network), and (2) a ready node may facilitate the joining of at most one new node at a time.

The main dynamic aspect of the protocol is the join process, explained as follows (see also Fig. 2). In LuPastry, each node is either *Dead* (not shown), *Waiting* (white), *OK* (gray) or *Ready* (black). Only Ready nodes facilitate the joining of new nodes into the network. A Dead node i that decides to join the network turns to Waiting and sends a *join request* to a Ready node j that it

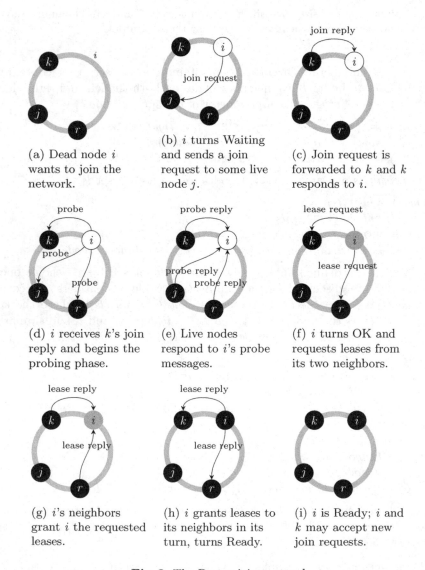

(a) Dead node i wants to join the network.

(b) i turns Waiting and sends a join request to some live node j.

(c) Join request is forwarded to k and k responds to i.

(d) i receives k's join reply and begins the probing phase.

(e) Live nodes respond to i's probe messages.

(f) i turns OK and requests leases from its two neighbors.

(g) i's neighbors grant i the requested leases.

(h) i grants leases to its neighbors in its turn, turns Ready.

(i) i is Ready; i and k may accept new join requests.

Fig. 2. The Pastry join protocol

knows about. The request is forwarded to the Ready node k that covers key i. Node k responds to i's request when it is free for handling a new join request, and communicates to i its own leaf set. Node i receives k's reply and, in order to construct its proper leaf set, sends *probe* messages to the nodes in the leaf set received from k. All non-Dead nodes that receive the probe add i to their leaf set if appropriate (*i.e.*, if node i is among the L closest live neighbor nodes), and send a *probe reply* to i with their own leaf set information. This process continues until i has probed all nodes it has heard about and that are close enough to i to be in i's leaf set, then i becomes OK. In order for i to become Ready and eventually serve the IDs closest to i, node i has to exchange *leases* with both its left and right neighbors (one of which must be k). Node i sends out *lease request* messages to both its neighbors. If i's neighbor is Ready or OK, and also considers i to be its neighbor, it grants i the lease in a *lease reply* message. When i has received lease replies from both its neighbors, it switches to Ready, and grants its neighbors leases in turn. When k receives i's lease, it may help other new nodes join the ring.

The global state of the LuPastry network is represented as the tuple *vars* of state variables.[2]

$$vars \triangleq \langle Messages, Status, LeafSets, Probing, Leases, Grants, ToJoin \rangle$$

Messages represents the set of messages currently in transmission. Variables *Status* and *LeafSets* are arrays whose i-th entries are the current status and leaf set of node i. Similarly, $Probing[i]$ is the set of nodes that node i has probed but has not heard back from yet, $Leases[i]$ and $Grants[i]$ are the set of nodes i has acquired leases from, and granted leases to, respectively. Lastly, $ToJoin[i]$ designates the node that is currently joining through i, if any, otherwise $ToJoin[i] = i$.

The TLA$^+$ specifications of the initial state and of the next-state relation are defined by the operators *Init* and *Next* shown in Fig. 3; they use auxiliary operators that represent elementary operations on leaf sets and routing tables, and that define the individual transitions of the Pastry protocol. The constant A is a parameter of the specification designating the nodes that are live (and Ready) initially. Note that, because LuPastry is a distributed system, each action may modify the local variables of at most one node i, the node executing it, besides the set *Messages*. The TLA$^+$ specification of LuPastry is defined as the formula $Spec \triangleq Init \wedge \square[Next]_{vars}$.

$EmptyLS(i)$ is a leaf set owned by i with no nodes in the right and left sides. $AddToLS(a, ls)$ is the leaf set obtained by adding the set of nodes a to ls. In case of an overflow, the new right (resp., left) leaf set consists of the L closest nodes to i from the right (resp., left); nodes that are farther away are discarded. Finally, $LeafSetContent(ls) \triangleq \{ls.node\} \cup ls.right \cup ls.left$ is the set of nodes in leaf set ls.

---

[2] For compactness, we omit parts of the specification irrelevant to the discussion. TLA$^+$ functions and operators have been given new names for better readability.

$Init \triangleq$
$\quad \wedge\ Messages = \{\}$
$\quad \wedge\ Status = [i \in I \mapsto$ IF $i \in A$ THEN "Ready" ELSE "Dead"$]$
$\quad \wedge\ ToJoin = [i \in I \mapsto i]\ \wedge\ Probing = [i \in I \mapsto \{\}]$
$\quad \wedge\ Leases = [i \in I \mapsto$ IF $i \in A$ THEN A ELSE $\{i\}]$
$\quad \wedge\ Grants = [i \in I \mapsto$ IF $i \in A$ THEN A ELSE $\{i\}]$
$\quad \wedge\ LeafSets = [i \in I \mapsto$ IF $i \in A$ THEN $AddToLS(A, EmptyLS(i))$
$\qquad\qquad\qquad\qquad$ ELSE $EmptyLS(i)]$

$Next \triangleq \exists i, j \in I :$
$\quad \vee\ Lookup(i,j) \qquad\qquad \vee\ RouteLookup(i,j)$
$\quad \vee\ DeliverLookup(i,j) \qquad \vee\ Join(i,j)$
$\quad \vee\ RouteJoinRequest(i,j) \quad \vee\ ReceiveJoinRequest(i)$
$\quad \vee\ ReceiveJoinReply(i) \qquad \vee\ ReceiveProbe(i)$
$\quad \vee\ ReceiveProbeReply(i) \qquad \vee\ RequestLease(i)$
$\quad \vee\ ReceiveLeaseRequest(i) \quad \vee\ ReceiveLeaseReply(i)$

Fig. 3. Initial condition and next-state relation specified in TLA$^+$.

3   The LuPastry$^+$ Model and Proof

Our contribution is illustrated in Fig. 4, and can be divided into two parts: (1) changes to the TLA$^+$ specification of LuPastry, and (2) the first complete proof of correctness.

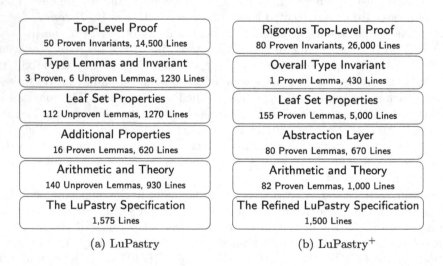

Top-Level Proof	Rigorous Top-Level Proof
50 Proven Invariants, 14,500 Lines	80 Proven Invariants, 26,000 Lines
Type Lemmas and Invariant	Overall Type Invariant
3 Proven, 6 Unproven Lemmas, 1230 Lines	1 Proven Lemma, 430 Lines
Leaf Set Properties	Leaf Set Properties
112 Unproven Lemmas, 1270 Lines	155 Proven Lemmas, 5,000 Lines
Additional Properties	Abstraction Layer
16 Proven Lemmas, 620 Lines	80 Proven Lemmas, 670 Lines
Arithmetic and Theory	Arithmetic and Theory
140 Unproven Lemmas, 930 Lines	82 Proven Lemmas, 1,000 Lines
The LuPastry Specification	The Refined LuPastry Specification
1,575 Lines	1,500 Lines
(a) LuPastry	(b) LuPastry$^+$

Fig. 4. Structure of the original LuPastry proof versus the rigorous version.

3.1   Changes to the LuPastry Specification

At the bottom layer, we refine LuPastry [6] as follows. In order for the proof to gain in modularity, readability, and simplicity, we introduce additional operators that abstract away from arithmetic calculations and reduce the use of TLA$^+$'s CHOOSE operator, which is difficult for back-end provers to reason about and hence restrains automation. This also makes the specification more concise.

Arithmetic calculations, which mainly involve comparisons between distances between nodes on the ring, appeared so extensively in LuPastry that arithmetic reasoning was frequently needed at the top level of the original proof. For example, typical subformulas $ClockwiseDistance(i, j) \leq ClockwiseDistance(i, k)$ require that the definition of $ClockwiseDistance$ be unfolded. Instead, in LuPastry$^+$ we define a predicate $ClockwiseArc(i, j, k)$, which holds if j lies on the clockwise path from i to k.

$$ClockwiseArc(i, j, k) \triangleq$$
$$ClockwiseDistance(i, j) \leq ClockwiseDistance(i, k)$$

We then prove once and for all the necessary properties of this relation in TLAPS, using the SMT backend for arithmetical reasoning, so that unfolding of the definition is no longer needed in the top-level proof. The following are some examples of these properties.

THEOREM $ArcAntiSymmetry \triangleq \forall x, y, z \in I :$
$\wedge\ ClockwiseArc(x, y, z) \wedge ClockwiseArc(x, z, y) \Rightarrow y = z$
$\wedge\ ClockwiseArc(x, y, z) \wedge ClockwiseArc(y, x, z) \Rightarrow x = y$

THEOREM $ArcRotation \triangleq \forall x, y, z \in I :$
$\wedge\ x \neq y \wedge ClockwiseArc(x, y, z) \Rightarrow ClockwiseArc(y, z, x)$
$\wedge\ y \neq z \wedge ClockwiseArc(x, y, z) \Rightarrow ClockwiseArc(z, x, y)$

This abstraction helps automate the proof process since now automatic back-ends like Zenon or Spass without native support for integer arithmetic are able to solve larger steps.

A related issue is the extensive use of Hilbert's ε-operator for definite choice. The TLA$^+$ expression CHOOSE $x \in S : P(x)$ denotes some fixed but arbitrary element x in set S for which the property P holds, if some such x exists. If P holds for no $x \in S$, as in CHOOSE $x \in Nat : x * 0 = 1$, the result of the CHOOSE expression is not specified.

The LuPastry definition of the operator $RightNeighbor$, shown in the previous section, uses CHOOSE. It is unwieldy to reason about operator $RightNeighbor$ by unfolding its definition because we would invariably have to show the existence of a node contained in $ls.right$ and whose distance to $ls.node$ is minimal among all these nodes. Formally, we need the following lemma:

LEMMA $\forall x \in I, S \in (\text{SUBSET } I) \setminus \{\{\}\} : \exists y \in S : \forall z \in S :$
$ClockwiseDistance(x, y) \leq ClockwiseDistance(x, z)$

In LuPastry$^+$, we perform three changes. First, we redefine $RightNeighbor$ to abstract away from the CHOOSE expression. The new term $ClosestFromTheRight$

is itself defined in terms of CHOOSE, but generalizes the previous operator and reuses the operator *ClockwiseArc* introduced above.

$RightNeighbor(ls) \triangleq ClosestFromTheRight(ls.node, ls.right)$

$ClosestFromTheRight(x, a) \triangleq$
    IF $a = \{\}$ THEN x
    ELSE CHOOSE $y \in a : \forall z \in a : ClockwiseArc(x, y, z)$

Second, since *ClosestFromTheRight* is defined using a CHOOSE expression, we add a lemma that guarantees the existence of a value satisfying the characteristic predicate.

LEMMA $choose\_ClosestFromTheRight \triangleq$
    $\forall x \in I, a \in$ SUBSET $I :$
       $a \neq \{\} \Rightarrow \exists y \in a : \forall z \in a : ClockwiseArc(x, y, z)$

Third, we add *type* and *expansion* lemmas that respectively provide type information and the relevant properties of *ClosestFromTheRight*. We introduce similar lemmas for operator *RightNeighbor*.

LEMMA $type\_ClosestFromTheRight \triangleq$
    $\forall x \in I, a \in$ SUBSET $I : ClosestFromTheRight(x, a) \in I$

LEMMA $def\_ClosestFromTheRight \triangleq$
    $\forall x \in I, a \in$ SUBSET $I :$
       $\wedge\ a = \{\} \Rightarrow ClosestFromTheRight(x, a) = x$
       $\wedge\ a \neq \{\} \Rightarrow ClosestFromTheRight(x, a) \in a$
       $\wedge\ \forall y \in a : ClockwiseArc(x, ClosestFromTheRight(x, a), y)$

We illustrate the effect of these abstractions using a simple lemma about adding new nodes to the leaf set data structure, that we prove once with and once without the use of the new operators.

LEMMA $\forall ls \in LeafSet, a \in$ SUBSET $I : IsProper(AddToLS(a, ls))$

Basically, the lemma says that the leaf set obtained by adding some new nodes to a leaf set is "proper", where "proper" is defined as follows.

$IsProper(ls) \triangleq$
    $\wedge\ \forall x \in ls.left \setminus ls.right,\ y \in ls.right : ClockwiseArc(y, x, ls.node)$
    $\wedge\ \forall x \in ls.right \setminus ls.left,\ y \in ls.left : ClockwiseArc(ls.node, x, y)$

The proof P_1 of this lemma according to the original definition of *IsProper* consists of 23 interactive proof steps that generate 64 proof obligations. With our new abstractions, the new proof P_2 consists of only 12 interactive proof steps (40 proof obligations). This significant difference comes from the fact that the new operators allow back-end provers to succeed directly on some steps in P_2, which have to be broken down into further substeps in the original proof P_1. Already for this simple example we observe a 50 % reduction in the number of steps, *i.e.*, user interactions.[3]

---

[3] Both examples can be found in our proof files in module ProofCorrectness.

Additionally, we fix some corner cases in the original specification. We modify the probing process so that the node does not probe itself. This is clearly unnecessary, and removing it simplifies some parts of the proof. We also add a missing border case to the TLA$^+$ formula $FindNext(i, j)$ that computes the next hop on the route from node i to node j.

3.2   New Proof of Correctness

The upper layers of Fig. 4 compare our new proof to Lu's original proof. The original LuPastry proof relied on a large number of unproven assumptions. In attempting to prove these assumptions, we found counter-examples to many of them, such as arithmetic assumptions ignoring border cases. Moreover, several assumptions were not actually used in the proof. For example, Lu's proof relied on 112 unproven assumptions about the leaf set data structure. Upon examining these assumptions, we could prove only 21 directly. We discovered that more than 30 were unused in Lu's proof. The rest of the assumptions were incorrect. Our analysis of Lu's assumptions led us to reformulate those that were needed for the top-level proof. This was possible for all but 6 of the incorrect assumptions. The following assumption, used by Lu as an unproven TLA$^+$ lemma, states that after adding some set of nodes a to a leaf set $ls1$, the right neighbor of the resulting leaf set $ls2$ can only be closer to the leaf set owner i than the original right neighbor of $ls1$.

LEMMA $\forall\, ls1,\ ls2 \in LeafSet,\ i \in I,\ a \in$ SUBSET I :
$i = ls1.node \land ls2 = AddToLS(a, ls1) \Rightarrow$
$ClockwiseDistance(i, RightNeighbor(ls2)) \leq$
$ClockwiseDistance(i, RightNeighbor(ls1))$

This lemma does not hold if the right-hand part of the leaf set is empty (*i.e.*, $ls1.right = \{\}$), because in this case $RightNeighbor(ls1) = i$ and i is closer to itself than to any other node. The lemma was therefore reformulated as follows.

LEMMA $\forall\, ls1,\ ls2 \in LeafSet,\ i \in I,\ a \in$ SUBSET I :
$i = ls1.node \land ls1.right \neq \{\} \land ls2 = AddToLS(a, ls1) \Rightarrow$
$ClockwiseDistance(i, RightNeighbor(ls2)) \leq$
$ClockwiseDistance(i, RightNeighbor(ls1))$

Other assumptions had to be eliminated entirely. For example,

LEMMA $\forall\, ls \in LeafSet,\ k \in I$:
$LeafSetContent(AddToLS(\{k\}, ls)) \setminus \{k\} = LeafSetContent(ls)$

states that the leaf set obtained by adding a node k to some leaf set ls, contains the same nodes in ls, and possibly also k. This is not true: if ls is full, adding a new node k to it will generally result in some other node being removed from the leaf set, invalidating the claimed equality.[4]

---

[4] See *AddToLSetInvCo* and *AddAndDelete* in LuPastry module *ProofLSetProp*.

Aside from reformulating and proving some assumptions from the original proof, we also added and proved many new facts that were helpful for the proof, resulting in more lemmas in the "Leaf Set Properties" layer.

In the top level of the proof, Lu proves correct delivery by reducing the property to 50 other properties and proves that these properties are invariants of LuPastry, based on the (partly wrong) assumptions made in the lower levels of the proof. Since some assumptions were discovered to fail, the following property *SemJoinLeafSet* is, in fact, not an invariant.

$$SemJoinLeafSet \ \stackrel{\Delta}{=}$$
$$\forall m \in \ Messages : m.content.type = \text{``JoinReply''} \Rightarrow$$
$$\text{LET} \ n \ \stackrel{\Delta}{=} \ m.content.ls.node$$
$$\text{IN} \quad \land \ ClockwiseDistance(LeftNeighbor(LeafSets[n]), n)$$
$$\leq ClockwiseDistance(LeftNeighbor(m.content.ls), n)$$
$$\land \ ClockwiseDistance(n, RightNeighbor(LeafSets[n]))$$
$$\leq ClockwiseDistance(n, RightNeighbor(m.content.ls))$$

The predicate asserts that if some node n sends a *JoinReply* message then n's current neighbors are closer to it than its neighbors were at the time when the message was sent. This is not true, however, if n's leaf set was empty at the time the message was sent. As mentioned earlier, in case of an empty leaf set, the left and right neighbors of node n are n itself. Any new neighbors of n will be farther away from n than n itself.

We have written a new, complete correctness proof for correct delivery of the revised protocol specification that does not rely on any unproven assumptions (see Fig. 4(b)). At the lowest level of the proof we have 82 lemmas about arithmetic. The abstraction layer provides some 80 lemmas relating to our new operators. The leaf set layer consists of 155 lemmas about the leaf set data structure. On top of this basis are 80 correctness invariants. All lemmas have fully machine-checked proofs.

Because our proof is rigorous, there was a need for a larger number of invariants than in Lu's proof. Also, some of the more involved invariants were split into several invariants in order to facilitate their proof. While our new proof LuPastry$^+$ shares some invariants with the original proof of LuPastry, there is no one-to-one correspondence between the two sets of invariants. In particular, while Lu's original proof depends more on the *lease exchange* phase of the protocol, our own correctness invariants focus more on *probing*.

4   A Proof of Correctness for LuPastry$^+$

Our main TLA$^+$ theorem *LuPastryCorrectness* proves correct delivery, as expressed by the following predicate [6], as an invariant of LuPastry$^+$. The full TLA$^+$ specification and proof are available online [1].[5]

---

[5] Currently, TLAPS requires that ENABLED be unfolded manually in the machine-checked proof.

$CorrectDelivery \triangleq$
  $\forall\, i,\, k \in I : \text{ENABLED } DeliverLookup(i,\, k) \Rightarrow$
    $\land\, \forall\, n \in I : n \neq i \land Status[n] = \text{``Ready''}$
      $\Rightarrow AbsoluteDistance(i,\, k) \leq AbsoluteDistance(n,\, k)$
    $\land\, \forall\, j \in I \setminus \{i\} : \neg\text{ENABLED } DeliverLookup(j,\, k)$

THEOREM $LuPastryCorrectness \triangleq Spec \Rightarrow \Box CorrectDelivery$

Action $DeliverLookup$ is defined as follows.

$DeliverLookup(i,\, j) \triangleq$
  $\land Status[i] = \text{``Ready''} \land Covers(LeafSets[i],\, j)$
  $\land \exists\, m \in Messages :$
    $\land m.content.type = \text{``Lookup''}$
    $\land m.destination = i$
    $\land m.content.node = j$
  $\land Messages' = (Messages \setminus \{m\})$
  $\land \text{UNCHANGED } \langle Status,\, LeafSets,\, Probing,\, Leases,\, Grants,\, ToJoin \rangle$

In what follows, we use shorthand notation instead of the full names of TLA$^+$ functions/operators for compactness. $RN(i) = RightNeighbor(LeafSets[i])$, and $CR(i) = ClosestFromTheRight(i,\, ReadyNodes \setminus \{i\})$ is the closest Ready/OK node to node i from the right. $LN(i)$ and $CL(i)$ are defined analogously. We use $i_1 \rightharpoonup \ldots \rightharpoonup i_n$ to denote a clockwise path on the ring; this is similar to the TLA$^+$ operator $ClockwiseArc$, but extended to an arbitrary number of nodes. The shortest "absolute" path between two nodes i and j may be $i \rightharpoonup j$ or $j \rightharpoonup i$; we denote this shortest path by $i \rightleftharpoons j$. In a ring of 16 nodes, for example, $(3 \rightleftharpoons 5) = (3 \rightharpoonup 5)$, but $(3 \rightleftharpoons 15) = (15 \rightharpoonup 3)$. We write $|p|$ for the length of the path p.

The idea of our proof is intuitive. As pointed out in Sect. 2, we basically need to prove non-overlapping coverage, $i.e.$, that the coverage of any ready node r_2 starts strictly after the coverage of any other ready node r_1 ends.

$NonOverlappingCoverage \triangleq \forall\, r1,\, r2 \in ReadyNodes : r1 \neq r2 \Rightarrow$
  $ClockwiseDistance(r1,\, RightCoverage(LeafSets[r1]))$
    $< ClockwiseDistance(r1,\, LeftCoverage(LeafSets[r2]))$

It is easy to prove non-overlapping coverage if we prove the following property, which (adapting notation) was already pointed out by Lu [6] as a main invariant.

$CloseNeighbors \triangleq \forall\, r1,\, r2 \in ReadyNodes : r1 \neq r2 \Rightarrow$
  $\land ClockwiseArc(r1,\, RightNeighbor(LeafSets[r1]),\, r2)$
  $\land ClockwiseArc(r2,\, LeftNeighbor(LeafSets[r1]),\, r1)$

We prove $CloseNeighbors$ by proving a stronger property which we call $stable$ $network$. A node i is $stable$ if $CR(i)$ and $CL(i)$ are in i's leaf set (Fig. 5a). A Pastry ring is $stable$ if all Ready or OK nodes are stable. It is clear that in a stable ring, the properties $CloseNeighbors$, and consequently $NonOverlappingCoverage$ hold. For a minimum leaf set size $L = 3$, stable network is an invariant of LuPastry$^+$. Let a $participating$ $node$ be a node that is either Ready or OK, or is the to-join node of a Ready node. Essentially, a participating node is any node that is known to some Ready or OK node. Let i and j be two consecutive Ready or OK nodes on the ring (see Fig. 5b). There can be at most two participating nodes k_1, k_2 between i and j: the to-join nodes of i and j. Any other non-Dead node between i and j must be a Waiting node whose join request has not been picked up by i or j (since they are busy facilitating the joins of k_1 and k_2), and so it can not be in the leaf sets of i or j. For a minimum leaf set size $L = 3$, we can ensure that stable i and j remain stable even if new nodes are added to their leaf sets.

We observe that the ring has the following properties, which we have proven in TLA$^+$.

P1 The coverage of a node is computed based on half the distance to its neighbors. A key k covered by a node i lies in either the *right* or *left* coverage regions of i (see Fig. 1). If i and j are leaf set neighbors, i.e., $i = LN(j)$ and $j = RN(i)$, their coverage regions cannot overlap.

P2 If k is in i's right (left) coverage region, $i \rightharpoonup k \rightharpoonup RN(i)$ $(LN(i) \rightharpoonup k \rightharpoonup i)$.

P3 If i is a stable node and $r \neq i$ is some Ready or OK node, then $i \rightharpoonup RN(i) \rightharpoonup r$ and $r \rightharpoonup LN(i) \rightharpoonup i$.

P4 Because we exclude node failure, all protocol actions that modify a node's leaf set do so through the operation *AddToLS*. Therefore, nodes are not purposely removed from a leaf set, but a node j may only be evicted from the leaf set for node i through an *AddToLS* operation that results in an overflow; i.e., if the leaf set of i becomes full and j is replaced by another node that is closer to i.

P5 A new node k joins the network through a Ready node i that initially covers it, and so k will remain closest to i on one side (right or left) until it finishes its join process. Only after k has finished joining and turned Ready can other nodes join the network between k and i. Therefore, any participating node between i and $CR(i)$ is either $ToJoin[i]$ or $ToJoin[CR(i)]$. That is, there can never be three different participating nodes k_1, k_2, k_3 such that $i \rightharpoonup k_1 \rightharpoonup k_2 \rightharpoonup k_3 \rightharpoonup CR(i)$, or dually, $CL(i) \rightharpoonup k_1 \rightharpoonup k_2 \rightharpoonup k_3 \rightharpoonup i$.

P6 If the leaf set size $L \geq 3$, no action can cause $CR(i)$ or $CL(i)$ to be removed from i's leaf set due to an overflow (see Fig. 5b).

P7 At any point in time, a participating node i is either probing $CR(i)$ (resp., $CL(i)$), or $CR(i)$ (resp., $CL(i)$) is in i's leaf set.

P8 At any point in time, a participating node i is either probing $CR(i)$ (resp., $CL(i)$), or i is in the leaf set of $CR(i)$ (resp., $CL(i)$).

Using these properties, our proof can be outlined in two theorems.

Theorem 1. *In any stable LuPastry$^+$ network, correct delivery holds.*

Proof (Outline). The action *DeliverLookup*(i, k) is enabled only if i is a Ready node that thinks it covers key k. Assume for the sake of contradiction that *DeliverLookup*(i, k) and *DeliverLookup*(j, k) are enabled where $j \neq i$. Nodes i and j are Ready, and both think they cover k. W.l.o.g., $i \rightharpoonup k \rightharpoonup j$; k is in i's

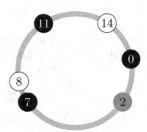

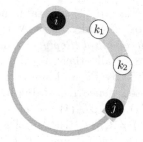

(a) This ring is stable if nodes $0, 2, 7, 11$ are stable. Node 0 is stable if its left leaf set contains 11 and its right leaf set contains 2.

(b) For $L \geq 3$, i's leaf set can always accommodate $k_1 = ToJoin[i]$, $k_2 = ToJoin[j]$, and i's Ready/OK neighbor j.

Fig. 5. Network stability

right coverage region and j's left coverage region. Therefore, $i \rightharpoonup k \rightharpoonup RN(i) \rightharpoonup j$ and $i \rightharpoonup LN(j) \rightharpoonup k \rightharpoonup j$, by P2, P3. In order for both to hold, it must be that $i = LN(j)$ and $j = RN(i)$. Therefore, the coverage regions of i and j cannot overlap (contradiction, by P1). Therefore, in a stable LuPastry$^+$ network, two Ready nodes i and j cannot both think they cover the same key k, and so at most one of $DeliverLookup(i, k)$ and $DeliverLookup(j, k)$ is enabled.

It remains to show that if $DeliverLookup(i, k)$ is enabled then i is closer to k than any other Ready node r is in terms of absolute distance on the ring. Assume again that k is in i's right coverage region, and so $i \rightharpoonup k \rightharpoonup RN(i) \rightharpoonup r$ (by P3). Because k lies within half the distance from i to $RN(i)$ (by P1), k must lie within half the distance from i to r. Therefore, $|i \rightharpoonup k| = |i \rightleftharpoons k| \leq |k \rightharpoonup r|$. If $|r \rightleftharpoons k| = |k \rightharpoonup r|$, we are done. Alternatively, assume $|r \rightleftharpoons k| = |r \rightharpoonup k|$. Because of the ordering of the nodes on the ring $r \rightharpoonup i \rightharpoonup k$, we have that $|r \rightharpoonup k| = |r \rightharpoonup i| + |i \rightharpoonup k|$. Since path lengths are non-negative, $(i \rightleftharpoons k) \leq (r \rightleftharpoons k)$. □

Theorem 2. *For $L \geq 3$, the network is always stable.*

Proof (Outline). The definition of *Init* implies that the LuPastry$^+$ ring is stable in the initial state: nodes in A are Ready and all other nodes are Dead. The leaf set of each A-node i is composed by adding all other A-nodes to i's empty leaf set. Consequently, the leaf set of i will contain its closest right and left A-neighbors. All A-nodes are stable, and so the network is stable. Now consider a stable Pastry network N. For the induction step, we need to show that N remains stable after executing any sub-action e of *Next*. We use N'_e to refer to the new state of N after the execution of e, and $CR'_e(i)$ and $CL'_e(i)$ the next values of $CR(i)$ (resp., $CL(i)$) for a node i. Note that for a node i that is not Ready or OK, the only action that can change i into a Ready or OK node is *ReceiveProbeReply(i)*: i receives the last probe reply message it was waiting for, its probing set becomes empty, and i becomes OK. We need to show that (1) if e results in some unstable node i in N to become Ready or OK, then

i is stable in N'_e, and (2) all stable nodes in N remain stable in N'_e. (1) Let i be an unstable N-node. Since N is stable, i is not Ready or OK. It must be that $e = ReceiveProbeReply(i)$. Since e can only change the local variables of i, the status of all other nodes remains unchanged; $CR'_e(i) = CR(i)$ and $CL'_e(i) = CL(i)$. By P7 and since i's probing set is empty, $CR(i)$ and $CL(i)$ are in i's leaf set, hence i is stable in N'_e. (2) Let i be a stable node in N. If $CR'_e(i) = CR(i)$ and $CL'_e(i) = CL(i)$, then i remains stable in N'_e by P6. Now suppose $CR'_e(i) \neq CR(i)$ (the proof is similar for $CL'_e(i) \neq CL(i)$). Therefore, $e = ReceiveProbeReply(j)$ and $CR'_e(i) = j$. Now, $i = CL'(j)$. By P8 and since j's probing set is empty, j is in i's leaf set. Therefore, i remains stable.   □

The total TLA$^+$ proof consists of more than 30,000 lines. The time taken to run the proof manager on the entire proof is 8 h and 57 min on a single Intel Xeon(R) CPU E5-2680 core running at 2.7 GHz with 256 GB RAM.

5   Conclusion

This paper presents LuPastry$^+$: the first completely machine-checked proof of correct delivery for the variant of Pastry introduced by Lu [6]. Like Lu's proof, our proof has been mechanized in the TLA$^+$ proof system. Compared to Lu's specification, we introduce some new TLA$^+$ operators that abstract away from arithmetic reasoning and other troublesome TLA$^+$ constructs, and this helps avoid low-level arithmetic reasoning at higher levels of the proof. Most importantly, our proof no longer relies on any unproven assumptions. Because we filled all the holes, our overall proof is longer (more than 30,000 lines overall) than Lu's original proof. Nevertheless, our new operators helped significantly reduce the number of steps in the main proof.

Lu [6] shows that correct delivery does not hold for the original published specification of Pastry, in particular in the case of node failure. Like Lu's original proof, our proof assumes that nodes never fail and focuses only on the join protocol. An interesting future work would be to identify assumptions on node failures that maintain the correctness of the protocol, or of a suitable variant, in the presence of nodes joining and leaving the ring.

Our experience shows that TLA$^+$ is well-suited for modeling concurrent and distributed algorithms such as Pastry. In particular, the set-theoretic nature of the specification language encourages the user to model the algorithm at a suitably high level of abstraction. Moreover, TLA$^+$'s hierarchical proof language lets a user focus on parts of the proof without having to remember details about unrelated parts of the proof. Like Lu, we rely on a large invariant for proving the main safety property of the protocol "in one shot", rather than proceeding by refinement from a high-level model where nodes join atomically, down to a detailed model of the real protocol. TLA$^+$ has a notion of refinement based on trace inclusion, and it would be interesting to develop a refinement-based proof and compare its complexity to that of our proof. The difficulty is that parts of the state, such as contents of leaf sets under construction, become visible when some node completes its join protocol, revealing information about other nodes joining concurrently that would have to be anticipated in a refinement-based development.

On a technical level, TLAPS includes facilities for checking the status of a proof that can identify which steps are affected by a change in the specification or the proof. While TLAPS can manage a proof of the size reported here, it is barely able to do so. For example, the Java heap size allotted to Eclipse has to be increased to several gigabytes, and status checking currently takes almost as much time as rerunning the proof. We are in contact with the TLAPS developers who are investigating solutions to these bottlenecks. Although users of a mechanical theorem prover always dream of better automation, the main difficulty in the formal verification of algorithms is in fact finding sufficiently strong inductive invariants that underpin the correctness argument, and any assistance in this task would be most welcome.

References

1. LuPastry+: Specification and Proof Files. http://www.mpi-inf.mpg.de/departments/automation-of-logic/people/noran-azmy/
2. Bakhshi, R., Gurov, D.: Verification of peer-to-peer algorithms: a case study. Electron. Notes Theor. Comput. Sci. **181**, 35–47 (2007)
3. Borgström, J., Nestmann, U., Onana, L., Gurov, D.: Verifying a structured peer-to-peer overlay network: the static case. In: Priami, C., Quaglia, P. (eds.) GC 2004. LNCS, vol. 3267, pp. 250–265. Springer, Heidelberg (2005)
4. Cousineau, D., Doligez, D., Lamport, L., Merz, S., Ricketts, D., Vanzetto, H.: TLA+ proofs. In: Méry, D., Giannakopoulou, D. (eds.) FM 2012. LNCS, vol. 7436, pp. 147–154. Springer, Heidelberg (2012)
5. Lamport, L.: Specifying Systems: The TLA+ Language and Tools for Hardware and Software Engineers. Addison-Wesley, Boston (2002)
6. Lu, T.: Formal verification of the pastry protocol using TLA+. In: Li, X., Liu, Z., Yi, W. (eds.) SETTA 2015. LNCS, vol. 9409, pp. 284–299. Springer, Heidelberg (2015). doi:10.1007/978-3-319-25942-0_19
7. Rowstron, A., Druschel, P.: Pastry: scalable, decentralized object location, and routing for large-scale peer-to-peer systems. In: Guerraoui, R. (ed.) Middleware 2001. LNCS, vol. 2218, pp. 329–350. Springer, Heidelberg (2001)
8. Stoica, I., Morris, R., Karger, D., Kaashoek, M.F., Balakrishnan, H.: Chord: a Scalable Peer-to-peer lookup service for internet applications. In: SIGCOMM 2001, pp. 149–160. ACM (2001)
9. Yu, Y., Manolios, P., Lamport, L.: Model checking TLA+ specifications. In: Pierre, L., Kropf, T. (eds.) CHARME 1999. LNCS, vol. 1703, pp. 54–66. Springer, Heidelberg (1999)
10. Zave, P.: Using lightweight modeling to understand chord. ACM SIGCOMM Comput. Commun. Rev. **42**(2), 49–57 (2012)
11. Zave, P.: How to Make Chord Correct (Using a Stable Base). CoRR abs/1502.06461 (2015)

Enabling Analysis for Event-B

Ivaylo Dobrikov$^{(\boxtimes)}$ and Michael Leuschel

Institut Für Informatik, Heinrich-Heine Universität Düsseldorf,
Universitätsstr. 1, 40225 Düsseldorf, Germany
{dobrikov,leuschel}@cs.uni-duesseldorf.de

Abstract. In this paper we present a static analysis to determine how
events influence each other in Event-B models. The analysis, called an
enabling analysis, uses syntactic and constraint-based techniques to compute the effect of executing one event on the guards of another event.
We describe the foundations of the approach along with the realisation in
PROB. The output of the analysis can help a user to understand the control flow of a formal model. Additionally, we discuss how the information
of the enabling analysis can be used to obtain a new optimised model
checking algorithm. We evaluate both the performance of the enabling
analysis and the new model checking technique on a variety of models.
The technique is also applicable to B, TLA$^+$, and Z models.

Keywords: Model comprehension · Model checking · Static analysis ·
Event-B · Constraint-based analysis

1 Introduction

In Event-B [2] the dynamic behaviour of a system is described by (atomic)
events and a system model is often composed of various components affecting
each other and also possibly having each their own control flow. In this context,
it can be very interesting to infer which events enable or disable which other
events, i.e., to infer the control flow inherent in a model. This information can be
useful to better understand the model, to find hidden control flow dependencies
and, in general, to validate the model. Moreover, this information can be highly
beneficial for other analyses, such as model checking or test-case generation.
For the latter, we have presented an application in [25], where a considerable
reduction in test-case generation time was achieved.

In this paper, we present the foundation of a static, constraint-based analysis
to infer the control-flow for Event-B models. The analysis, which we denote in
this work as *enabling analysis*, is implemented in PROB[21] and can be applied
also for B [1], TLA$^+$ [18], and Z [27] since PROB supports also these formalisms
[16,24]. We validate the performance of the enabling analysis on a variety of
models and present the results of the analysis on one particular model [28],
presented within the landing gear case study track of ABZ 2014, and discuss the
possible representations of the enabling information.

© Springer International Publishing Switzerland 2016
M. Butler et al. (Eds.): ABZ 2016, LNCS 9675, pp. 102–118, 2016.
DOI: 10.1007/978-3-319-33600-8_6

In addition, a new technique for state space exploration for B and Event-B is introduced, which makes use of the enabling information during model checking. The new method of state space exploration is implemented in ProB, which comprises a model checker for automatic verification of specifications written in B, Event-B, TLA+, Z, and CSP. The technique is thoroughly discussed and evaluated on various B and Event-B models. For simplicity of the presentation, we will concentrate mainly on Event-B models in this paper.

In the next section, we give a brief overview of the Event-B formalism. The foundations of the enabling analysis are introduced in Sect. 3. In Sect. 4 we present and evaluate our new state space exploration technique for Event-B. The related work is outlined in Sect. 5. Finally, we discuss future directions and draw the conclusions of our work.

2 Preliminaries

The static parts of an Event-B model, such as carrier sets, constants, axioms, and theorems, are contained in contexts, whereas the dynamic parts of the model are included in machines. A machine comprises variables, invariants, and events. We denote the conjunction of all invariants by Inv and the set of all events by $Events$. The variables make up the states of the model. An event consists of two main parts: guards and actions. Formally, an event has the following form:

$$\textbf{event } e \mathrel{\widehat{=}} \textbf{any } t \textbf{ when } G(x, t) \textbf{ then } S(x, t, x') \textbf{ end}$$

The symbols x and x' stand for the evaluation of the machine variables before and after the execution of the event e, and t for the parameters of the respective event. The parameters t in the **any** clause are typed and restricted in the guard $G(x, t)$ of the event. The action part $S(x, t, x')$ of an event is comprised of a list of assignments to machine variables. When the event is executed, all assignments in the action part are completed simultaneously. All variables that have not been assigned to remain unchanged. In case an event has no parameters we will denote the guard of the event by $G(x)$ and the action part by $S(x, x')$. Note that the guard of an event can simply be truth ($\top$) and the action part can be empty (also known as skip).

Every machine contains the special initialisation event, which has no guards and whose actions are not allowed to refer to the current variable values x of the machine. A state of an Event-B machine with variables x is a vector of values of the correct type. We write $s \models p(x)$ to denote that the predicate p over variables x is true in s.

Definition 1 *(Event Enabledness). For an event e we define grd_e by*

$$grd_e(x) = \begin{cases} G(x), & \text{if } e \text{ has no parameters} \\ \exists t \cdot\ G(x, t), & \text{otherwise} \end{cases}$$

An event e is said to be **enabled** (respectively **disabled**) in a state s iff $s \models grd_e$ (respectively $s \models \neg grd_e$). Further, we define

$$enabled(s) = \{e \in Events \mid s \models grd_e\}$$

to denote the set of all enabled events in state s.

By $s \xrightarrow{e} s'$ we will denote the transition that goes from s to s' by executing the event e from s, where s and s' are states of the machine to which e belongs and further $s \models grd_e$. As a running example throughout the next section we will use the Event-B machine in Example 1.

Example 1 (Example of an Event-B Machine).

machine $M_{v,w}$
variables v, w
invariants $v \geq 0 \wedge w \geq v$
events
    initialisation $\hat{=} v := 0 \parallel w := 1$
    event $vinc \hat{=}$ **when** $v < w$ **then** $v := v + 1$ **end**
    event $w2inc \hat{=}$ **when** $v = w$ **then** $w := w + 2$ **end**
end

Referring to Example 1 and Definition 1, we can observe that in the state $\langle v = 0, w = 1 \rangle$ the event $vinc$ is enabled and $w2inc$ is disabled, whereas in $\langle v = 1, w = 1 \rangle$ the event $vinc$ is disabled but $w2inc$ is enabled.

3   Enabling Analysis

In this section, we study the effect of executing one event on the status of the guard of another event. At first, we introduce the definition of the *before-after* predicate of an event e which expresses a logical statement relating the values of the variables before e (also denoted by x) to the values of the variables after e (also denoted by x').

Definition 2 *(Before-After Predicate BA_e). We define the before-after predicate BA_e of an event e by $BA_e(x, x') = \exists t.G(x, t) \wedge S(x, t, x')$ in case the event has parameters. If e has no parameters, then $BA_e(x, x') = G(x) \wedge S(x, x')$.*

The next definition captures whether—provided the invariant holds[1] and a pre-condition P— an event can be executed and make a post-condition Q true.

---

[1] Note: we include the invariant Inv here, meaning that all results are only valid so-long as the invariant remains true. In practice, this is usually ok: animation and model checking with PROB will detect invariant violations. Adding the invariant is often important to help the constraint solver. On the other hand, it is possible to remove the invariant from Definition 3 and one would then obtain an analysis that is also valid for states which do not satisfy the invariant.

Definition 3 *(Conditional Event-Feasibility $\leadsto_e$). For an event e and the predicates P and Q, we say that an event e is feasible under the conditions P and Q, denoted by $P \leadsto_e Q$, iff there exists a state s such that $s \models Inv \wedge P$ and $s \models \exists s'.(BA_e(s, s') \wedge Q)$. If there is no such a state s, then we write $P \not\leadsto_e Q$ to denote that e is not conditionally feasible under P and Q.*

For the machine $M_{v,w}$ in Example 1, we have that $(v = 10) \leadsto_{w2inc} (w = 12)$, $(v = w) \not\leadsto_{w2inc} (v = w)$, $(v = w) \not\leadsto_{vinc} (v = w)$, or $(v < w) \leadsto_{vinc} (v = w)$. To establish that $(v < w) \leadsto_{vinc} (v = w)$ is satisfied according to Definition 3, we can find for $M_{v,w}$, for example, the state $s = \langle v = 1, w = 2 \rangle$ satisfying $v < w$ and the solution for s' with $\langle v = 2, w = 2 \rangle$ for the conditional feasibility of $vinc$. Note that in contrast to the Hoare triple $\{P\}S\{Q\}$, $P \not\leadsto_e Q$ does not ensure that Q holds after e, only that Q may hold, for some parameter values and nondeterministic execution. Based on Definition 3 we can already characterise certain possible effects of the execution of an event e_1 on the status of another event e_2 as given in Definition 4.

Definition 4 *(feasible, guaranteed, impossible). An event e is feasible if there exists a state s such that $s \models Inv \wedge grd_e$. A feasible event e is guaranteed if there exists no state s such that $s \models Inv \wedge \neg grd_e$.*
Event e_2 is impossible after a feasible event e_1 iff $\top \not\leadsto_{e_1} grd_{e_2}$. Event e_2 is guaranteed after a feasible event e_1 iff $\top \not\leadsto_{e_1} \neg grd_{e_2}$.

In our example Example 1, both events are feasible, and $vinc$ is guaranteed after $w2inc$ but $w2inc$ itself is impossible after $w2inc$. After $vinc$ neither event $w2inc$ nor $vinc$ is impossible or guaranteed.

We now want to obtain a more precise characterisation of the effect of an event e_1 on the enabling condition of another event e_2. We say that an event e_2 is **enabled** by some event e_1 if there is a transition $s \xrightarrow{e_1} s'$ such that $s \models \neg grd_{e_2}$ and $s' \models grd_{e_2}$. Similarly, we say that e_2 is **disabled** by e_1 if there is a transition $s \xrightarrow{e_1} s'$ such that $s \models grd_{e_2}$ and $s' \models \neg grd_{e_2}$.

In the Definition 5 below we check four different conditions: can e_1 enable e_2, can e_1 disable e_2, can e_1 keep e_2 enabled, and can e_1 keep e_2 disabled. The answer to each of these questions can be true or false, giving rise to 16 different combinations. We can view the above four conditions as possible edges in graph, consisting of possible values of the guards before and after an execution. This leads to the following definition of an *enabling relation*.

Definition 5. *Let e_1, e_2 be events. By $\mathcal{ER}(e_1, e_2)$, the **enabling relation** for e_2 via e_1, we denote the binary relation over $\{\top, \bot\}$ defined by*

1. $\bot \mapsto \top \in \mathcal{ER}(e_1, e_2)$ *iff* $\neg grd_{e_2} \leadsto_{e_1} grd_{e_2}$ *(e_1 can enable e_2)*
2. $\top \mapsto \bot \in \mathcal{ER}(e_1, e_2)$ *iff* $grd_{e_2} \leadsto_{e_1} \neg grd_{e_2}$ *(e_1 can disable e_2)*
3. $\top \mapsto \top \in \mathcal{ER}(e_1, e_2)$ *iff* $grd_{e_2} \leadsto_{e_1} grd_{e_2}$ *(e_1 can keep e_2 enabled)*
4. $\bot \mapsto \bot \in \mathcal{ER}(e_1, e_2)$ *iff* $\neg grd_{e_2} \leadsto_{e_1} \neg grd_{e_2}$ *(e_1 can keep e_2 disabled)*

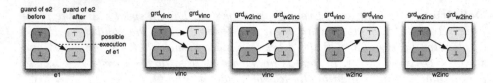

Fig. 1. Graphical representation of the effect $\mathcal{ER}(e_1, e_2)$, illustrated on Example 1

We provide a graphical representation of enabling relations, explained and illustrated on Example 1 in Fig. 1.

Providing the user with a table containing enabling diagrams will probably turn out to be overwhelming. We have therefore tried to group the 16 possibilities into concepts which can be more easily grasped by users. Earlier in Definition 4 we have already introduced the concepts of guaranteed and impossible. Three further concepts are those introduced in Definition 6.

Definition 6. *We say that e_1 **keeps** e_2 if e_2 remains enabled respectively disabled after e_1, i.e. we have: $(grd_{e_2} \not\leadsto_{e_1} \neg grd_{e_2}) \wedge (\neg grd_{e_2} \not\leadsto_{e_1} grd_{e_2})$.*
*We say that e_2 **can enable** e_1 if e_2 cannot disable e_1 and may enable e_1, i.e. the following constraints are fulfilled: $(grd_{e_2} \not\leadsto_{e_1} \neg grd_{e_2})$ and $(\neg grd_{e_2} \leadsto_{e_1} grd_{e_2})$.*
*We say that e_2 **can disable** e_1 if e_2 cannot enable e_1 and may disable e_1, i.e. the following constraints are fulfilled: $(\neg grd_{e_2} \not\leadsto_{e_1} grd_{e_2})$ and $(grd_{e_2} \leadsto_{e_1} \neg grd_{e_2})$.*

Figure 2 shows all possible enabling relations, and shows how we have grouped them using the concepts from Definitions 4 and 6. These concepts are also presented to the user in our implementation of the enabling analysis in PROB, either as a table or a graph. All combinations in Fig. 2 can actually arise in practice.[2]

Implementation and Empirical Evaluation

We have implemented the enabling analysis within the PROB toolset, also to answer the questions whether the analysis can provide interesting feedback to the user and whether the analysis can scale up despite the inherent quadratic complexity and the possibly complex constraints. Indeed, for any given e, P, Q we use PROB's constraint solver to determine whether $P \leadsto_e Q$ or $P \not\leadsto_e Q$ holds. For example, for $(w > v) \leadsto_{vinc} (v = w)$ the constraint solver would find a solution state s satisfying $v > w$ from which after executing $vinc$ at s a solution state s' will be found that fulfils $v = w$. Possible solution states could be, for instance, $s = \langle v = 1, w = 2 \rangle$ and $s' = \langle v = 1, w = 2 \rangle$. In case a time-out occurs during constraint solving, we have no information about whether $P \leadsto_e Q$ or $P \not\leadsto_e Q$ holds. An occurrence of a time-out during constraint solving means that the solver could not find a solution for the constraints in the given time from the user. In the graphical representation, an occurrence of a time-out could

---

[2] In addition, we illustrate some of the enabling relations on concrete examples in
    https://www3.hhu.de/stups/prob/index.php/Tutorial_Enabling_Analysis.

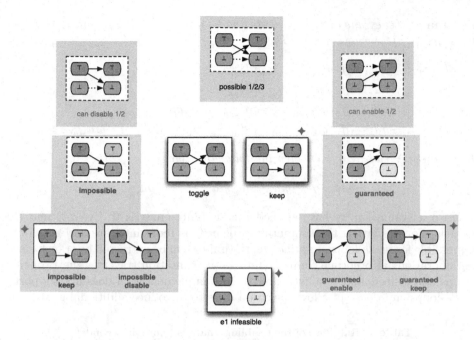

Fig. 2. Classification of the possible effects of an event e_1 on the guards of e_2. Dotted-edges mean the absence or presence of the edge does not influence the classifcation.

be visualised by having a dashed edge. This would also mean that considering a time-out for the definition of the different types of enabling relations gives rise of $3^4 = 81$ combinations rather than 16.

In addition to determining $\mathcal{ER}(e_1, e_2)$ for each pair of events, including $e_1 = e_2$, we also compute for every event e_2 the possible status after the initialisation event. That is, we compute also $\top \leadsto_{INIT} grd_e$ and $\top \leadsto_{INIT} \neg grd_e$, the conditional feasibility operator in Definition 3, where $BA_{INIT}(s, s')$ is the after-predicate of the initialisation event when determining the respective $\leadsto_{INIT}$ conditional-feasibility of $INIT$ in regard to some event e.

Syntactic conditions: The enabledness of an event e is determined by the values of the variables that are read in the guard of e (grd_e). The set of read machine variables in grd_e will be denoted by $read_G(e)$. Accordingly, by $read_S(e)$ and $write(e)$ we will denote the variables that are read and the variables that are written in the action part of e, respectively.

In our implementation we have used syntactic conditions to avoid calling the constraint solver as much as possible. The following lemma captures this optimisation. It decomposes the guard of the second event into two parts: those conjuncts that cannot be influenced by the execution of the first event (grd_{static}) and those that can (grd_{dyn}).

Lemma 1. *Let e_1 and e_2 be two events and let $grd_{e_2} = (grd_{static} \wedge grd_{dyn})$ where $vars(grd_{static}) \cap write(e_1) = \varnothing$. Then*

$$grd_{e_2} \leadsto_{e_1} grd_{e_2} \quad \textit{iff} \quad grd_{e_2} \leadsto_{e_1} grd_{dyn}$$

$$grd_{e_2} \leadsto_{e_1} \neg grd_{e_2} \quad \textit{iff} \quad grd_{e_2} \leadsto_{e_1} \neg grd_{dyn}$$

$$\neg grd_{e_2} \leadsto_{e_1} grd_{e_2} \quad \textit{iff} \quad (grd_{static} \wedge \neg grd_{dyn}) \leadsto_{e_1} grd_{dyn}$$

$$\neg grd_{e_2} \leadsto_{e_1} \neg grd_{e_2} \quad \textit{iff} \quad (grd_{static} \wedge \neg grd_{dyn}) \leadsto_{e_1} \neg grd_{dyn} \textit{ or } \neg grd_{static} \leadsto_{e_1} \top$$

If $write(e_1) \cap read_G(e_2) = \varnothing$, then the following hold:

$$\neg grd_{e_2} \not\leadsto_{e_1} grd_{e_2} \textit{ and } grd_{e_2} \not\leadsto_{e_1} \neg grd_{e_2}$$

In our evaluation we have applied the enabling analysis to Event-B and B models. For the latter, the computation of grd_e is more intricate and actually impossible for *while*-loops. In principle, the guard can be extracted by transforming operations into normal form [1]. Our implementation traverses the B operations and collects and combines guards. The implementation does not support *while*-loops and does not allow operation calls to introduce additional guards.

Table 1. Runtimes of the enabling analysis (times in seconds)

Benchmark	# Events	# Pairs	# Timeouts	Analysis Time
CAN BUS	21	462	2	1.565
Cruise Control	26	702	14	9.480
DeMoney	8	72	0	0.685
DeMoney Refl	8	72	11	5.503
Scheduler	5	30	0	0.687
USB 4 Endpoints	28	1482	0	5.708
Travel Agency	10	110	77	36.123
Landing Gear v1	16	272	0	0.599
Landing Gear v4	32	1056	208	151.622
LandingGear_Abrial3_m0	6	42	2	3.502
LandingGear_Abrial3_m1	7	56	7	5.537
LandingGear_Abrial3_m2	11	132	8	7.951
LandingGear_Abrial3_m3	21	462	27	29.996
LandingGear_Abrial3_m4	26	702	138	96.190

In Table 1 we present some timing results, showing that the technique can scale to interesting B and Event-B models.[3] In column "**# Pairs**" we have listed the number of pairs of events (e_1, e_2) needed to determine all possible enabling

---

[3] The models and the results of the enabling analysis can be obtained from the following web page http://nightly.cobra.cs.uni-duesseldorf.de/enabling_analysis/.

relations. Note that for the enabling analysis we also determine $\mathcal{ER}(INIT, e)$ for each event e of the respective machine. The table also shows the number of time-outs that occurred: a time-out means that one of the four edges had a time-out. We have used a time-out of at most 300 ms for every solver call $P \leadsto_e Q$. (See [10]) All measurements were made on an Intel Xeon Server, 8 x 3.00 GHz Intel(R) Xeon(TM) CPU with 8 GB RAM running Ubuntu 12.04.3 LTS.

At the lower part of the Table 1 we have listed the results of the enabling analysis of two Event-B models describing a landing gear system: *Landing Gear* [15] and *LandingGear_Abrial3* [28]. Both models were developed using refinement. In the case of the *Landing Gear* model we have listed the first and fourth refinement of the model, whereas for *LandingGear_Abrial3* we have given the abstract model and its four consecutive refinements from [28]. In the upper part of the table the following specifications are given: *CAN BUS* represents an Event-B model specifying a controller area network bus. *Cruise Control* is a model written in B representing a case study at Volvo on a typical vehicle function. *DeMoney* and *Demoney Ref1* present the first two levels of an electronic purse used to demonstrate GeneSyst in [6]. *Scheduler* is the model of a process scheduler from [19] and *USB 4 Endpoints* a B specification of a USB protocol, developed by the French company ClearSy. The *Travel Agency* model is a 296 lines B specification of a distributed online travel agency, through which users can make hotel and car rental bookings. Some operations of the *Travel Agency* specification are very complex consisting up to 98 lines of nested conditionals and any statements.

We have found the analysis to be very useful on many practical problems. For example, on the *CAN BUS* the analysis clearly shows that the system cycles through three distinct phases. It can help getting an understanding of models written by somebody else, or even confirm one's intuition about the control flow of a model. Our technique can be applied to classical B, TLA, Z models, but is probably most useful for Event-B and Event-B style models. Indeed, due to the lack of constructs such as sequential composition, conditional statements or loops, Event-B models tend to have many relatively simple events. Also, the control flow tends to be encoded using explicit program counters, which can be dealt with quite well by our constraint solver. So, the enabling analysis is more useful to the user, and scales better due to the events being much simpler.

For the *LandingGear_Abrial3* specifications one can observe in Table 1 the increasing number of events for the advancing refinement levels. Another interesting fact is the increasing number of time-outs in each further refinement of *LandingGear_Abrial3*. This could be explained by the fact that at each next refinement level of *LandingGear_Abrial3* the invariant and the guards of the events are getting more involved and thus the constraint solver needs to handle more complex constraints. In our opinion, the only disappointing result in Table 1 is the *Travel Agency* model, for which a large number of time-outs occurs. This is due to the use of complicated substitutions in the machine, where there are some operations consisting of up to 98 lines with nested conditionals and any statements. The enabling analysis, however, still provides some useful insights for the *Travel Agency* model. For instance, there are various operations in the model such as *bookRoom* or *bookCar* that can possibly disable themselves.

Below we show one particular result, the model m0 of the third landing gear model in [28] (LandingGear_Abrial3_m0.mch). The results of the enabling analysis can be exported as table, which we reproduce here. It uses the classification from Fig. 2, rather than showing the individual enabling graphs (but which are also available if the user wishes to inspect them).

Origin	act1	X_act	X_chg	act6	beg_X	end_X
INIT.	guaranteed	impossible	impossible	impossible	impossible	impossible
act1	impossible	impossible*	guaranteed	impossible*	guaranteed	impossible*
X_act	impossible*	impossible	guaranteed	possible_enable	can_enable	can_enable
X_chg	impossible*	possible	guaranteed	possible	possible	keep
act6	guaranteed	impossible*	impossible	impossible	impossible*	impossible*
beg_X	unchanged	impossible*	unchanged	impossible*	impossible	guaranteed
end_X	unchanged	can_enable	unchanged	can_enable	impossible*	impossible

unchanged = syntactic_unchanged,
impossible* = impossible_keep (must be disabled before)

Instead of a tabular representation, we have also provided a graph representation of the enabling information in Fig. 3. It contains the events as nodes and

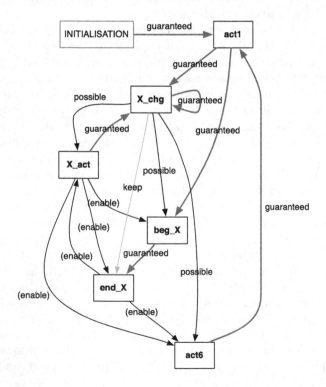

Fig. 3. Enabling results for LandingGear_Abrial3_m0.mch

the above classification as edges, with the exception that combinations marked as impossible and unchanged are not shown in the graph. One can clearly see the control flow of the model in Fig. 3, e.g. that beg_X is guaranteed to enable end_X.

4   Optimising the Model Checker

Consistency Checking Algorithm. The results of the enabling analysis can be used for optimising a model checker for B and Event-B. In particular, one can use the outcome of the enabling analysis to improve model checking in PROB [20,21]. The optimisation, designated as partial guard evaluation (PGE), uses the event relations *impossible* (see Definition 4) and *keep* (see Definition 6) for improving the process of consistency checking of Event-B models.

When checking an Event-B model for consistency (e.g., invariant satisfaction and deadlock freedom) the PROB model checker (see also Algorithm 5.1 in [21]) traverses the state space of the model beginning at the initial states and checking each reachable state for errors. If no error is detected in the currently explored state, then its successor states are computed. When computing the successors of a state s the guard of each event of the machine is tested in s. If an event is enabled, then the actions of the event are applied to s, which results in various (possibly new) successor states. The search for errors proceeds until an error state is found or all possible states of the model are visited and checked. The effort for checking a state amounts to checking the state for errors (testing for invariant violation, assertion violations, etc.) plus the computation of the successors.

Checking the guard of every event in each reachable state can sometimes be a time-consuming task. In some cases one can examine whether an event is disabled by just observing the incoming transitions. If, for example, event e_2 is impossible after e_1 (i.e. $\top \not\leadsto_{e_1} grd_{e_2}$) and the currently explored state s is reachable by e_1, then we can safely skip the evaluation of the guard of e_2 in s when computing the enabled events in s. The information for the impossibility of e_2 to be enabled after e_1 can be obtained from our enabling analysis.

Especially, when the model checker has to check exhaustively Event-B models with large state spaces and a large number of events, the effort of testing the guards in every state may be considerable. The idea of our PGE optimisation is to identify a set of disabled events in each visited state, using the information of the enabling analysis. The set of disabled events in each state s is determined with respect to the predecessor states and the incoming events of s.

The Optimisation. The algorithm of the PGE optimisation is outlined in Algorithm 1 and can be described as follows. Consider a state s that is currently explored and has a set of disabled events $s.disabled$. Each event evt that is not an element of $s.disabled$ is tested for being enabled in s (**for**-loop condition in line 7). If evt is enabled at s, then its actions are applied at s. The effect of executing evt at s results in a set of successor states $Succ$ (line 9). Subsequently, the set of disabled events $Disabled$ in the successors of s is computed (line 10). All events that are asserted by the enabling analysis to be *impossibly* enabled after *evt* are considered to be disabled at each $s' \in Succ$. Further, each event *e*, regarded as disabled in *s*, will be included into the set of disabling events of each $s' \in Succ$ if *evt* cannot influence the guard of *e*, i.e. *evt* keeps *e* disabled.

Once the set of disabled events with respect to *s* and *evt* is computed, we should initialise $s'.disabled$ for each $s' \in Succ$. This depends on whether the respective successor state s' was already generated ($s' \in Visited$) or it occurs for the first time during the state space exploration ($s' \notin Visited$). If s' has not yet been visited, then we assign to *s.disabled* the set of disabled events *Disabled*. Otherwise, if s' has already been visited, then we update the set of disabled events $s'.disabled$ for s' (line 17) as there could be "new" events that have not yet been added to $s'.disabled$. In the latter case we update the set $s'.disabled$ by adding *Disabled* in order to increase the possibility for saving more unnecessary guards evaluations when s' is explored later. As a consequence, the optimisation in Algorithm 1 can sometimes perform differently for different exploration strategies.

For each state *s* the set of disabled events *s.disabled* comprises the *obviously* disabled events in *s*. Usually, there are events in the model which are not enabled in *s* and their disabledness in *s* cannot be determined by means of the information provided by the enabling analysis. These events are also included to the set of disabled events *s.disabled* in the process of exploration of *s* (line 21). In this way, the possibility is increased for adding more events to the sets of disabled events of the successor states by means of the *keep* relation.

Evaluation. For evaluating the approach we focussed on Event-B models with large state spaces, so that a large number of skipped guard evaluations allows us to recover the cost of the static enabling analysis. The performance of Algorithm 1 does not depend solely on the overall number of skipped guard evaluations, but also on the guard complexity of the events whose enabledness tests are omitted in the various states. The detection of the redundant guard evaluations depends also on how the events influence each other in the respective model, as well as on the accuracy of the results of the enabling analysis. We have evaluated the optimisation on various Event-B models[4]. In Table 2 we list a part of the results of the evaluation.

For every benchmark we carried out three types of performance comparisons: *mixed breadth- and depth-first* (**BF/DF**) search, *breadth-first* (**BF**) search, and *depth-first* (**DF**) search. For each of the three search strategies we analysed the performance of checking by means of PROB's original algorithm (Algorithm 5.1 in [21]) and Algorithm 1. The search strategy has an impact on the overall number of skipped guard evaluations when exploring the state space of a model by means of Algorithm 1. This is due to the fact that when we explore a state, say *s*, in Algorithm 1 the set of the "obviously" disabled events in *s* is determined by the predecessor states of *s* or, more precisely, by the incoming transitions of *s* and the sets of disabled events in the predecessor states of *s*. If a state *s* has multiple predecessor states, then at the moment of exploring *s* the number of

---

Algorithm 1. Consistency Checking with Partial Guard Evaluation

---

1 $Queue := \langle root \rangle$; $root.disabled := \varnothing$; $Visited := \{\}$; $Graph := \{\}$;
2 **while** $Queue$ *is not empty* **do**
3      $s :=$ get_state($Queue$);
4   **if** $error(s)$ **then**
5   **return** counter-example trace in $Graph$ from $root$ to s
6   **else**
7   **foreach** $evt \in EVENTS$ **such that** $evt \notin s.disabled$ **do**
8   **if** evt *is enabled at* s **then**
9                  $Succ := \{s' \mid BA_{evt}(s, s')\}$ /* compute successors of s for evt */ ;
10                 $Disabled := \{e \in Events_M \mid e \text{ impossible after } evt\}$
                        $\cup \{e \in s.disabled \mid evt \text{ keeps } e\}$;
11   **foreach** $s' \in Succ$ **do**
12                     $Graph := Graph \cup \{s \xrightarrow{evt} s'\}$;
13   **if** $s' \notin Visited$ **then**
14   push_to_front(s', $Queue$) ; $Visited := Visited \cup \{s'\}$;
15                         $s'.disabled := Disabled$
16   **else**
17                         $s'.disabled := s'.disabled \cup Disabled$
18   **end if**
19   **end foreach**
20   **else**
21                 $s.disabled := s.disabled \cup \{evt\}$
22   **end if**
23   **end foreach**
24   **end if**
25 **end while**
26 **return** ok

---

predecessor states may depend on the exploration strategy. The rule of thumb is then the more predecessor states are explored before s is being processed, the higher is the possibility that more events that are disabled in s are determined without testing their guards for enabledness.

All tests with the option PGE (BF/DF+PGE, BF+PGE and DF+PGE) in Table 2 use Algorithm 1 with the respective search strategy. All other entries used PROB's original consistency algorithm. The model checking times as well as the times for performing the enabling analysis (in case the PGE optimisation is used) are given in the table. We also report the number of the overall and the skipped guard evaluations. Other statistics like number of states, transitions, and events of every Event-B model are shown in the first column of Table 2. Each of the experiments was performed ten times and the geometric mean of the model checking and enabling analysis times are reported. All measurements were made on an Intel Xeon Server, 8 x 3.00 GHz Intel(R) Xeon(TM) CPU with 8 GB RAM running Ubuntu 12.04.3 LTS.

Table 2. Part of the model checking experimental results (times in seconds)

Model & State Space Stats.	Algorithm	Analysis Time	Skipped/Total Guard Tests	Model Checking Time
Complex Guards	BF/DF	-	0/2,099,622	1350.641
(Best-Case)	BF/DF+PGE	8.040	1,899,629/2,099,622	620.662
# Events: 21	BF	-	0/2,099,622	1343.740
States: 99,982	BF+PGE	8.079	1,899,629/2,099,622	609.669
Transitions: 99,984	DF	-	0/2,099,622	1337.831
	DF+PGE	8.105	1,899,629/2,099,622	621.547
CAN BUS	BF/DF	-	0/2,784,600	496.922
	BF/DF+PGE	0.682	2,257,505/2,784,600	251.275
# Events: 21	BF	-	0/2,784,600	487.146
States: 132,600	BF+PGE	0.673	2,284,693/2,784,600	230.327
Transitions: 340,267	DF	-	0/2,784,600	496.389
	DF+PGE	0.660	2,242,223/2,784,600	268.185
Lift	BF/DF	-	0/1,257,986	390.713
	BF/DF+PGE	5.508	783,429/1,257,986	364.272
# Events: 21	BF	-	0/1,257,986	382.276
States: 58,226	BF+PGE	5.554	793,256/1,257,986	350.986
Transitions: 357,147	DF	-	0/1,257,986	407.274
	DF+PGE	5.693	788,464/1,257,986	369.104
Cruise Control	BF/DF	-	0/35,282	11.168
	BF/DF+PGE	2.220	16,846/35,282	11.727
# Events: 26	BF	-	0/35,282	10.498
States: 1,361	BF+PGE	2.199	15,192/35,282	11.656
Transitions: 25,697	DF	-	0/35,282	10.925
	DF+PGE	2.173	16,839/35,282	11.564
All Enabled	BF/DF	-	0/600,012	218.063
(Worst-Case)	BF/DF+PGE	0.287	0/600,012	252.213
# Events: 6	BF	-	0/600,012	211.401
States: 100,002	BF+PGE	0.285	0/600,012	254.937
Transitions: 550,003	DF	-	0/600,012	198.161
	DF+PGE	0.282	0/600,012	252.321

The models *Complex Guards* and *All Enabled* are toy examples created in order to show the best and worst case when model checking Event-B models using partial guard evaluation, respectively. The best case example, *Complex Guard*, constitutes a model with 21 events in which only one event is enabled per state and in addition each event has a guard which is relatively expensive to be checked. On the other hand, the worst case example, *All Enabled*, represents a simple model for which all events are enabled in each state of the model and thus no event can be disabled after the execution of some of the other events. *CAN BUS* and *Cruise Control* are the models that we have introduced in Table 1 in Sect. 3. *Lift* represents an Event-B model of a lift.

In almost all experiments, except for *Cruise Control* and *All Enabled*, the new PGE consistency checking algorithm (Algorithm 1) is faster than the original one. For *CAN BUS* and *Lift*, the breadth-first search strategy works best with

PGE; indeed, in breadth-first mode a node is more likely to already have more incoming edges when being processed as compared to depth-first. The number of spared guard evaluations varies for the different search strategies for the cases where the PGE optimisation is used.

In the worst case (*All Enabled*), the performance of Algorithm 1 is not significantly different from the performance of the ordinary search. In this case no guard evaluation has been skipped since all six events are always enabled. No performance improvement was obtained in the *Cruise Control* experiment, although a considerable number of guard evaluations were removed. However, the guards are probably too simple (involving many boolean variables) and the additional bookkeeping of Algorithm 1 seems more expensive than the guard evaluations.

5 Related Work

Another approach for determining how events can influence each other was presented in [4]. It annotates the edges in a graph by predicates, which are derived by proof and predicate simplification. Our approach is constraint-based and provides a new, more fine-grained way of presenting and visualising the enabling information. Another related work is the GeneSyst system [6] which is also semi-automatic and proof based, and tries to generate an abstract state space representation. However, it does also support linking refinements with abstractions. It focusses more on the set of reachable states, not on enabling and disabling of events as in this paper. The techniques from [17,22] use the explicitly constructed state space. This is more precise, but cannot be applied for infinite or large state spaces and obviously cannot be used to optimise model checking. An approach like UML2B [26] works the other way to our enabling analysis: the B model is generated from a control flow description, rather than the other way around. The works [12,14] try to generate UML state charts from B models, but do not specify how this is to be done ([14] refers to [5], a precursor to [6] described above). Maybe, our enabling analysis or alternatively [4] could be used to generate UML state charts rather than, e.g., the graphs in Fig. 3.

The enabling analysis in the present paper was adapted in [9] for computing the independence and enabling relations between events.[5] Both types of relations, the independence and enabling relations, are used in the process of model checking Event-B machines using partial order reduction. This technique is orthogonal to the PGE evaluation presented here and [9] does not discuss the use of the enabling analysis for model comprehension. The enabling analysis result has also been used to considerably improve the performance of test-case generation [25], by pruning infeasible paths.

Model checking is a practical technique allowing an automatic formal verification of various properties on finite-state models. The state space explosion impedes in most cases the formal verification via model checking. As a consequence, various techniques have been proposed for combating the state space

---

[5] Ideally the present paper should have been published before [9].

explosion problem: partial order reduction [8,13], symbolic model checking [23], symmetry reduction [7,11], directed model checking, and etc. A lot of work has been devoted to optimising the ordinary model checker of PROB for B and Event-B in order to tackle the state space explosion problem. Besides partial order reduction [9], several symmetry reduction techniques such as [29] were developed for B. Another optimisation of model checking was presented in [3], where proof information is used to optimise invariant preservation checking. This optimisation was used in the experiments in this paper.

6   Conclusion and Future Work

We have described a new static analysis for computing enabling relations of events for B and Event-B using syntactic and constraint-based analyses. The information of the enabling analysis can contribute to better understanding of a model, as well as to identify the program flow of it. We have shown that enabling analysis is not only beneficial for the better understanding, but also that it delivers a valuable information for the model checker. We have presented a more elaborated state space exploration that makes use of the results of the enabling analysis. We have demonstrated that the new state space exploration technique performs considerably better for very large state models than the ordinary state space exploration.

Further work needs to be done in investigating whether other relations, besides *impossible* and *keep*, can be used to optimise model checking of Event-B models. Another interesting avenue of research would be to generate UML state diagrams from Event-B models, possibly taking the refinement structure into account.

Acknowledgements. We would like to thank the reviewers of ABZ'16 for their very useful suggestions, e.g., concerning Fig. 2. We also thank Jens Bendisposto for very useful feedback and ideas.

References

1. Abrial, J.-R.: The B-Book: Assigning Programs to Meanings. Cambridge University Press, New York (1996)
2. Abrial, J.R.: Modeling in Event-B: System and Software Engineering, 1st edn. Cambridge University Press, New York (2010)
3. Bendisposto, J., Leuschel, M.: Proof assisted model checking for B. In: Breitman, K., Cavalcanti, A. (eds.) ICFEM 2009. LNCS, vol. 5885, pp. 504–520. Springer, Heidelberg (2009)
4. Bendisposto, J., Leuschel, M.: Automatic flow analysis for Event-B. In: Giannakopoulou, D., Orejas, F. (eds.) FASE 2011. LNCS, vol. 6603, pp. 50–64. Springer, Heidelberg (2011)
5. Bert, D., Cave, F.: Construction of finite labelled transition systems from B abstract systems. In: Grieskamp, W., Santen, T., Stoddart, B. (eds.) IFM 2000. LNCS, vol. 1945, pp. 235–254. Springer, Heidelberg (2000)

6. Bert, D., Potet, M.-L., Stouls, N.: Genesyst: a tool to reason about behavioral aspects of B event specifications. application to security properties. In: ZB , pp. 299–318 (2005)
7. Clarke, E., Enders, R., Filkorn, T., Jha, S.: Exploiting symmetry in temporal logic model checking. Formal Methods Syst. Des. **9**(1–2), 77–104 (1996)
8. Clarke, E., Grumberg, O., Minea, M., Peled, D.: State space reduction using partial order techniques. Int. J. STTT **2**(3), 279–287 (1999)
9. Dobrikov, I., Leuschel, M.: Optimising the ProB model checker for B using partial order reduction. In: Giannakopoulou, D., Salaün, G. (eds.) SEFM 2014. LNCS, vol. 8702, pp. 220–234. Springer, Heidelberg (2014)
10. Dobrikov, I., Leuschel, M.: Enabling analysis for Event-B (technical report). Technical report, Institut für Informatik, University of Düsseldorf (2016). http://stups.hhu.de/w/Special:Publication/LeuschelDobrikov-EnablingTR
11. Donaldson, A.F., Miller, A.: Exact and approximate strategies for symmetry reduction in model checking. In: Misra, J., Nipkow, T., Sekerinski, E. (eds.) FM 2006. LNCS, vol. 4085, pp. 541–556. Springer, Heidelberg (2006)
12. Fekih H, Ayed LJ, Merz S.: Transformation of B specifications into UML class diagrams and state machines. ACM Symposium on Applied Computing - SAC 2006, vol. 2, pp. 1840–1844. Dijon, France (Apr. 2006)
13. Godefroid, P. (ed.): Partial-Order Methods for the Verification of Concurrent Systems. LNCS, vol. 1032. Springer, Heidelberg (1996)
14. Hammad, A., Tatibouët, B., Voisinet, J.-C., Wu, W.-P.: From a B specification to UML statechart diagrams. In: George, C.W., Miao, H. (eds.) ICFEM 2002. LNCS, vol. 2495, pp. 511–522. Springer, Heidelberg (2002)
15. Hansen, D., Ladenberger, L., Wiegard, H., Bendisposto, J., Leuschel, M.: Validation of the ABZ landing gear system using ProB. In: Boniol, F., Wiels, V., Ait Ameur, Y., Schewe, K.-D. (eds.) ABZ 2014. CCIS, vol. 433, pp. 66–79. Springer, Heidelberg (2014)
16. Hansen, D., Leuschel, M.: Translating TLA$^+$ to B for validation with PROB. In: Derrick, J., Gnesi, S., Latella, D., Treharne, H. (eds.) IFM 2012. LNCS, vol. 7321, pp. 24–38. Springer, Heidelberg (2012)
17. Ladenberger, L., Leuschel, M.: Mastering the visualization of larger state spaces with projection diagrams. In: Butler, M., Conchon, S., Zaïdi, F. (eds.) Formal Methods and Software Engineering. LNCS, pp. 153–169. Springer, Switzerland (2015)
18. Lamport, L.: Specifying Systems: The TLA+ Language and Tools for Hardware and Software Engineers. Addison-Wesley Longman Publishing Co., Inc, Boston (2002)
19. Legeard, B., Peureux, F., Utting, M.: Automated boundary testing from Z and B. In: Eriksson, L.-H., Lindsay, P.A. (eds.) FME 2002. LNCS, vol. 2391, pp. 21–40. Springer, Heidelberg (2002)
20. Leuschel, M., Butler, M.: ProB: a model checker for B. In: Araki, K., Gnesi, S., Mandrioli, D. (eds.) FME 2003: Formal Methods. LNCS, pp. 855–874. Springer, Heidelberg (2003)
21. Leuschel, M., Butler, M.: ProB: an automated analysis toolset for the B method. STTT **10**(2), 185–203 (2008)
22. Leuschel, M., Turner, E.: Visualising larger state spaces in Pro **B**. In: Treharne, H., King, S., C. Henson, M., Schneider, S. (eds.) ZB 2005. LNCS, vol. 3455, pp. 6–23. Springer, Heidelberg (2005)

23. McMillan, K.L.: Symbolic Model Checking: An Approach to the State Explosion Problem. Ph. D. thesis, Carnegie Mellon University, Pittsburgh, PA, USA, UMI Order No. GAX92-24209 (1992)
24. Plagge, D., Leuschel, M.: Validating Z specifications using the PROB animator and model checker. In: Davies, J., Gibbons, J. (eds.) IFM 2007. LNCS, vol. 4591, pp. 480–500. Springer, Heidelberg (2007)
25. Savary, A., Frappier, M., Leuschel, M., Lanet, J.-L.: Model-based robustness testing in Event-B using mutation. In: Calinescu, R., Rumpe, B. (eds.) SEFM 2015. LNCS, vol. 9276, pp. 132–147. Springer, Heidelberg (2015)
26. Snook, C., Butler, M.: Verifying Dynamic Properties of UML Models by Translation to the B Language and Toolkit. In: UML 2000 WORKSHOP Dynamic Behaviour in UML Models: Semantic Questions, October 2000
27. Spivey, J.M.: The Z Notation: A Reference Manual. Prentice-Hall Inc, Upper Saddle River (1989)
28. Su, W., Abrial, J.-R.: Aircraft landing gear system: approaches with Event-B to the modeling of an industrial system. In: Boniol, F., Wiels, V., Ait Ameur, Y., Schewe, K.-D. (eds.) ABZ 2014. CCIS, vol. 433, pp. 19–35. Springer, Heidelberg (2014)
29. Turner, E., Leuschel, M., Spermann, C., Butler, M.: Symmetry reduced model checking for B. In: Proceedings TASE, pp. 25–34. IEEE (2007)

A Compact Encoding of Sequential ASMs in Event-B

Michael Leuschel[1](✉) and Egon Börger[2]

[1] Institut für Informatik,
Universität Düsseldorf, Universitätsstr. 1, 40225 Düsseldorf, Germany
`leuschel@cs.uni-duesseldorf.de`
[2] Dipartimento di Informatica, University of Pisa, Pisa, Italy

Abstract. We present a translation of sequential ASMs to Event-B specifications. The translation also addresses the partial update problem, and allows a variable to be updated (consistently) in parallel. On the theoretical side, the translation highlights the intricacies of ASM rule execution in terms of Event-B semantics. On the practical side, we show on a series of examples that the Event-B encoding remains compact and is amenable to proof within Rodin as well as animation and model checking using PROB.

Keywords: ASM · B-method · Model checking · Constraint-solving · Tools

1 Motivation

ASMs have been used since 1989 to model computational systems of different kinds; various tools have been built to simulate such models [5,12,13] and various theorem proving systems have been adopted to prove properties of ASMs, for references see [8, Chap. 8.1] and more recent publications in the ABZ Conference Proceedings. Given the close relationship between ASMs and B and Event-B [2] models we want to investigate more closely the relation between ASMs, B and Event-B concepts. In particular we want to clarify whether and how one can translate ASMs to Event-B in a reasonable fashion, without blow-up, such that the translation can be effectively applied and permits the use of the animation, model checking, and constraint solving tool PROB[17] as well as of provers [3,10,11] to mechanically support proving properties for ASMs.

2 Background

2.1 ASM Background

The syntax and semantics of ASMs is defined in [8, Chap. 2.4]. We recapitulate the part considered in this paper.

Part of this research by the second author was funded by a renewed Forschungspreis grant of the Humboldt Foundation in the summer of 2015.

© Springer International Publishing Switzerland 2016
M. Butler et al. (Eds.): ABZ 2016, LNCS 9675, pp. 119–134, 2016.
DOI: 10.1007/978-3-319-33600-8_7

ASMs are defined as rules which transform structures of a given signature, their 'states'. Without loss of generality one can consider a signature as a family of function symbols f^n of arity n; predicates (relations) can be dealt with in terms of their characteristic Boolean-valued functions. Thus a state over such a signature is the mathematical structure formed by a set U (called the (super-) universe of the state) together with an interpretation of each function symbol f^n by an n-ary function with arguments and values in U; specific domains can be defined as subsets of the superuniverse U.

There are various constructs which can appear in ASM rules to modify a state by changing the value of some of its functions at some of their arguments, essentially using updates of form $f(s_1, \ldots, s_n) := t$ where s_i, t are terms (functional expressions). In particular the following recursively defined set of ASM rules turned out to be sufficient to describe any sequential algorithm at whatever level of abstraction[1]

> **skip**   // empty action
> $f(s_1, \ldots, s_n) := t$ with terms s_i, t // assignment rule
> R **par** S // simultaneous parallel execution
> **if** ϕ **then** R **else** S with formula ϕ // case distinction
> **let** $x = t$ **in** R // call by value
> **choose** x **with** ϕ **do** R // nondeterministic choice

Notably the **forall** construct for universal quantification and the two clauses for submachines and sequential functional composition in [8, Table 2.2] are missing.

Usually R_1 **par** R_2 **par** $\ldots$ **par** R_n is written in vertical notation without **par**. Below we also write $S \mid S'$ for S **par** S'.

The parallel composition operator **par** is considered to be characteristic for ASMs. The semantics of $R = R_1$ **par** R_2 **par** $\ldots$ **par** R_n is to execute all component rules R_i simultaneously, atomically. This means that for one (an atomic) step of R in a given state s, the updates of each R_i are applied together in one step to s. This yields the next state s' (successor state of s) if the computed set of updates is consistent; otherwise the step cannot be computed and the successor state of s is undefined.

Updates are pairs $((f, (a_1, \ldots, a_n)), val)$ of so-called locations $(f, (a_1, \ldots, a_n))$ and a newly to be assigned (not necessarily new) value for that location; here f is a function symbol (of arity n), $(a_1, \ldots, a_n)$ an argument tuple of elements of some set in the given state s—typically the interpretation in state s of some terms $s_1, \ldots, s_n$ in an assignment rule $f(s_1, \ldots, s_n) := t$—and val the interpretation of t in state s. A set of updates (usually called an 'update set') is **consistent** if it does not contain two pairs $(l, v), (l, v')$ with $v \neq v'$. Thus parallel updates to a same variable (location) are allowed, but they must be consistent. In case the update set is inconsistent (i.e., contains two updates $(l, v), (l, v')$ for the same location l with different values v, v') the next state s' is not defined. Note that 0-ary locations $(f, ())$ are just variables (written f as usual).

---

[1] This fact is known as the sequential ASM Thesis Gurevich proved in [14] from three natural postulates which axiomatize the underlying concept of 'sequential algorithm'.

For the deterministic form of the sequential ASM Thesis it turned out that the following sequential normal form ASMs suffice:

if ϕ_1 **then** $f_1(s_{1,1}, \ldots, s_{1,n_1}) := t_1$

$\vdots$

if ϕ_k **then** $f_k(s_{k,1}, \ldots, s_{k,n_k}) := t_k$

where the guards ϕ_i are boolean combinations of equations between terms.
Note that in fact

- **if** ϕ **then** R **else** S is equivalent to

  (**if** ϕ **then** R) **par** (**if not** ϕ **then** S)

- rules of the form **let** $x = t$ **in** R exist only for practical purposes, a hidden form of sequentialization as used for example in initializations. The rule is equivalent to **choose** x **with** $x = t$ **do** R, though **choose** is not needed for the deterministic case. Note that the **let** rule could also be programmed (without using **choose**) by mere **if then else** rules.

- occurrences of **choose** can be moved to the outside, using renaming to avoid clashes between variables. Under the assumption that x is not used in B and S, the construct **if** B **then** (**choose** x **with** ϕ **do** R) **else** S becomes

  choose x **with** $(B \Rightarrow \phi$ **and not** $(B) \Rightarrow x = default)$ **do**
      if B **then** R **else** S

  where *default* is some value compatible with the type of x. Indeed, when B is false, the variable x is not used and the role of the assignment $x = default$ is simply to prevent unnecessary non-determinism in the else branch.

In the non-deterministic case of the sequential ASM Thesis one has to consider the bounded choices an algorithm can make—not only the environment—using the **choose** operator. Gurevich's sequential ASM Thesis remains provable but with a slight restatement of one of the three postulates (namely the abstract-state postulate) and the following extension of the normal form to:

choose $i \in \{1, 2, \ldots, k\}$ **do**
    if $i = 1$ **then** R_1

    $\vdots$

    if $i = k$ **then** R_k

Here the transitions rules R_i consist of a finite set of parallel updates. For details see [8, pp. 306–7].

2.2 Event-B Background

Both the B-method [1] and its successor Event-B [2] are state-based formal methods rooted in set theory. Event-B has a richer refinement notion, with the aim of systems modelling rather than software development. On the other hand, Event-B has a much simpler structure for statements: notably there are no conditionals and no let statements. The static parts of an Event-B model, such as carrier sets, constants, axioms, and theorems are contained in contexts, whereas the dynamic parts of the model are contained in machines. A machine comprises variables, invariants, and events. An event consists of two main parts: guards and actions. Formally, an event has the following form:

$$\textbf{event } e = \textbf{any } t \textbf{ when } G(x,t) \textbf{ then } S(x,t,x') \textbf{ end}$$

Here, t are the parameters of the event and the guard $G(x,t)$ can be an arbitrarily complex predicate over the state variables x and the parameters t. The statements $S(x,t,x')$, however, are very restricted and consist of parallel assignments of the form

- $v := E(x,t)$ *(deterministic assignment)*
- $v :\in E(x,t)$ *(non-deterministic assignment from a set of values)*
- $v : |P(x,v',t)$ *(non-deterministic assignment using a predicate)*

The statement list can also be empty, which corresponds to **skip**. It is not allowed to assign to the same variable v twice within the same event.

An event is *enabled* if there exists a value for the parameters t which makes the guard $G(x,t)$ true. If no such value exists, the event is *disabled*. Let us present a small example. The following event decrements a variable x by the amount a. In case $x \leq 0$, the event is disabled, as no solution for $a \in 1..x$ exists.

$$\textbf{event } decrement = \textbf{any } a \textbf{ when } a \in 1..x \textbf{ then } x := x - a \textbf{ end}$$

All machines also contain a special event, the *initialisation* which is not allowed to refer to the current state of the variables. The way events are executed in Event-B is somewhat different to ASMs. First, the initialisation event is executed to generate an initial state of the model. In any given state, any enabled event e can be executed atomically, resulting in a new state. All the actions of e are executed in parallel. So, in contrast to ASMs, events are executed in isolation; events cannot be executed in parallel together (but the individual actions of an event are).

Event-B machines contain an *invariant*, which is a predicate over the variables (and constants) of the machine. In order to establish that the invariant is indeed true in all reachable states, proof obligations are generated: one has to prove that the initialisation establishes the invariant and that each event— when enabled— preserves the invariant.

The Event-B language supports a rich set of datatypes, encompassing integers, booleans, user-defined types, sets, relations, and (higher-order) functions. For this paper one notation for functions will be important:

$$\lambda x.P \mid E$$

Given a predicate P and an expression E, this represents the function whose domain is all those values for x which make P true, in which case the function returns the value of E. For example,

$$\lambda x.x > 0 \mid x * x$$

is the squaring function defined for strictly positive integers. In B, functions are just seen as relations which in turn are sets of pairs. The above function could also have been written as

$$\{x \mapsto y \mid x > 0 \wedge y = x * x\}$$

The function $\lambda x.x \in 1..3 \mid x*x$ is a finite function and could thus also have been written as a set extension $\{1 \mapsto 1, 2 \mapsto 4, 3 \mapsto 9\}$. Set operators can also be used for functions, e.g., the predicate $\{2 \mapsto 4\} \subset \lambda x.x > 0 \mid x * x$ is true.

In summary, the differences with ASMs are:

1. richer actions are allowed in ASM rules (conditionals, let, ...),
2. parallel updates to the same variable are allowed in ASMs and not in Event-B (we return to this in Sect. 4),
3. the way rules are fired (all enabled ones are fired simultaneously in ASMs as opposed to one enabled event in Event-B).

3   Translating Conditional and Parallel Statements

Both Event-B and ASM allow parallel updates but differ in a quite fundamental way. Indeed, as mentioned in Sect. 2.2, Event-B only allows parallel updates of disjoint variables, such as $x := 1 \| y := 2$. However, the following is not allowed in Event-B (or classical B):

$$x := 1 \| x := y$$

ASMs, however, do allow this parallel assignment but then impose a consistency condition on all parallel updates.[2] For this example, this implies that the update is allowed in case $y = 1$ and considered inconsistent otherwise. If the update set to be applied to a state s is inconsistent, the next state s' is not defined. When combined with conditional statements — which Event-B does not support— the translation becomes even more intricate. Let us take an ASM machine with a variable x initialised to 0 and the rule depicted in Fig. 1.

Suppose we are interested in proving that this rule preserves the invariant $x \in \mathbb{N}$. How can we translate this to Event-B, without using conditional actions and parallel assignments to the same variable x?

---

[2] This is related to the circumstance that for reasons of generality arbitrary terms s, not only variables, are permitted on the left side of an assignment statement $s := t$. Therefore a machine may contain assignment statements $t_1 := t$ and $t_2 := t'$ with syntactically different $t_i = f(t_{i,1}, \ldots, t_{i,n})$ for which however in some state S their evaluation may yield the same arguments $eval_S(t_{1,j}) = eval_S(t_{2,j})$ for all $1 \le j \le n$ resulting in two updates $(l, eval_S(t))$ and $(l, eval_S(t'))$ to the same location $l = (f, (eval_S(t_{1,1}), \ldots, eval_S(t_{1,n})))$ so that the consistency condition $eval_S(t) = eval_S(t')$ is required.

if $x > 10$ then $x := x - 1$ |if $x < 5$ then $x := x + 1$

Fig. 1. Simple ASM rule

3.1   A Simple Translation Using Case Distinctions

The first solution that comes to mind is to encode every possible path through the rule as one event in Event-B. In this case there are in principle four possible paths: each of the two if-conditions can be either true or false. A translation of this ASM machine using four events is presented in Fig. 2.

```
machine ASM_4
variables x
invariants @inv x∈N
events
  event INITIALISATION begin @ini x=0
  end
  event asmif_tt when @gtt x>10 ∧ x<5 theorem @thm x-1=x+1 then
     @act  x = x-1
  end
  event asmif_tf when @gtf x>10 ∧ ¬(x<5) then
     @dec  x = x-1
  end
  event asmif_ft when  @gft ¬(x>10) ∧ x<5  then
     @inc  x = x+1
  end
  event asmif_ff when @gff ¬(x>10) ∧ ¬(x<5) end
end
```

Fig. 2. Non-linear translation of if $x > 10$ then $x := x - 1$ |if $x < 5$ then $x := x + 1$ from Fig. 1

On the positive side, this machine can be animated and model checked with PROB and proven fully automatically in Rodin with the standard autotactics. In other words, we have established that our ASM rule preserves the invariant $x \in \mathbb{N}$. Note that, for the case that both assignments are triggered (asmif_tt), we have added the guard theorem $x + 1 = x - 1$ to encode that these two assignments yield the same result. In Event-B this gives rise to the following proof obligation (where the guards and invariants are in the hypotheses):

$$x \in \mathbb{N} \land x > 10 \land x < 5 \models x + 1 = x - 1$$

As the hypotheses are unsatisfiable, the theorem can be proven. We have thus also proven that no conflict between the parallel assignments to x can occur. If we change the test $x < 5$ to, e.g., $x < 15$ in Figs. 1 and 2 the corresponding guard theorem proof obligation $x \in \mathbb{N} \land x > 10 \land x < 15 \models x + 1 = x - 1$ can no longer be discharged.

On the negative side, however, this translation can lead to a combinatorial blow up in the number of events. The many events will share many common predicates (i.e., $x > 10$ or $x < 5$ above) which will be re-evaluated by tools such as a model checker or even the provers. In this paper, we try to find a translation into Event-B which does *not* lead to such an explosion of the events, but which remains linear in the size of the original ASM.

3.2 Translation Using Update Functions

One idea of our solution is to encode the updates to variables into composable *update functions*. Basically, an update function u for the variable x will be used to construct an Event-B assignment of the form

$$x := u(x)$$

This solution can later be extended (in Sect. 4) to deal with the important issue of partial updates, but it also solves the problem of conditional total updates. Let us return to our ASM rule from Fig. 1, where we now add the implicit else branches:

if $x > 10$ **then** $x := x - 1$ **else skip** $|$ **if** $x < 5$ **then** $x := x + 1$ **else skip**

The idea is that every branch First, the update function for the (implicit) else branches is the identity function $\mathrm{id}_{\mathbb{Z}}$ for the type $\mathbb{Z}$ of x, defined by:

$$\mathrm{id}_T = \lambda v.v \in T \mid v$$

Second, the update function for the assignment $x := x - 1$ is the function $cst_{\mathbb{Z}}(x - 1)$ defined by (where x' does not occur in C):[3]

$$cst_T(C) = \lambda v.v \in T \mid C$$

Similarly, the update function for the assignment $x := x + 1$ is the function $cst_{\mathbb{Z}}(x+1)$. While id_T copies the old value v of a variable, $cst_T(C)$ simply ignores it and overwrites it with C.

For conditionals, we construct update functions by inserting conditions into the update functions of each branch. This scheme is defined formally below. First, we need the following two auxiliary definitions (where v is a variable not occurring in u):

- $cond(P, u) = \lambda v.P \mid u(v)$
- $if(P, u_1, u_2) = cond(P, u_1) \cup cond(\neg P, u_2)$

We can now formally define the ternary relation $S \leadsto_x u$, denoting that the ASM statement S results in the update function u for the variable x. We assume that, like above, a conditional statement without else branch is first translated into a conditional statement whose else branch is **skip**.

We also have the set of location (entry) names *LocEntry*, which are the names of functions/variables which are updated in the ASM machine under consideration.

---

[3] An overwrite particle in the terminology of [15].

From the point of view of the Event-B translation, these will be the variable names of the B machine. Below, $type(E)$ refers to the type of an expression.

$$\frac{}{\mathbf{skip} \leadsto_y \mathrm{id}_{type(y)}} \; y \in LocEntry$$

$$\frac{}{x := E \leadsto_x cst_{type(x)}(E)}$$

$$\frac{}{x := E \leadsto_y \mathrm{id}_{type(y)}} \; y \in LocEntry \setminus \{x\}$$

$$\frac{S \leadsto_x u \wedge S' \leadsto_x u'}{\mathbf{if} \; B \; \mathbf{then} \; S \; \mathbf{else} \; S' \leadsto_x if(B, u, u')}$$

$$\frac{S_1 \leadsto_x u_1 \wedge S_2 \leadsto_x u_2}{S_1 \mid S_2 \leadsto_x u_1 \circ u_2}$$

Rule 1 above stipulates that **skip** does not modify any location, i.e., we always obtain the identity function as update function. Rule 2 says that an assignment to x does indeed modify location x, while Rule 3 stipulates that it does not influence locations $y \neq x$, i.e., we obtain the identity update function $\mathrm{id}_{type(y)}$ for y. In the case of the parallel composition in the fifth rule, how do we know that the update functions u_1 and u_2 are compatible ? Also, the last rule has two obvious solutions: $u_1 \circ u_2$ and $u_2 \circ u_1$. Which one should be chosen? Basically, we will add proof obligations to ensure that $u_1 \circ u_2 = u_2 \circ u_1$, which in turn guarantees that the update functions are compatible. Indeed, commutativity is one way of defining compatible updates (see the functional applicative algebras in [15]). As such we will add a guard theorem $u_1 \circ u_2 = u_2 \circ u_1$ to the translated events; this theorem will result in a proof obligation but does not influence the event execution as such.[4] Actually, for technical reasons we will add the following weaker guard theorem, which only requires commutativity for the actual values of x encountered:

$$u_1(u_2(x)) = u_2(u_1(x))$$

This theorem can be dealt with more easily by the Rodin provers and by tools such as PROB. The precise computation of these guard theorems, also encompassing partial updates, will be formalised later in Sect. 4.

Note that the above rules distinguish between **skip** and $x := x$: the former has the update function $\mathrm{id}_{\mathbb{Z}}$ while the latter has the update function $cst_{\mathbb{Z}}(x) = \lambda v.v \in \mathbb{Z} \mid x$. This is very important. Indeed, $x := x \mid x := x - 1$ is inconsistent while $\mathbf{skip} \mid x := x - 1$ is valid and equivalent to $x := x - 1$. Indeed, for the latter we have commutativity of the update functions:

$$\mathrm{id}_{\mathbb{Z}} \circ cst_{\mathbb{Z}}(x - 1) = cst_{\mathbb{Z}}(x - 1) \circ \mathrm{id}_{\mathbb{Z}} = cst_{\mathbb{Z}}(x - 1)$$

but not for the former:

$$cst_{\mathbb{Z}}(x) \circ cst_{\mathbb{Z}}(x - 1) = cst_{\mathbb{Z}}(x) \neq cst_{\mathbb{Z}}(x - 1) \circ cst_{\mathbb{Z}}(x)$$

---

[4] See http://handbook.event-b.org/current/html/theorems.html.

We can now use the following inference rules to construct the action parts of an Event-B event for an ASM rule R as follows

$$\frac{R \leadsto_x u}{R \leadsto_{act} x := u(x)} \quad x \in LocEntry \wedge u \neq \mathrm{id}_{type(x)}$$

To construct the Event-B action we simply take all solutions A for $R \leadsto_{act} A$ and put them into parallel and then add the commutativity theorems.

Let us return to our ASM rule R from Fig. 1. We have $R \leadsto_x u_1 \circ u_2$ with

- $u_1 = if(x > 10, cst(x - 1), id)$ and
- $u_2 = if(x < 5, cst(x + 1), id)$.

We can compute these update functions and their composition as follows:

- $cst(x - 1) = \lambda v.v \in \mathbb{Z} \mid x - 1$
- $u_1 = (\lambda v.x > 10 \mid x - 1) \cup (\lambda v.\neg(x > 10) \mid v)$
- $u_2 = (\lambda v.x < 5 \mid x + 1) \cup (\lambda v.\neg(x < 5) \mid v)$
- $u_1 \circ u_2 = (\lambda v.x > 10 \mid x - 1) \cup (\lambda v.\neg(x > 10) \wedge x < 5 \mid x + 1) \cup$
  $\qquad (\lambda v.\neg(x > 10) \wedge \neg(x < 5) \mid v)$
- $u_2 \circ u_1 = (\lambda v.x < 5 \mid x + 1) \cup (\lambda v.\neg(x < 5) \wedge x > 10 \mid x - 1) \cup$
  $\qquad (\lambda v.\neg(x > 5) \wedge \neg(x > 10) \mid v) =$
  $u_1 \circ u_2$ because $x > 10 \wedge x < 5$ is unsatisfiable.

```
machine ASM1
variables x
invariants @inv x∈ℕ
events
    event INITIALISATION begin @ini x=0
    end
    event asmifs // if x>10 then x:=x-1 | if x<5 then x:=x+1
        any u1 u2 where
        @g1 u1 = (λv·v∈ℤ ∧ x>10 | x-1)  ∪  (λv·v∈ℤ ∧ ¬(x>10)| v)
        @g2 u2 = (λv·v∈ℤ ∧ x<5 | x+1)  ∪  (λv·v∈ℤ ∧ ¬(x<5)| v)
        theorem @comm u1(u2(x)) = u2(u1(x))
        then
            @a   x = u1(u2(x))
        end
end
```

Fig. 3. Event-B translation of an ASM rule with parallel conditional update on the same variable

Figure 3 shows the complete Event-B translation of the above example with the commutativity theorem. We use the syntax of the text editor Camille [6]. The labels such as @g1, @g2 and @a only play a role for user feedback during proof; they have no attached semantics. For readability we have extracted the two composed update functions into event parameters. The machine has three proof obligations related to the translated event `asmifs`, all of which can be automatically discharged using the SMT prover plugin [10,11]. However, the standard autotactics of Rodin no longer discharge all of them. Indeed, proving commutativity typically requires case distinctions, which the SMT prover plugin is good at but not so much the Atelier-B provers ML and PP. PROB can animate and model check the model; it has 6 distinct states ($x \in 0..5$).

4   Partial Update Problem

In Event-B a function—just like a relation— is seen like a set of tuples. An array is just a special case of a function, where the domain is a contiguous set of indexes. We will adopt the same view for our ASM machines. Below we also use $\lambda v.E$ as a shortcut for $\lambda v.v \in type(E) \mid E$.

The parallel update problem becomes more interesting when partial updates are concerned. An assignment like $f(2) := 3$ is a partial update in the sense that the function f is only changed at the position 2. A possible value for f would be $\{1 \mapsto 0, 2 \mapsto 0\}$ and the effect of the above assignment would be to change f into $\{1 \mapsto 0, 2 \mapsto 3\}$. Furthermore, in Event-B the above assignment is seen as syntactic sugar for $f := f \mathbin{\lhd\mkern-9mu-} \{2 \mapsto 3\}$, where the override operator $\mathbin{\lhd\mkern-9mu-}$ can be defined as follows:

$$r \mathbin{\lhd\mkern-9mu-} s = \{x \mapsto y \mid x \mapsto y \in s \vee (x \mapsto y \in r \wedge x \notin dom(s))\}$$

As such Event-B does not allow the parallel updates:

$$f(1) := 2 \mid f(2) := 3$$

as indeed they get translated to the following parallel total updates of f:

$$f := f \mathbin{\lhd\mkern-9mu-} \{1 \mapsto 2\} \mid f := f \mathbin{\lhd\mkern-9mu-} \{2 \mapsto 3\}$$

First, parallel updates of the same variable are not allowed in Event-B. Second, even if they were, these two total updates are not consistent, as $f \mathbin{\lhd\mkern-9mu-} \{1 \mapsto 2\} = \{1 \mapsto 2, 2 \mapsto 0\} \neq \{1 \mapsto 0, 2 \mapsto 3\}$ for our value of $f = \{1 \mapsto 0, 2 \mapsto 0\}$ above. In ASMs, however, these parallel partial updates are allowed (and are not seen as equivalent to a total update) and result in $f = \{1 \mapsto 2, 2 \mapsto 3\}$.

The situation is somewhat similar to the conditional updates in Sect. 3, and our solution is the same: we represent partial updates also as update functions which can be composed and which have to be checked for commutativity. So, in the example above we would generate two update functions:

1. $u_1 = \lambda v.v \mathbin{\lhd\mkern-9mu-} \{1 \mapsto 2\}$
2. $u_2 = \lambda v.v \mathbin{\lhd\mkern-9mu-} \{2 \mapsto 3\}$

We have commutativity $u_1 \circ u_2 = u_2 \circ u_1 = \lambda v.v \mathbin{\lhd\mkern-9mu-} \{1 \mapsto 2, 2 \mapsto 3\}$ and we can achieve the combined parallel effect of the assignments using the single assignment

$$f := u_1(u_2(f))$$

Let us consider again the assignment $x := x + 1$ from Sect. 3. There we have used the update function $u_c = cst_\mathbb{Z}(x+1)$. However, an alternate update function would have been $u_{cum} = \lambda v.v \in \mathbb{Z} \mid v + 1$. Note that u_c is idempotent and does not commute with $u'_c = cst_\mathbb{Z}(x - 1)$. u_{cum} on the other hand is not idempotent

$$u_{cum} \circ u_{cum} = \lambda v.v \in \mathbb{Z} \mid v + 2$$

but does commute with $u'_{cum} = \lambda v.v \in \mathbb{Z} \mid v - 1$:

$$u_{cum} \circ u'_{cum} = u'_{cum} \circ u_{cum} = \mathrm{id}_\mathbb{Z}$$

So, the update function u_{cum} corresponds to another, *cumulative* interpretation of $x := x + 1$; when executed in parallel with itself it increments x by 2. u_{cum} is an example of what are called particles in [15].

Figure 4 contains a few more example update functions. For queues we assume the representation of a sequence of length n as a total function with domain $1..n$. Within the update function for pop, $(\{1\} \lhd v)$ removes the first element from the sequence v without adjusting the indices of the other elements; that is achieved by using relational composition with the successor function succ. Our translation to Event-B can in principle cope with any of these interpretations, cumulative or not. In the remainder of the paper we will only focus on total assignment and function update, though.

Description	Syntax	Update function u
Regular total assignment	$x := y$	$\lambda v.y$
Cumulative integer addition	$x + = \Delta$	$\lambda v.v \in \mathbb{Z} \mid v + \Delta$
Function update	$x(s) := t$	$\lambda v.v \lhd \{s \mapsto t\}$
Function update level 2	$x(s_1)(s_2) := t$	$\lambda v.v \lhd \{s_1 \mapsto (v(s_1) \lhd \{s_2 \mapsto t\})\}$
Cumulative set addition	$x \cup = y$	$\lambda v.v \cup y$
Cumulative set removal	$x \setminus = y$	$\lambda v.v \setminus y$
Queue push	$push(x, e)$	$\lambda v.v \cup \{(card(v) + 1) \mapsto e\}$
Queue pop	$pop(x)$	$\lambda v.(\mathrm{succ} \,;(\{1\} \lhd v))$

Fig. 4. Example update functions ($\lambda v.E$ is a shortcut for $\lambda v \in type(E) \mid E$)

The core idea is that two partial updates are consistent if they commute and the practical question is how can we check this, in particular when multiple update functions are combined. We could try out all permutations, but that quickly blows up. Note that the composition of multiple update functions is always associative, as function composition $\circ$ is associative. We return to this issue later.

5 Translation Scheme

We now finish the translation scheme started in Sect. 3.2, adding rules for partial updates and describing how the guard theorems are generated. Basically, we have already described the rule $ASM \leadsto_{act} x := u(x)$ and below we describe the relation $ASM \leadsto_{thm} G$. Let $\{A_1, \ldots, A_k\} = \{A \mid ASM \leadsto_{act} A\}$ and let $\{G_1, \ldots, G_m\} = \{G \mid ASM \leadsto_{thm} G\}$. Then we generate the Event-B event:

event ASM = **when theorem** G_1 ... **theorem** G_m **then** $A_1 \mid \ldots \mid A_k$ **end**

Observe that we have no parameters and no proper guard, just guard theorems. The guard theorems just give rise to proof obligations which ensure that the update functions commute and we have consistent updates. We could (and actually do so in the implementation) lift the individual update functions u in the A_j's to be parameters and reuse them within the G_i's. We already did so in Fig. 3.

In addition to the basic update functions id, $cst(C)$, and $if(P, u_1, u_2)$ we now need one more construct defined as follows:

- $upd(s, t) = \lambda v.v \lessdot \{s \mapsto t\}$

We extend the inference rules for $\leadsto$ from Sect. 3.2 by the two following ones:

$$\overline{x(E) := F \leadsto_x upd(E, F)}$$

$$\frac{}{x(E) := F \leadsto_y \mathrm{id}_{type(y)}} \quad y \in LocEntry \setminus \{x\}$$

The guard theorems are generated using this inference rule for $\leadsto_{thm}$:

$$\frac{ASM \leadsto_x u_1 \circ \ldots u_k}{ASM \leadsto_{thm} perm_x((u_1 \circ \ldots u_k))}$$

We define $perm_x(u)$ to be true if u is a basic update function not of the form $u_1 \circ u_2$. Otherwise we have:

$$perm_x(u_1 \circ u_2) =_{def} (u_1(u_2(x)) = u_2(u_1(x))) \wedge perm_x(u_1) \wedge perm_x(u_2)$$

Take the following ASM rule $f(x) := 1 \mid \mathbf{if}\ y \neq x\ \mathbf{then}\ f(y) := -1$ where we wish to establish that the invariant $f(x) \geq 0$ is preserved by the rule. The result of our translation can be found in Fig. 5.

For this example the proofs no longer go through automatically in Rodin: we have to do manual case distinctions on $x = y$. In future, one should probably generate a library of "update function proof rules"; we return to this issue below. We can animate and model check the system using PROB without problem though.

Some Optimisations. In order to make the translation more amenable to proof and animation, we suggest a few optimisation rules (some of which we have implemented in our prototype). Let us first define this notation:

$$u_1 \diamond u_2 =_{def} (u_1 \circ u_2) = (u_2 \circ u_1)$$

```
event asm // f(x) := 1 || if y/= x then f(y) := -1
  any u1 u2 where
    @g1 u1 =   (λg·g∈z↔z| g◁{x↦1})
    @g2 u2 =   (λg·g∈z↔z ∧ y≠x| g◁{y↦-1}) ∪
               (λg·g∈z↔z ∧ y=x| g)
    theorem @commutativity u1(u2(f)) = u2(u1(f))
  then
    @a  f = u1(u2(f))
  end
```

Fig. 5. Event-B translation of an ASM rule with partial updates and conditionals

Note that $u_1 \diamond u_2$ is symmetric. The following selection of results hold for commutativity of our update functions and can ease proving our guard theorems:

$$id_T \diamond u \Leftrightarrow \top$$

$$cst_T(C_1) \diamond cst_T(C_2) \Leftrightarrow (C_1 = C_2)$$

$$cst_T(C) \diamond upd(s, t) \Leftrightarrow (C(s) = t)$$

$$upd(s, t) \diamond upd(s', t') \Leftrightarrow (s \neq s' \vee t = t')$$

It would make sense to add these as theorems to our translation, to ease automatic proving.

Initialisation. The Event-B initialisation allows no parameters. In case parameters are used in an ASM initialisation, we either need to translate these into a second initialisation event (along with a boolean variable *isInitialised*). Alternatively, one can use Event-B constants to represent the parameters.

Well-Definedness. The ASM statement `f(x) := undef` is translated to domain subtraction in Event-B $f := \{x\} \lhd f$. Apart from that we suppose well-definedness in Event-B style.

Translating Choose and Forall. The proof of the Parallel ASM Thesis in [7] yields the following normal form for parallel ASMs R (i.e., sequential ASMs with the additional **forall** x **with** ϕ **do** M construct):

forall $x \in U_R$
  if $Cond_1(x)$ **then** $upd_1(x)$
  $\vdots$
  if $Cond_r(x)$ **then** $upd_r(x)$

for a multiset term U_R, single assignments $upd_i(x)$ of the form $f(t_1, \ldots, t_n) := t$ and some r depending on the given ASM program R. Essentially the term U_R represents the multiset U of updates to be applied by R—the empty set in case R computes an inconsistent multiset of updates so that if $U \neq \varnothing$ the assignments $upd_i(x)$ are consistent.

Suppose now we have a parallel ASM rule **forall** i **with** $\Phi(i)$ **do** $S(i)$ in normal form (i.e., where **forall** appears only as the outer constructor of the ASM).

If the set $\{i \mid \Phi(i)\}$ is static, the machine can be translated to a sequential machine as follows. One can transform $\Phi(i)$ into $i \in \{K_1, \ldots, K_n\} \wedge \Phi'(i)$, where the expressions $K_1, \ldots, K_n$ are known statically. Then we can generate the following sequential ASM which is equivalent to the given ASM:

if $\Phi'(K_1)$ **then** $S(K_1)$
if $\Phi'(K_2) \wedge K_2 \notin \{K_1\}$ **then** $S(K_2)$
$\vdots$
if $\Phi'(K_n) \wedge K_n \notin \{K_1, \ldots, K_{n-1}\}$ **then** $S(K_n)$

to which one can apply the translation rules described above.

6   Prototype, Discussions and Future Work

Prototype Implementation and Experiments

A prototype translator has been implemented within PROB, using Prolog to implement our inference rules. It uses the classical B syntax to express ASMs. To avoid generating errors for parallel assignments to the same variable, a preference within PROB has been added to turn off a variety of static checks. The translator generates the update functions and theorems as described above. We have then manually copied and pasted the result into Rodin.

However, while writing the paper we have streamlined and improved the notations and also the transformation process. Initially we experimented with various other approaches, notably collecting the updates in sets. All of these were more cumbersome than the solution presented here and actually less amenable to automated proof.

The prototype implementation should now be rewritten using the simpler concepts and should also work on "real" ASM input files rather than use classical B syntax. We plan to do this in future work. Still, we were able to experiment with some non-trivial ASM rules and animate, model check, and prove them.

Discussion and Future Work

Translating Monitored Variables. ASMs foresee certain variables (locations) to be monitored and implicitly modified by the environment (between two internal ASM steps, see [8, Definition 2.4.22]). We can translate this behaviour by adding a "turn" variable and the respective guards to alternate between executing the ASM rules proper and the environment which modifies the monitored variables.

Translating Derived Functions. ASMs also foresee derived functions such as:

```
derived isMaster(m) = (index(m)=0)
```

There are various ways these could be translated. One solution would be to add new definition using the Event-B Theory Plugin [9]. Alternatively, one could add a new function in a context which takes all variables as arguments. For this example that would be: $isMaster = \lambda m.m \in Agents \mid bool(index(m)) = 0$.

Translating Submachine Calls. These use "call by reference" in ASM. In our experiments we have thus expanded such rule calls like macros (possibly with renaming to avoid clashes and variable capture).

Bounded Exploration Postulate. Sequential ASMs satisfy Gurevich's bounded exploration postulate so that there can only be finitely many changes to locations in one step. In B (as in parallel ASMs) you can do infinitely many changes, e.g., the assignments $f := \mathbb{N} \times \{1\}$ or $f := \{d \mapsto r \mid d \in \mathbb{N} \wedge r = d * x\}$ set the value of f in infinitely many locations, as do the non-sequential parallel ASMs **forall** $n \in \mathbb{N}$ $f(n) := 1$ resp. **forall** $n \in \mathbb{N}$ $f(n) := n * x$.

Validation. To prove our translation correct we would need an expression of ASM semantics in B; but this is what we are trying to develop in the first place. However, once we have a tool in place we could cross check the results of various tools, such as [12] or [4].

By combining all updates for a variable, our translation avoids re-evaluating the same predicate multiple times (unlike the naive translation in Fig. 2). However, the same predicate can still appear in different update functions for different variables. We propose to solve this issue by using PROB's common-sub-expression elimination.[5]

To what extent our technique can really scale to bigger ASM machines is still open. As far as proving is concerned, the commutativity theorems usually require a series of case distinctions. As such, provers such as the SMT prover plugin [10, 11] or the PROB Disprover [16] are probably essential for our translated models. Theorems about applicative algebras in [15] may be of help. Simulation and model checking can also be done, but it still remains open what the performance will be compared to running CoreASM [12] or using AsmetaSMV [4].

Outside of tooling support for ASMs, we hope that our work has also clarified the intricacies of ASM rules to Event-B researchers, and also shown the ASM researchers the power of B's predicates and expressions.

In summary, we have developed a translation from sequential ASMs to Event-B machines, which prevents a blow-up of the number of events. The translated machines can be validated using the Event-B tools. We hope that the efforts help in bringing the ASM and B communities closer to together.

---

[5] A recent feature of PROB; it needs to be explicitly enabled via the **CSE** preference; it does not work across multiple events and thus cannot be applied to Fig. 2.

Acknowledgement. We would like to thank the reviewers of ABZ'16 for very useful feedback. The second author thanks Laurent Voisin for discussing with him during the Dagstuhl seminar *Integration of Tools for Rigorous Software Construction and Analysis* (September 8–13, 2013) the problem of an Asm2EventB translation. (http:// drops.dagstuhl.de/opus/volltexte/2014/4358/.)

References

1. Abrial, J.-R.: The B-Book. Cambridge University Press, New York (1996)
2. Abrial, J.-R.: Modeling in Event-B: System and Software Engineering. Cambridge University Press, New York (2010)
3. Abrial, J.-R., Butler, M., Hallerstede, S., Voisin, L.: An open extensible tool environment for Event-B. In: Liu, Z., Kleinberg, R.D. (eds.) ICFEM 2006. LNCS, vol. 4260, pp. 588–605. Springer, Heidelberg (2006)
4. Arcaini, P., Gargantini, A., Riccobene, E.: AsmetaSMV: a way to link high-level ASM models to low-level NuSMV specifications. In: Frappier, M., Glässer, U., Khurshid, S., Laleau, R., Reeves, S. (eds.) ABZ 2010. LNCS, vol. 5977, pp. 61–74. Springer, Heidelberg (2010)
5. The Abstract State Machine Metamodel Website (2006). http://asmeta. sourceforge.net
6. Bendisposto, J., Fritz, F., Jastram, M., Leuschel, M., Weigelt, I.: Developing Camille, a text editor for Rodin. Softw. Pract. Exp. **41**(2), 189–198 (2011)
7. Blass, A., Gurevich, Y.: Abstract state machines capture parallel algorithms: correction and extension. ACM Trans. Comput. Log. **8**(3), 19:1–19:32 (2008)
8. Börger, E., Stärk, R.F.: Abstract State Machines. A Method for High-Level System Design and Analysis. Springer, Heidelberg (2003)
9. Butler, M., Maamria, I.: Practical theory extension in Event-B. In: Liu, Z., Woodcock, J., Zhu, H. (eds.) Theories of Programming and Formal Methods. LNCS, vol. 8051, pp. 67–81. Springer, Heidelberg (2013)
10. Déharbe, D.: Automatic verification for a class of proof obligations with SMT-solvers. In: Proceedings ASM, pp. 217–230 (2010)
11. Déharbe, D., Fontaine, P., Guyot, Y., Voisin, L.: SMT solvers for Rodin. In: Derrick, J., Fitzgerald, J., Gnesi, S., Khurshid, S., Leuschel, M., Reeves, S., Riccobene, E. (eds.) ABZ 2012. LNCS, vol. 7316, pp. 194–207. Springer, Heidelberg (2012)
12. Farahbod, R., et al.: The CoreASM Project. http://www.coreasm.org and https:// github.com/coreasm/
13. Foundations of Software Engineering Group, Microsoft Research. AsmL (2001). http://research.microsoft.com/foundations/AsmL/
14. Gurevich, Y.: Sequential abstract state machines capture sequential algorithms. ACM Trans. Comput. Log. **1**(1), 77–111 (2000)
15. Gurevich, Y., Tillmann, N.: Partial updates. Theor. Comput. Sci. **336**(2-3), 311–342 (2005)
16. Krings, S., Bendisposto, J., Leuschel, M.: From failure to proof: the ProB disprover for B and Event-B. In: Calinescu, R., Rumpe, B. (eds.) SEFM 2015. LNCS, vol. 9276, pp. 199–214. Springer, Heidelberg (2015)
17. Leuschel, M., Butler, M.J.: ProB: an automated analysis toolset for the B method. STTT **10**(2), 185–203 (2008)

Proof Assisted Symbolic Model Checking
for B and Event-B

Sebastian Krings$^{(\boxtimes)}$ and Michael Leuschel

Institut für Informatik, Universität Düsseldorf,
Universitätsstr. 1, 40225 Düsseldorf, Germany
{krings,leuschel}@cs.uni-duesseldorf.de

Abstract. We have implemented various symbolic model checking algorithms, like BMC, k-Induction and IC3 for B and Event-B. The high-level nature of B and Event-B accounts for complicated constraints arising in these symbolic analysis techniques. In this paper we suggest using static information stemming from proof obligations to simplify occurring constraints. We show how to include proof information in the aforementioned algorithms. Using different benchmarks we compare explicit state to symbolic model checking as well as techniques with and without proof assistance. In particular for models with large branching factor, e.g., due to complicated data values being manipulated, the symbolic techniques fare much better than explicit state model checking. The inclusion of proof information results in further clear performance improvements.

Keywords: B-Method · Event-B · Proof · Symbolic model checking

1 Introduction and Motivation

Model checking is one of the key techniques used in formal software development. Two variants are currently in use: explicit state model checking and symbolic model checking. In explicit state model checking, every state is computed, the invariant is verified and discovered successor states are queued to be analyzed themselves. Symbolic model checking on the other hand tries to represent the state space and possible paths through it by predicates representing multiple states at once. Instead of stepwise exploration of the state space graph, the model checking problem is encoded as a formula and given to a constraint solver.[1]

So far, existing model checkers for B and Event-B like PROB [20,21], Eboc [23], pyB [30] or TLC [31] (via [17]) rely on explicit state model checking. PROB features some symbolic techniques for error detection [14] and test-case generation [27], but not full-blown symbolic model checking. This is mostly due

---

S. Krings and M. Leuschel—Part of this research has been initially sponsored by the EU funded FP7 project 287563 (ADVANCE).

[1] BDD-style model checking [10] is also called symbolic model checking. In recent work PROB has been integrated with LTSMin for such kind of model checking.

© Springer International Publishing Switzerland 2016
M. Butler et al. (Eds.): ABZ 2016, LNCS 9675, pp. 135–150, 2016.
DOI: 10.1007/978-3-319-33600-8_8

to the high-level nature of B and Event-B. Both the usage of higher-order constructs and the underlying non-determinism accounts for complicated constraints during symbolic model checking. Some complexity can be coped with by relying on SMT solvers [12] or SAT solvers [25], but this is not always the case.

In this paper we have implemented various symbolic model checking algorithms for B and then study various ways to use proof information to optimize them. The proof information is used to strengthen constraints and reduce the counterexample search space. For Event-B, the information stems from discharged proof obligations exported from Rodin [1]. For classical B, no automatic proof information is available at the moment within PROB.[2] However, we can recompute the information using PROB's proof capabilities as outlined in [18]. Essentially, for a B operation with before-after-predicate BA we search for a solution to

$$invariant(x) \land BA(x, x') \land \neg conjunct\_of\_invariant(x').$$

If PROB reports a contradiction, we know that the operation can not lead to a violation of the particular conjunct. In addition to the technique in [18], we developed a bridge to the Atelier B provers ml and pp.

All techniques used in this paper have been implemented both for classical B and Event-B. Both languages will be used in our empirical evaluation. For the sake of brevity we will only talk about Event-B events in the following sections instead of distinguishing events and operations.

We will introduce the model checking algorithms BMC, k-Induction and IC3 in Sects. 2.1, 2.2, and 2.3. For each of them we will show how to include proof information into the occurring constraints. Following, in Sect. 3, we will empirically compare symbolic model checking to explicit state model checking and model checking with and without proof assistance. Discussion and conclusions will be presented in Sect. 4.

2   Proof Assisted Symbolic Model Checking

When using the B method to develop a software or system, one often alternates between different phases. Among those are writing and adapting the specification, manual and automated proof efforts as well as model checking. Yet, the different steps are only loosely coupled when it comes to tool support.

In [4] the authors have shown how to augment explicit state model checking with proof information. In the following, we introduce three symbolic model checking techniques for B and incorporate proof information in a similar fashion.

We will use the running example in Fig. 1 to illustrate various concepts in our paper. First, let us introduce the notation we will be using. By x we will denote a vector of state variables. x' denotes the state variables in the successor state. A predicate p over the state variables x is denoted by $p(x)$. The same predicate over the successor state is written as $p(x')$. By $Events$ we denote the set of events of a B machine. By Inv we denote its invariant.

---

[2] In theory, one could export proof information from Atelier B.

```
MACHINE Counter
CONSTANTS m
PROPERTIES m : {127,255}
VARIABLES c
INVARIANT c>=0 & c<=m
INITIALISATION c:=0
OPERATIONS
   incby(i) = PRE i:1..64 THEN c := c+i END
END
```

Fig. 1. A simple, erroneous B machine

Definition 1. *For an event* $evt \in Events$ *let* $BA_{evt}(x, x')$ *denote the before-after-predicate connecting state variables in* x *to their successors in* x'.

Definition 2. *For a predicate* $p = \bigwedge_{i \in I} p_i$ *and event* evt *let* $proven_{evt,p}$ *denote a set of conjuncts* p_i *that are proven to hold after the execution of* evt *on a* p*-state, i.e., we have* $proven_{evt,p} = \bigwedge_{i \in J} p_i$ *for some* $J \subseteq I$ *such that*

$$\forall x, x'.p(x) \wedge BA_{evt}(x, x') \Rightarrow proven_{evt,p}(x').$$

Let $unproven_{evt,p} = \bigwedge_{i \in I \setminus J} p_i$ *denote the complement of* $proven_{evt,p}$*, i. e., all the conjuncts of* p *that are not in* $proven_{evt,p}$*. We also define* $proven_{evt} = proven_{evt,Inv}$ *and* $unproven_{evt} = unproven_{evt,Inv}$.

For the example in Fig. 1, we have $BA_{incby}(c, c') = \exists i \in 1..64 \wedge c' = c + i$. Furthermore, $proven_{incby}(c) = c \geq 0$; the invariant $c \geq 0$ is preserved by $incby$. This implies $unproven_{incby}(c) = c \leq m$.

In our current implementation, we have that $proven_{evt,p} = \bigwedge_{j \in J} p_j$ with $J \subseteq I$. This however is not a strict limitation. One could add other predicates discovered to be implied to $proven_{evt,p}$ so as to further strengthen the predicates given below.

Lemma 1. *A valid solution for Definition 2 is always* $proven_{evt,p} = true$*, meaning that nothing is proven for the event* evt*. At the other extreme, if all conjuncts of* p *are proven to hold after the execution of* evt *then* $proven_{evt,p} = p$ *and* $unproven_{evt,p} = true$.

Definition 3. *By* T *we refer to a monolithic transition predicate, i. e., the disjunction of all before-after-predicates:* $T(x, x') = \bigvee_{e \in Events} BA_e(x, x')$*. By* I *we denote the after predicate of the initialization; including the properties about the constants.*

For Fig. 1 we have $I(c) = m \in \{127, 255\} \wedge c = 0$.

In the following sections, we show how proof information can be embedded in the queries of bounded model checking (Sect. 2.1), k-Induction based model checking (Sect. 2.2) and IC3 (Sect. 2.3). An empirical evaluation of the algorithms and the influence of using proof information will be performed in Sect. 3.

2.1   BMC — Bounded Model Checking

BMC [5] has been suggested by Armin Biere et al. in 1999 [6]. One of the main goals is to avoid the blowup and resulting slow down of BDDs-based model checking algorithms. This is achieved by replacing the BDDs by a SAT solver.

The basic idea is as follows: For an initial state relation I, a transition relation T and a property p and a bound k starting with $k = 0$, a sequence of propositional formulas is generated. Each of the formulas is satisfiable if and only if there exists a counterexample to the property with length $\leq k$. This can be expressed as:

$$BMC(p, k) = I(s_0) \wedge \bigwedge_{i=0}^{k-1} T(s_i, s_{i+1}) \wedge \bigvee_{i=0}^{k} \neg p(s_i)$$

In order to include proof information we have to rewrite the predicate. First of all, if we increase k step-by-step as done in PROB's implementation of BMC, it is sufficient to check only the last state for a violation of p:

$$BMC(p, k) = I(s_0) \wedge \bigwedge_{i=0}^{k-1} T(s_i, s_{i+1}) \wedge \neg p(s_k)$$

For the example machine in Fig. 1, we have:

$$BMC(Inv, 0) = m \in \{127, 255\} \wedge c_0 = 0 \wedge \neg(c_0 \geq 0 \wedge c_0 \leq m) \tag{1}$$

$$BMC(Inv, 1) = m \in \{127, 255\} \wedge c_0 = 0 \wedge \exists i.(i \in 1..64 \wedge c_1 = c_0 + i)$$
$$\wedge \neg(c_1 \geq 0 \wedge c_1 \leq m) \tag{2}$$

$$BMC(Inv, 2) = m \in \{127, 255\} \wedge c_0 = 0 \wedge \exists i.(i \in 1..64 \wedge c_1 = c_0 + i)$$
$$\wedge \exists i.(i \in 1..64 \wedge c_2 = c_1 + i) \wedge \neg(c_2 \geq 0 \wedge c_2 \leq m) \tag{3}$$

PROB's constraint solver finds no solution for Eqs. (1) and (2), but does so for Eq. (3): $m = 127, c_0 = 0, c_1 = 64, c_2 = 128$. One can see that the constraint solver has instantiated the parameter of the event in such a way as to violate the invariant. PROB's classical model checker on the other hand "blindly" enumerates all 64 possible successor states. Using breadth-first search, the counterexample is found after having generated 325 states and 12420 transitions; taking $\sim 1.5s$ whereas BMC finds the counterexample with $k = 2$, i.e., after three calls to the constraint solver and $\sim 1s$. A depth-first search may generate a long counterexample of up to 127 steps, depending in which order the successors are processed. PROB in this case actually processes the successors with the larger i values first; leading to a counterexample of length 4 after generating 324 states and 323 transitions. The state space is shown in Fig. 2, the corresponding counterexample is shown in Fig. 3. The larger the branching-factor, the better BMC becomes as compared to explicit state model checking. When the number

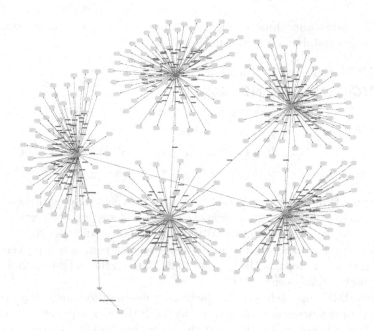

Fig. 2. State space of explicit state model checking

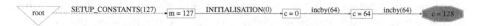

Fig. 3. Counterexample found by BMC

of possible parameter values becomes unbounded, e.g., supposing the incby event had no upper bound on i, BMC is often the only practical solution.[3]

Next, we extend the transition relation to either assert a property after every step or assert its negation:

Definition 4. *For a predicate p we define T_p and T_p^- by*

$$T_p(x, x') = \bigvee_{e \in Events} (BA_e(x, x') \wedge p(x'))$$

$$T_p^-(x, x') = \bigvee_{e \in Events} (BA_e(x, x') \wedge proven_{evt,p}(x') \wedge \neg unproven_{evt,p}(x'))$$

For $k \geq 1$, the proven conjuncts of p can be used to strengthen the constraint:

$$BMC(p, k) = I(s_0) \wedge p(s_0) \wedge \bigwedge_{i=0}^{k-2} T_p(s_i, s_{i+1}) \wedge T_p^-(s_{k-1}, s_k) \qquad (4)$$

---

[3] PROB gives the user the opportunity to set an upper-bound on the number of successor states per event for the explicit model checker; exhaustive model checking is then not possible but counterexamples can still be found.

For the example machine in Fig. 1, we have that for $k = 0$ the constraint remains unchanged, but for $k = 1$ and $k = 2$ we obtain:

$$BMC(Inv, 0) = m \in \{127, 255\} \wedge c_0 = 0 \wedge \neg(c_0 \leq m)$$

$$BMC(Inv, 1) = m \in \{127, 255\} \wedge c_0 = 0$$
$$\wedge \exists i.(i \in 1..64 \wedge c_1 = c_0 + i) \wedge c_1 \geq 0 \wedge \neg(c_1 \leq m)$$

$$BMC(Inv, 2) = m \in \{127, 255\} \wedge c_0 = 0$$
$$\wedge \exists i.(i \in 1..64 \wedge c_1 = c_0 + i) \wedge c_1 \geq 0 \wedge c_1 \leq m$$
$$\wedge \exists i.(i \in 1..64 \wedge c_2 = c_1 + i) \wedge c_2 \geq 0 \wedge \neg(c_2 \leq m)$$

Remember that $unproven_{evt,p}(s_k)$ evaluates to true if all conjuncts of p have been proven to hold after the execution of evt. Hence, for completely proven events $\neg unproven_{evt,p}(s_k)$ is false and the corresponding disjunct in $T_p^\neg$ is obviously unsatisfiable. However, we can not remove such completely proven events from the first $k - 1$ steps as they might contribute to the path to a violation of p, using another final event.

Another BMC approach is to use the test-case generation algorithm from [27], using $\neg p$ as target predicate. In contrast to the BMC technique above, the transition predicate is not monolithic, and the algorithm builds up a tree of feasible paths. We have extended the algorithm from [27] to also use $\neg unproven_{evt,p}$ instead of $\neg p$, where evt is the last event of any given path. The algorithm optionally uses a static enabling analysis to filter out infeasible paths before calling the solver. In the remainder of the paper we refer to this algorithm as BMC*.

2.2   k-Induction

k-Induction [29] is a mixture of BMC and proof by induction. For the method to be complete, one has to avoid getting stuck in loops. Hence, the constraints are strengthened to avoid a state occurring twice on a given path.[4]

The base condition is encoded in Eq. (5); it is basically a BMC step and tries to find a counterexample of length k starting from the initialization. Like in Sect. 2.1 we assume that we gradually increase the value of k starting from 0, as shown in Algorithm 1. The inductive step, including the uniqueness of states, is expressed in Eq. (6), where Axm are the axioms on the constants of the model (e.g., $m \in \{127, 255\}$ in our running example).

$$Base(p, k) = I(s_0) \wedge \bigwedge_{i=0}^{k-1} T(s_i, s_{i+1}) \wedge \neg p(s_k) \tag{5}$$

$$Step(p, k) = Axm \wedge \bigwedge_{0 \leq i < j \leq k} s_i \neq s_j \wedge \bigwedge_{i=0}^{k} T(s_i, s_{i+1}) \wedge \bigwedge_{i=0}^{k} p(s_i) \wedge \neg p(s_{k+1}) \tag{6}$$

---

[4] We could have added theses constraints $s_i \neq s_j$ also in Sect. 2.1.

Data: Property P
Result: true iff P holds
1 **procedure boolean** k-induction(P)
2      $k := 0$
3   **while** *true* **do**
4   **if** $Base(P, k)$ satisfiable **then return** false
5   **elsif** $Step(P, k)$ unsatisfiable **then return** true
6   **else** $k := k + 1$ **end**
7   **end**

Algorithm 1. k-Induction

For $k = 0$ $Step(Inv, k)$ corresponds to trying to find counterexamples to the B invariant preservation proof obligations. In a similar fashion, $Base(Inv, 0)$ corresponds to finding initial states which violate the invariant. Hence, if $Base(p, 0)$ and $Step(T, p, 0)$ are unsatisfiable, we have found an inductive proof of the property p. However, the difference with B's approach to proving invariants does appear when $Step(T, p, 0)$ is satisfiable, i.e., there exists a state which satisfies p and a successor state violates p. The k-induction method tries to construct a real counterexample, starting from a valid initial state, not from *any* state satisfying p. Hence, the value of k is now increased and we try to find a real counterexample of length $k + 1$ using the BMC constraint Eq. (5).

Compared to BMC, k-Induction has the advantage of including an explicit termination condition. Suppose for example we take for p the predicate $c \geq -2 \wedge c \neq -1$ for Fig. 1. In this case BMC will never terminate, as for every value of k no counterexample can be found. k-Induction, however, can already stop with $k = 1$, as $Step(Inv, 1)$ is unsatisfiable. This is an interesting result, given that the state space of the model is infinite. The constraints are shown in Fig. 4 and are unsatisfiable except for $Step(Inv, 0)$. $Step(Inv, 0)$ corresponds to the B proof obligation for the event, checking that p is inductive. Hence, the B proof method is not able to prove that $c \geq -2 \wedge c \neq -1$ always holds. (A user would need to find an inductive invariant such as $c \geq 0$ implying the property. The IC3 algorithm in the next section will do just that automatically.)

The *Base* constraint is equal to the one in BMC: $Base(p, k) = BMC(p, k)$. Hence, we can include proof information in the same fashion and simply reuse the optimized constraint Eq. (4) from Sect. 2.1. For the inductive step, we can again use T_p^- for the last step, to only look for violations of *unproven* parts of p. Following Algorithm 1, we also know that all intermediate states must satisfy the property p; this we can encode using T_p, leading to the definition in Eq. (7). As in BMC, we can not remove before-after-predicates from the first steps. Constraints are simplified but the search space is not reduced.

$$Step(p, k) = Axm \wedge p(s_0) \wedge \bigwedge_{0 \leq i < j \leq k} s_i \neq s_j \wedge \bigwedge_{i=0}^{k-1} T_p(s_i, s_{i+1}) \wedge T_p^-(s_k, s_{k+1}) \quad (7)$$

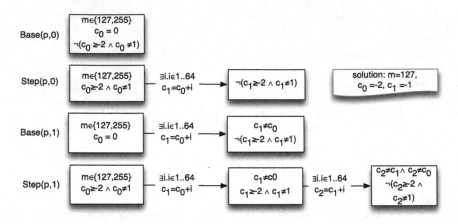

Fig. 4. Steps of k-Induction Algorithm 1 for Fig. 1 with property $c \geq -2 \wedge c \neq -1$

2.3 IC3

In contrast to BMC presented in Sect. 2.1 and k-Induction presented in Sect. 2.2 the IC3 algorithm does not use an unwinding ($\bigwedge_{i=0}^{k} T(s_i, s_{i+1})$) of the transition system. Instead, only single step queries are performed.

In order to verify a system, IC3 tries to *automatically* find an inductive invariant implying the property in question. To do so, it keeps a list of *frames* F_i over-approximating the set of states reachable in $\leq i$ steps. Counterexamples reachable in one or two steps are handled as a special case, as shown in line 2 of Algorithm 2. Afterwards, for each level k IC3 tries to find a property violation in a single step, i. e., a solution to $F_k \wedge T \wedge \neg p$.

If no solution exists, k is incremented and a new frame holding p is added. Otherwise, IC3 tries to show that the faulty state is in fact not reachable from the initialization. This is done by incrementally strengthening frames until F_k becomes strong enough to prevent the property violation from occurring. A partial outline is shown in procedure *strengthen* in Algorithm 2. The *Counterexample* exception is thrown by *inductivelyGeneralize* if generalization fails and the counterexample can not be proven spurious.

Afterwards, a new counterexample might be found and IC3 will start to iterate between finding counterexamples and strengthening frames. If strengthening the frames eventually fails, a counterexample to the property is found. Otherwise, an inductive invariant has been found.

In the following, we will only go into details of IC3 wherever proof information can be incorporated. For a complete overview, see Bradley's original paper [9] or the one by Een et al. in [13]. Algorithms 2 and 3 follow the implementation of [9].

The first change to incorporate proof support takes place in the main loop of IC3. When implemented as suggested by Bradley in [9], IC3 features a special case for 0-step and 1-step reachability of a property violation as explained above.

Data: Property p, Transition predicate T, Initial state predicate I
Result: true iff p holds
1 **procedure boolean** ic3(p, T, I)
2   **if** $sat(I(s_0) \wedge \neg p(s_0)) \vee sat(I(s_0) \wedge T(s_0, s_1) \wedge \neg p(s_1))$ **then return** false
    end
3      $F_0 := I,\ clauses(F_0) := \emptyset$
4      $F_1 := p,\ clauses(F_1) := \emptyset$
5      $k := 1$
6   **while** *true* **do**
7   **if** not strengthen(k, P, T) **then return** false **end**
8   propagate_clauses(k)
9   **if** $\exists i \in [1, k]: clauses(F_i) = clauses(F_{i+1})$ **then return** true **end**
10          $k := k + 1$
11   **end**

Algorithm 2. IC3: Main Loop

---

1 **procedure boolean** strengthen(k, p, T)
2   **try**
3   **while** $sat(F_k(s) \wedge T(s, s') \wedge \neg p(s'))$ **do**
4              $s :=$ the predecessor extracted from the witness
5              $n := inductivelyGeneralize(s, k - 2, k)$
6              $pushGeneralization((n + 1, s), k)$
7   **end**
8   **return** *true*
9   **catch** *Counterexample*
10   **return** *false*

Algorithm 3. IC3: Strengthen

This is shown in line 2 of Algorithm 2. The query on line 2 can be changed in the same way we did for BMC and k-Induction. After splitting the transition relation and adding proof information we obtain:

$$sat(I(s_0) \wedge \neg p(s_0)) \vee sat(I(s_0) \wedge T_p^{\neg}(s_0, s_1))$$

The key point where adding proof assistance improves the performance however is inside the *strengthen* procedure of IC3. The original version is given in Algorithm 3. Inside, the algorithm tries to find a state included in F_k that has a successor violating the property. With the usual transformation, $F_k(s) \wedge T(s, s') \wedge \neg p(s')$ is transformed into $F_k(s) \wedge T_p^{\neg}(s, s')$.

In addition to simplifying the query itself, we can provide the sub routines *inductivelyGeneralize* and *pushGeneralization* with the event that lead to the violating state. This enables simplifying the respective predicates considerably.

In IC3, adding proof information has more benefits than just simplifying the occurring constraints. Due to the one-step nature of queries, constraint solving can be skipped altogether if $unproven_{evt,p} = true$. As no paths are built up explicitly, fully proven events have to be considered only during strengthening.

They can safely be omitted during the counterexample search. Thus, including proof information leads to a reduction of the search space.

3   Empirical Results

The four algorithms described above have been implemented and are available in the nightly builds of PROB[5]. For the empirical evaluation we want to focus on two questions:

- Does the usage of proof information considerably improve the performance of symbolic model checking algorithms for B and Event-B?
- Can symbolic model checking algorithms compete with explicit state model checking (MC) as done by PROB?

We apply both the algorithms introduced in Sect. 2 as well as PROB's explicit state model checker (MC) to a selection of models, including artificial and real benchmarks. We use the explicit state model checker with and without proof support as outlined in [4]. The following models were used:

- *LargeBranching*, a crafted benchmark featuring a counterexample reachable in two steps. However, the initialization has numerous outgoing edges. Discovering the counterexample thus heavily relies on picking the right transitions to follow. The model is included to show that the symbolic algorithms are not influenced by this fact.
- *Search*, a classical B model of a binary search algorithm. *SearchEvents* models the same algorithm, but is written in Event-B style with simpler events. While this leads to simpler constraints, it increases the number of conjuncts due to the increased number of events.
- *TravelAgency*, a classical B model of a travel agency system storing and managing car and room rentals. The model includes an invariant violation.
- *Coloring*, a model of a graph coloring algorithm by Andriamiarina and Méry. In this particular model, the algorithm works on a concrete graph of 40 nodes.
- *f_m0* and *f_m1*, two hybrid models taken from [2].
- *Counters(Wrong)*, two artificial benchmarks featuring two independent counters, one of them bounded and one counting up infinitely. Both models feature an infinite state space. *CountersWrong* has a finite counterexample.
- *R0_Gear_Door*, *R1_Valve*, *R2_Outputs*, *R3_Sensors* and *R4_Handle* are the first refinement levels of our model [15] for the ABZ 2014 landing gear case study [8].

All benchmarks were run on a MacBook Pro featuring a 2.6 GHz i7 CPU and 8 GB of RAM. We did not run anything in parallel in order to avoid issues due to hyper-threading or scheduling. For each benchmark, a number of different results can occur:

---

[5] Available at http://stups.hhu.de/ProB. Information on how to use the new algorithms can be found on the PROB wiki: For the BMC* algorithm see http://stups.hhu.de/ProB/Bounded_Model_Checking. The other algorithms are documented at http://stups.hhu.de/ProB/Symbolic_Model_Checking.

Table 1. Runtimes (in seconds) and speedup (in percent)

Model	MC		BMC		BMC*		k-Induction		IC3	
Use proof info	No	Yes	No	Yes	No	Yes	No	Yes	No	Yes
Models with invariant violations										
LargeBranching	-	-	1.18	0.99 (16.1%)	0.97	0.98 (−1.03%)	-	1.1 (∞)	1.0	1.0 (0.0%)
Search	-	-	-	-	-	-	-	-	-	-
SearchEvents	-	-	1.12	1.12 (0.0%)	1.08	1.05 (2.78%)	-	-	1.06	1.07 (−0.94%)
TravelAgency	1.11	1.09 (1.8%)	-	-	32.18	21.64 (32.75%)	-	-	-	-
CountersWrong	0.83	0.83 (0.0%)	0.9	0.96 (−6.67%)	0.94	0.88 (6.38%)	0.95	0.99 (−4.21%)	0.97	0.96 (1.03%)
Correct Models										
Coloring	-	-	-	-	-	-	-	1.22 (∞)	1.44	1.44 (0.0%)
Counters	-	-	-	-	-	-	1.0	0.92 (8.0%)	0.87	0.87 (0.0%)
f_m0	0.82	0.88 (−7.32%)	-	-	-	-	-	-	0.84	0.83 (1.19%)
f_m1	0.81	0.8 (1.23%)	-	-	-	-	-	-	0.9	0.89 (1.11%)
R0_GearDoor	0.78	0.79 (−1.28%)	-	-	-	-	0.96	0.86 (10.42%)	0.94	0.91 (3.19%)
R1_Valve	0.92	0.92 (0.0%)	-	-	-	-	7.24	0.85 (88.26%)	1.01	0.96 (4.95%)
R2_Outputs	1.84	1.78 (3.26%)	-	-	-	-	0.84	0.9 (−7.14%)	0.89	0.9 (−1.12%)
R3_Sensors	3.06	2.85 (6.86%)	-	-	-	-	-	0.93 (∞)	1.21	0.96 (20.66%)
R4_Handle	33.19	27.61 (16.81%)	-	-	-	-	-	-	-	-

- *verified*, i.e., the model could be model checked exhaustively without an invariant violation being detected.
- *counterexample found*, i.e., a state violating the invariant was found in the model and a trace to it has been computed.
- *incomplete*, i.e., no invariant violation has been found but model checking was not exhaustive. This could be due to timeouts or due to PROB being unable to solve occurring constraints. Currently, we do not try to recover. In case of BMC or k-Induction one could for instance try to increase k anyway.

The results are given in Table 1 showing the runtimes on successful benchmarks as well as the speedup achieved by using proof information.

The state space of the *Search* model is too large to be traversed by PROB's explicit state model checker. Unfortunately, the involved substitutions result in complex constraints that cannot be checked by the symbolic algorithms. The effect is increased by the unwinding of the transition system, as complicated constraints start to occur multiple times.

The *SearchEvents* model features simpler substitutions and is thus more suited for symbolic analysis. Using proof information, all symbolic algorithms are able to find the counterexample. Without proof information, k-Induction is not able to check the model anymore. *LargeBranching* paints a similar picture.

The *TravelAgency* model on the other hand has a relatively small state space and can easily be verified using MC. However, it features involved constructs like sequences resulting in complicated constraints. BMC* is the only symbolic technique to find the counterexample, albeit taking much longer than MC.

The *Coloring* model is quite big and can not be checked exhaustively by MC in the given time. Only IC3 and k-Induction with proof information are able to do so. For IC3 this is due to its focus on one step reachability in combination with the model being correct: Only a small amount of counterexample candidates are discovered by IC3 and are immediately detected as spurious.

Abrial's hybrid models can be verified by MC and IC3. Here, constraints become considerably more involved with each unwinding of the transition relation done in BMC and k-Induction. IC3 is again able to verify the model thanks to its local search for counterexamples.

The infinite counters show one of the key limitations of explicit state model checking. Once a state space is infinite, exhaustive analysis is obviously impossible. For the correct model, BMC reaches its iteration limit without detecting an error. Both k-Induction and IC3 are able to analyze the models.

The landing gear model shows that the benefit of using proof information increases with the complexity of the model. As can be seen in Table 1 computation times go down once proof information is used. As for *SearchEvents* and *LargeBranching*, for *R3_Handle* k-Induction can only successfully be used if proof information is considered. For the first refinement steps, IC3 is quicker than explicit state model checking with PROB. However, once the fourth refinement level is reached, none of the symbolic algorithms can handle the occurring constraints anymore.

Regarding speedup, we can report from ~7 % (*CountersWrong* with BMC) up to ~88 % (*R1_Valve* with k-Induction). For most of the models, incorporating proof information leads to a speedup. Using IC3, our approach leads to a performance decline for some models. We suspect it is because adding additional constraints is not necessarily beneficial for a constraint solver.

Summarizing, we can answer the two questions stated at the beginning:

- The inclusion of proof information into the symbolic model checking algorithms does improve the performance most of the time. Furthermore, some models can only be checked if proof information is used.
- For some, albeit small, models symbolic techniques can compete with explicit state model checking. Symbolic model checkers allow to verify infinite state spaces which are beyond the scope of ProB's classical model checker.
- Among the symbolic techniques, BMC* was the best for erroneous models, while IC3 was best for correct models.
- However, existing solvers for B and Event-B are still too weak to handle the constraints occurring in larger or more involved models. This currently hinders symbolic model checking efforts.

4 Discussion, Related Work and Conclusion

In [4] the authors presented a similar integration of proof information into explicit state model checking algorithms. As is the case with our implementation, the authors report a speedup by not checking invariants known to be true. In contrast to our approach, the use of proof information never slowed down the model checking process.

Compared with [4], we have added a way to construct proof information within ProB itself, using a bridge to the Atelier B provers and using ProB's proving capabilities [18]. Of course this takes time and does not always pay off.

In [4], as with BMC and k-Induction, the search space itself is never reduced. Search space reduction through using proof techniques is considered in [28] and [24]. For model checking CTL and LTL properties, proof information can be used as well. In [26] the model checker SMV is coupled with theorem proving techniques. In a similar fashion, [3] combines the Alloy Analyzer with the Athena theorem prover.

Instead of using theorem provers to support model checking, one can use model checkers for theorem proving. We have done so using ProB [18,22].

Our evaluation shows that using symbolic model checking techniques for B and Event-B models is beneficial: Several counterexamples could only be detected by the symbolic algorithms. Furthermore, some models could be model checked exhaustively. As already outlined in [14], symbolic techniques prove to be a valuable addition to explicit techniques. The techniques are actually also applicable to TLA$^+$, via ProB's translation from TLA$^+$ to B [16][6].

---

[6] For further information regarding TLA$^+$ support in ProBhave a look at http:// stups.hhu.de/ProB/TLA.

They key weakness of employing symbolic model checking techniques lies within the expressiveness of B and Event-B. Even though constraint solvers and SMT solvers have increased their efficiency by a huge margin, the constraints occurring during symbolic model checking of high-level languages like B are still too involved. Among other abstraction techniques, integrating static (proof) information into the constraints is one way to help. It brings down computation times and sometimes enables successful validation. We are also working on strengthening the underlying constraint solver, by integrating SMT solvers such as Z3 [19]. Still, more improvements need to be achieved until full symbolic verification of B and Event-B models becomes viable.

Regarding the different model checking algorithms, especially IC3 seems promising. In contrast to the other two algorithms, its focus on one step reachability keeps occurring constraints easier. This makes it more suited for symbolic model checking of high-level languages like B and Event-B. Additionally, the integration of proof information can lead to a reduced search space. As IC3 has originally been developed for hardware model checking, it is not trivial to lift it to the software world. To do so, we would like to investigate IC3 for B together with abstraction techniques as introduced in [11] or [7].

Another direction of future work could be to generate missing proof obligations from the model checking run. Analyzing predicates that lead to a timeout one could find problematic properties and try to prove them externally or in an independent run. Once the constraint solver gets stuck we could ask an external solver[7] to proof or disprove further invariants. Afterwords, one could extend the set of properties under consideration.

In summary, we have implemented four symbolic model checking algorithms for B and Event-B and have shown how to integrate proof information to improve the algorithms' performance. Our evaluation shows that bounded model checking can effectively find counterexamples in models with very large branching factors and that IC3 is capable of automatically proving models with infinite state spaces correct. Further research is, however, needed to scale up the symbolic techniques to models with more involved events.

Acknowledgements. We would like to thank the reviewers of ABZ'2016 for their useful feedback. We also thank Aymerick Savary for comments and ideas, in particular relating to BMC and test-case generation.

References

1. Abrial, J.-R., Butler, M., Hallerstede, S., Hoang, T., Mehta, F., Voisin, L.: Rodin: an open toolset for modelling and reasoning in Event-B. Int. J. Softw. Tools Technol. Transf. **12**(6), 447–466 (2010)
2. Abrial, J.-R., Su, W., Zhu, H.: Formalizing hybrid systems with event-B. In: Derrick, J., Fitzgerald, J., Gnesi, S., Khurshid, S., Leuschel, M., Reeves, S., Riccobene, E. (eds.) ABZ 2012. LNCS, vol. 7316, pp. 178–193. Springer, Heidelberg (2012)

---

[7] Like the Atelier B provers or the SMT solvers for Rodin.

3. Arkoudas, K., Khurshid, S., Marinov, D., Rinard, M.: Integrating model checking and theorem proving for relational reasoning. In: Berghammer, R., Möller, B., Struth, G. (eds.) RelMiCS 2003. LNCS, vol. 3051, pp. 21–33. Springer, Heidelberg (2004)

4. Bendisposto, J., Leuschel, M.: Proof assisted model checking for B. In: Breitman, K., Cavalcanti, A. (eds.) ICFEM 2009. LNCS, vol. 5885, pp. 504–520. Springer, Heidelberg (2009)

5. Biere, A.: Bounded model checking. In: Handbook of Satisfiability, pp. 457–481 (2009)

6. Biere, A., Cimatti, A., Clarke, E., Zhu, Y.: Symbolic model checking without BDDs. In: Cleaveland, W.R. (ed.) TACAS 1999. LNCS, vol. 1579, p. 193. Springer, Heidelberg (1999)

7. Birgmeier, J., Bradley, A.R., Weissenbacher, G.: Counterexample to induction-guided abstraction-refinement (CTIGAR). In: Biere, A., Bloem, R. (eds.) CAV 2014. LNCS, vol. 8559, pp. 831–848. Springer, Heidelberg (2014)

8. Boniol, F., Wiels, V.: The landing gear system case study. In: Boniol, F., Wiels, V., Ait Ameur, Y., Schewe, K.-D. (eds.) ABZ 2014. CCIS, vol. 433, pp. 1–18. Springer, Heidelberg (2014)

9. Bradley, A.R.: SAT-based model checking without unrolling. In: Jhala, R., Schmidt, D. (eds.) VMCAI 2011. LNCS, vol. 6538, pp. 70–87. Springer, Heidelberg (2011)

10. Burch, J.R., Clarke, E.M., McMillan, K.L., Dill, D.L., Hwang, L.J.: Symbolic model checking: 10^{20} states and beyond. Inf. Comput. **98**(2), 142–170 (1992)

11. Cimatti, A., Griggio, A.: Software model checking via IC3. In: Madhusudan, P., Seshia, S.A. (eds.) CAV 2012. LNCS, vol. 7358, pp. 277–293. Springer, Heidelberg (2012)

12. Déharbe, D., Fontaine, P., Guyot, Y., Voisin, L.: Integrating SMT solvers in Rodin. Sci. Comput. Program. **94**(P2), 130–143 (2014)

13. Een, N., Mishchenko, A., Brayton, R.: Efficient implementation of property directed reachability. In: Proceedings of the International Conference on Formal Methods in Computer-Aided Design (FMCAD 2011), pp. 125–134, Austin, TX, FMCAD Inc (2011)

14. Hallerstede, S., Leuschel, M.: Constraint-based deadlock checking of high-level specifications. Theor. Pract. Logic Program. **11**(4–5), 767–782 (2011)

15. Hansen, D., Ladenberger, L., Wiegard, H., Bendisposto, J., Leuschel, M.: Validation of the ABZ landing gear system using ProB. In: Boniol, F., Wiels, V., Ait Ameur, Y., Schewe, K.-D. (eds.) ABZ 2014. CCIS, vol. 433, pp. 66–79. Springer, Heidelberg (2014)

16. Hansen, D., Leuschel, M.: Translating TLA$^+$ to B for validation with PROB. In: Derrick, J., Gnesi, S., Latella, D., Treharne, H. (eds.) IFM 2012. LNCS, vol. 7321, pp. 24–38. Springer, Heidelberg (2012)

17. Hansen, D., Leuschel, M.: Translating B to TLA$^+$ for validation with TLC. In: Ait Ameur, Y., Schewe, K.-D. (eds.) ABZ 2014. LNCS, vol. 8477, pp. 40–55. Springer, Heidelberg (2014)

18. Krings, S., Bendisposto, J., Leuschel, M.: From failure to proof: the prob disprover for B and event-B. In: Calinescu, R., Rumpe, B. (eds.) SEFM 2015. LNCS, vol. 9276, pp. 199–214. Springer, Heidelberg (2015)

19. Krings, S., Leuschel, M.: SMT Solvers for Validation of B and Event-B models. In: Proceedings iFM'2016, LNCS. Springer (2016). to appear

20. Leuschel, M., Butler, M.: ProB: a model checker for B. In: Araki, K., Gnesi, S., Mandrioli, D. (eds.) FME 2003. LNCS, vol. 2805, pp. 855–874. Springer, Heidelberg (2003)
21. Leuschel, M., Butler, M.: ProB: an automated analysis toolset for the B method. Int. J. Softw. Tools Technol. Transf. **10**(2), 185–203 (2008)
22. Ligot, O., Bendisposto, J., Leuschel, M.: Debugging event-B models using the ProB disprover plug-in. In: Proceedings AFADL 2007, June 2007
23. Matos, P.J., Fischer, B., Marques-Silva, J.: A lazy unbounded model checker for EVENT-B. In: Breitman, K., Cavalcanti, A. (eds.) ICFEM 2009. LNCS, vol. 5885, pp. 485–503. Springer, Heidelberg (2009)
24. Müller, O., Nipkow, T.: Combining model checking and deduction for I/O-automata. In: Brinksma, E., Steffen, B., Cleaveland, W.R., Larsen, K.G., Margaria, T. (eds.) TACAS 1995. LNCS, vol. 1019, pp. 1–16. Springer, Heidelberg (1995)
25. Plagge, D., Leuschel, M.: Validating B, Z and TLA$^+$ using PROB and Kodkod. In: Giannakopoulou, D., Méry, D. (eds.) FM 2012. LNCS, vol. 7436, pp. 372–386. Springer, Heidelberg (2012)
26. Pnueli, A., Ruah, S., Zuck, L.D.: Automatic deductive verification with invisible invariants. In: Margaria, T., Yi, W. (eds.) TACAS 2001. LNCS, vol. 2031, p. 82. Springer, Heidelberg (2001)
27. Savary, A., Frappier, M., Leuschel, M., Lanet, J.-L.: Model-based robustness testing in event-B using mutation. In: Calinescu, R., Rumpe, B. (eds.) SEFM 2015. LNCS, vol. 9276, pp. 132–147. Springer, Heidelberg (2015)
28. Shankar, N.: Combining theorem proving and model checking through symbolic analysis. In: Palamidessi, C. (ed.) CONCUR 2000. LNCS, vol. 1877, p. 1. Springer, Heidelberg (2000)
29. Sheeran, M., Singh, S., Stålmarck, G.: Checking safety properties using induction and a SAT-solver. In: Johnson, S.D., Hunt Jr., W.A. (eds.) FMCAD 2000. LNCS, vol. 1954, pp. 108–125. Springer, Heidelberg (2000)
30. Witulski, J., Leuschel, M.: Checking computations of formal method tools - a secondary toolchain for prob. In: Proceedings of the 1st Workshop on Formal-IDE (EPTCS), Electronic Proceedings in Theoretical Computer Science, vol. 149 (2014)
31. Yu, Y., Manolios, P., Lamport, L.: Model checking TLA$^+$ specifications. In: Pierre, L., Kropf, T. (eds.) CHARME 1999. LNCS, vol. 1703, pp. 54–66. Springer, Heidelberg (1999)

On Component-Based Reuse for Event-B

Andrew Edmunds[1(✉)], Colin Snook[2], and Marina Waldén[1]

[1] Åbo Akademi University, Åbo, Finland
aedmunds@abo.fi
[2] University of Southampton, Southampton, UK

Abstract. Efficient reuse is a goal of many software engineering strategies and is useful in the safety-critical domain where formal development is required. Event-B can be used to develop safety-critical systems, but could be improved by a component-based reuse strategy. In this paper, we outline a component-based reuse methodology for Event-B. It provides a means for bottom-up scalability, and can also be used with the existing top-down approach. We describe the process of creating library components, their composition, and specification of new properties (involving the composed elements). We introduce Event-B component interfaces and propose to use a diagrammatic representation of component instances (based on iUML-B) which can be used to describe the relationships between the composed elements. We also discuss the specification of communication flow across component boundaries and describe the additional proof obligations that are required.

1 Introduction

Formal methods can play a useful role in the development of safety-critical systems. Having flexibility in the formal approaches will make them more useful in the development process. Event-B [3] is a formal method, with tool support [11], which has been used in industry. We are seeking to improve the re-use of Event-B artefacts, with the aim of increasing agility. The creation of a library of components and a way to assemble them would facilitate this. Our proposal is based on shared-event composition [23], since we believe that it provides an intuitive abstraction for the encapsulation that is often seen in object-oriented software components.

In its current form, the existing composition approach, and tools, give little guidance as to how machines and their elements should be combined. Components based on shared-event composition provide a useful encapsulation abstraction. The shared-event approach models the interactions between machines using event synchronization, we can view this as an abstraction of method calling in object-oriented components [6]. Since we are focussed on the potential for reuse, we need a way for developers to interpret the intended use of a component. Typically, this is achieved through the use of interfaces, in conventional software engineering practice. In our approach, we introduce *interface events* to make events 'available for use' by potential users of a machine. When considering the

© Springer International Publishing Switzerland 2016
M. Butler et al. (Eds.): ABZ 2016, LNCS 9675, pp. 151–166, 2016.
DOI: 10.1007/978-3-319-33600-8_9

design and reuse of components, we consider how a developer understands what a component does. The state updates are described by events, in the normal way, but to understand the flow of information across the interface boundary we need to introduce additional annotations to represent parameter directions.

Decomposition is a technique for simplifying complex developments or introducing structural partitions. A single machine is split into multiple sub-units, and the equivalence is maintained using a composition technique [22–24]. In this paper, we introduce Event-B components, interfaces, and composite components which builds on the existing composition techniques. To visualize developments, and assist with their specification, iUML-B [26] provides a graphical interface, with state-machines and class diagrams [21,25]. We propose an extension to iUML-B class diagrams to assist with the use of components. We introduce a composed machine diagram showing which machines and components to include in a composition; and we introduce a new component instance diagram to specify how machines and component instances are related. In addition to this, it may be desirable to specify properties involving the elements of newly composed components. We describe how we could extend the existing composition approach, by adding guards to a composed machine, to ensure that these properties are satisfiable.

In Sect. 2, we provide an overview of Event-B, and Sect. 3 describes Event-B composition. Section 4 introduces ideas for component composition and interfaces. Section 5 discusses use of *composition invariants*, and Sect. 6 introduces proof obligations showing that communication between assembled components is feasible. Section 7 shows an example of a Component. Section 8 discusses related work, and concluding remarks appear in Sect. 9. The work presented here was done as part of the ADVICeS[1] project [28].

2   Event-B

Event-B is a specification language and methodology [1,3] with tool support provided by the Rodin tool [11]. Event-B has received interest from industry, for the development of railway, automotive, and other safety-critical systems [20]. In Event-B, the system, and its properties, are specified using set-theory and predicate logic. It uses proof and refinement [19] to show that the properties hold as the development proceeds. Refinement iterations add detail to the development. Event-B tools are designed to reduce the amount of interactive proof required during specification and refinement steps [8]. Proof obligations in the form of sequents are automatically generated by the Rodin tool. The automatic prover can discharge many of the P.O.s, and the remainder can be tackled using the interactive prover. The basic Event-B elements are *contexts*, *machines* and *composed-machines*. Contexts define the static parts of the system using sets, constants and axioms which we denote by s, c, and a. Machines describe the dynamic parts of a system using variables and events: v and e, and use invariant

---

[1] The ADVICeS project is funded by Academy of Finland, grant No. 266373.

predicates I to describe the properties that should hold. We specify an event in the following way,

$$e \triangleq \textbf{ANY } p \textbf{ WHERE } G(p, s, c, v) \textbf{ THEN } A(p, s, c, v) \textbf{ END},$$

where e has parameter names p; a guarding predicate G; and actions A. State updates (described in the action) can take place only when the guard is true. Guards and actions can refer to the parameters, sets, constants and variables of the machine and seen contexts. For events to occur, the environment non-deterministically chooses an event from the set of enabled events. For clarity, in the remainder of the paper, we omit sets and constants from the description where possible; the discussion largely focusses on parameters and machine variables. As development proceeds, the models can become very detailed, these can be broken down into more tractable sub-units using decomposition [24].

iUML-B [26] is a graphical modelling approach, for Event-B, for specifying state-machines, and class diagrams [21,25]. The diagrams are linked to a parent machine and contribute to its content using automatic translation tools. State-machine diagrams impose an ordering on the machine's events, and the behaviour can be illustrated using a diagram animator. Class diagrams are used to define data entities and their relationships. We propose to extend class diagrams to expose component interfaces. An example of the extension is shown in Fig. 1, and described in more detail in Sect. 4.

3   Composition of Decomposed Machines

Previous work [23] describes the composition of events arising from the decomposition of one machine into multiple sub-units. We make use of the shared-event approach for decomposition, where variables are partitioned into different machines, and events can be combined. The multiple, decomposed sub-units and the composed-machine construct form a refinement of the abstract machine. The combined-events clause of the composed-machine refines an abstract event e. We write $e_a \parallel e_b$ to combine events e_a and e_b, where subscripts a and b also identify distinct sub-units (machines). These combined-events are said to *synchronize* (i.e., both of the events are enabled) when the conjunction of the guards are true. The combined actions are composed in parallel. The semantics of synchronizing events is inspired by the CSP semantics of synchronization [10], however (unlike CSP) matching event names are not required in the shared-event approach. This is due to one of the features of the composed-machine specification, which allows a developer to select which events to synchronize.

$$e_a \triangleq \textbf{ANY } p?_a, p!_a, x_a \textbf{ WHERE } G_a(p_a, x_a, v_a) \textbf{ THEN } A_a(p_a, x_a, v_a) \textbf{ END}$$

$$e_b \triangleq \textbf{ANY } p?_b, p!_b, x_b \textbf{ WHERE } G_b(p_b, x_b, v_b) \textbf{ THEN } A_b(p_b, x_b, v_b) \textbf{ END}$$

$$e_a \parallel e_b \triangleq \textbf{ANY } p, x_a, x_b \textbf{ WHERE } G_a(p, x_a, v_a) \wedge G_b(p, x_b, v_b) \tag{1}$$
$$\textbf{THEN } A_a(p, x_a, v_a) \parallel A_b(p, x_x, v_b) \textbf{ END}$$

Events e_a and e_b may have a set of parameters p in common, with parameters matched by name. Parameter sets are annotated with "!" and "?" to describe output sets and input sets respectively. The annotations are not part of the parameter name, but simply inform us about the direction of data flow into, and out of, events. The annotation might alternatively be written using the Ada parameter mode style 'p : *in*' for input, and 'p : *out*' for output. To account for multiple machines we use a machine name subscript; the set of output parameter names of an event in machine a is written $p!_a$. This is paired with a set of input parameter names in machine b written $p?_b$. Using syntactic sugar, we can write $e_a(p!_a)$ for $e_a \triangleq \mathbf{ANY} \; p!_a \ldots \mathbf{END}$. Events can have sets of uniquely named, non-shared parameters x_a and x_b, which consist of the local variables of the combined-event. The guards G_a and G_b, and actions A_a and A_b, range over the parameters of the event and the machine variables v_a and v_b.

The decomposed sub-units, together with the composed-machine construct, form a refinement of the abstract machine. The composed-machine and sub-units can be merged into a single, unifying machine without changing the composition's semantics. This can result in duplication of the events guards, and some simplification may be necessary. The set of communicating parameters of an event $(p!_a \parallel p?_b) \cup (p!_b \parallel p?_a)$ reduces to p when combined. This can be seen in the combined-events of Eq. 1. In an event, to pass a machine variable w as an output parameter $q!$, we add a guard $q = w$. To use an input parameter $q?$, we can assign it to a machine variable w in an action, using the assignment $w := q$.

Parameter names are not duplicated when merging input-output pairs. For each input parameter $q?$ that is paired with its output parameter $q!$, after merging we have only a single parameter q, so, $q = q! \parallel q?$.

4   Composition with Components

An important feature of a library component is its interface. It defines how the component reveals itself to the outside world. Since we intend to use the components in shared-event style composition, we need to reveal a set of events that can synchronize with some other machine. We mark the events on the class diagram with an annotation; interface events have the letter i next to the event name, see Fig. 1. The interface event may involve communication across the component boundary. This will involve parameter passing, so the interface event needs to reveal information about the names and ranges of the *communicating* parameters. Combined-events that communicate via parameters are required to do so through parameters that have the same name. Events that are not

Fig. 1. The FIFO buffer component

marked with the interface annotation may not synchronize: they are 'hidden' from other components. However, they may be non-deterministically selected by the environment, as usual.

4.1 Using Components in a Development

The composition diagram, shown in Fig. 2, is used to import components into a development. It is a new graphical representation of the existing composed machine, but, additionally, it makes use of pre-existing components, which is a new concept. The composed-machine *Cm* *includes* library machine components *L* and machines under construction *M*. In addition, the machines *M* and *L* may be associated with an existing refinement chain, or be used to specify a new one. In the diagram, combined-events are represented by dashed lines between the machines. The diagram would be similar to an iUML-B diagram, in that diagrammatic elements are added to the canvas, and the underlying Event-B can be generated, or existing elements linked to it. One shortcoming of the composition diagram is that it gives no information about the number of *instances* of each component. A user should be able to select a component and drop an *instance* onto a canvas. The diagram would be linked to a composed machine, in the style of iUML-B [26]. In this diagram, the number of instances and their relationships with other components and machines can be specified. A component instance diagram, showing this, is depicted in Fig. 4. However, the concepts are best explained using an example, which we defer until Sect. 7.2.

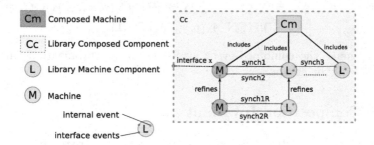

Fig. 2. Using components in a composition diagram

There are two scenarios for instance creation, one is where the library machine links to a machine that initially has no corresponding events. In that case, new events will be added to the machine under construction. The second case is where two existing events are to be synchronized, where a check for compatible parameter names and directions would be done. When no corresponding synchronizing event exists, event stubs can be added. An event stub is a concept taken from programming, where a partial implementation (usually of a method, operation, procedure or function) is generated automatically. This is illustrated in Eq. 2, e_a is an event in the library machine (annotated with **interface**) and e_b is the automatically generated stub. For each output parameter in $p!_a$, we generate an input parameter in $p?_b$ and vice versa. Order of declaration is not important since parameters are simply matched by name, regardless of the order in which they appear. Typing guards for parameters may be suggested at the time of instantiation, but no other event guards and actions are created automatically. The developer will complete the necessary details during further development.

$$\textbf{interface } e_a \triangleq \textbf{ANY } p?_a, p!_a, x_a \textbf{ WHERE } G_a(p_a, x_a, v_a)$$
$$\textbf{THEN } A_a(p_a, x_a, v_a) \textbf{ END}$$
$$e_b \triangleq \textbf{ANY } p?_b, p!_b \textbf{ WHERE } G_b(p_b) \textbf{ END} \tag{2}$$
$$e_a \parallel e_b \triangleq \textbf{ANY } p, x_a \textbf{ WHERE } G_a(p, x_a, v_a) \wedge G_b(p)$$
$$\textbf{THEN } A_a(p, x_a, v_a) \textbf{ END}$$

4.2 Composite Components

When a composed-machine is defined, it can be added as a library component. The system boundary is then represented by the outer, dashed box, see Fig. 2. We need to decide which of the events of the new component are revealed in the interface. We assume that, by default, all events of a composed-machine are hidden, in which case we would need to promote some new, or existing, events to the new interface. The parameters of the exposed events can be marked with the input/output annotations, ? or !. The composed event of Eq. 2 could be promoted to the composite component interface using the **interface** annotation, as follows,

$$\textbf{interface } e_a \parallel e_b \triangleq \textbf{ANY } p, \ x_a \textbf{ WHERE } G_a(p, x_a, v_a) \wedge G_b(p)$$
$$\textbf{THEN } A_a(p, x_a, v_a) \textbf{ END} \tag{3}$$

This would make the combined-event available for synchronization with some event outside of the component.

5   The Composition Invariant

5.1 Adding a Guard to Satisfy the Composition Invariant

The existing composed-machine CM is made up of the included machines M_0 .. M_m, a list of combined-events, and a composition invariant CI. Any of the machines M_0 .. M_m may be library machines. The CI can be used to specify properties relating the elements of separate components of a composition. These properties cannot be specified in machine invariants since the elements they refer to reside in separate machines. In the case where a top-down development introduces components in a refinement, and one finds that a particular invariant in the abstraction involves elements that reside in separate components, then, in the refinement, the CI in the composition will reproduce the invariant from the abstraction.

The *composition invariant*, CI(s, c, v), has visibility of *all* of the sets and constants of the included contexts, and variables of the composed-machines s, c, and v respectively. To identify the sets, constants and variables of the individual machines, in a composition of machines $M_0 .. M_m$, we write $s = s_0 .. s_m$ for sets, $c = c_0 .. c_m$ for constants and $v = v_0 .. v_m$ for variables. The composed-machine invariant CMI is a conjunction of the individual machine invariants $MI_0 .. MI_m$ and CI, where each machine invariant has visibility of its own variables and sets, and also the constants of its seen contexts as follows,

$$CMI(CM, M_0 .. M_m) = CI(s, c, v) \wedge MI_0(s_0, c_0, v_0) \wedge .. \wedge MI_m(s_m, c_m, v_m) \quad (4)$$

To ensure that the composition invariant CI is preserved, we need to add guards G_{CI}, but currently there is no mechanism in the existing tool that does this automatically, so this remains as future work. We would like G_{CI} to range over the whole of v, c and s. That is, the guard requires component-wide visibility of variables, and of the sets and constants of the seen contexts of the included machines. The intuitive place to do this is in the composed-machine, where we propose to add an additional guard clause to the combined-event clause. We extend the combined-event of Eq. 1 with G_{CI}, as follows,

$$e_a \parallel e_b \triangleq \textbf{ANY } p, x_a, x_b \textbf{ WHERE } \textbf{\textit{G}}_{CI}(v) \wedge G_a(p, x_a, v_a) \wedge G_b(p, x_b, v_b)$$
$$\textbf{THEN } A_a(p, x_a, v_a) \parallel A_b(p, x_b, v_b) \textbf{ END}$$

$$(5)$$

In the composed-machine, we should demonstrate that the invariants (including the CI) are preserved for all events of the included machines. The invariant preservation proof obligation $INV_{e_a \parallel e_b}$ follows, for each invariant i in I, where local variables x are omitted, and the remainder of the parameters refer to those in events before composition,

$$INV_{ea} : I_a(v_a) \wedge G_a(p_a, v_a) \wedge A_a(p_a, v_a, v_a') \vdash i_a(v_a') \quad (6)$$

$$INV_{eb} : I_b(v_b) \wedge G_b(p_b, v_b) \wedge A_b(p_b, v_b, v_b') \vdash i_b(v_b') \quad (7)$$

$$\begin{aligned} INV_{e_a \parallel e_b} : \ & CI(v) \wedge I_a(v_a) \wedge I_b(v_b) \\ & \wedge G_a(p_a, v_a) \wedge G_b(p_b, v_b) \wedge \textbf{\textit{G}}_{CI}(v) \\ & \wedge A_a(p_a, v_a, v_a') \wedge A_b(p_b, v_b, v_b') \\ & \vdash i_a(v_a') \wedge i_b(v_b') \wedge CI(v') \end{aligned} \quad (8)$$

As seen above, we are required to choose an appropriate guard $\textbf{\textit{G}}_{CI}$ to show that the invariant holds. It appears in the antecedent of the combined-event's invariant proof obligation.

5.2 Component Development

One of the benefits of the existing decomposition approach is that once decomposition has taken place, the individual machines can be refined independently. This is possible for the components that are used in compositions, too, and allows components/machines to be further refined, by a number of teams, independently. To see how this is possible we comment on the two scopes of visibility in a composition. The top-level scope is defined by the composed machine, which has visibility of all of the sets, constants and variables of the machines that it *includes*, and of the contexts that those machines *see*. The CI resides at the top-level in the composed machine, and can refer to variables of multiple machines. The composed machines can have guards added to combined-events, these also have visibility of the variables of the included machines. An important point here, is that the CI should only describe the properties relating to the composition (i.e. properties that cannot be described in a machine/component in isolation). Otherwise, those properties should reside in the normal machine invariants. Each included machine, and its refinement chain, in a composed machine, forms a lower-level scope of visibility. At the lower level scope, the included machines and refinements can be worked on independently since it contains no information about the composition.

The need to recompose components is a natural consequence of placing constraints on elements residing in different machines of a composition. For each of the included machines and their refinement chains, further refinements can be added independently by adding new variables, strengthening guards, and data refinement. If a composition is complex, it will be possible to add further composed-machines to the refinement chain (which may or may not *include* existing components), thereby allowing specification of emerging composition properties as development proceeds.

6 Proof Obligations

6.1 Feasibility of Inputs and Outputs

The use of components and their interfaces can be described using a contract with pre- and post-conditions. However, pre-condition semantics are missing in Event-B. So, how do we expect component users to understand what the interface provides? Since we propose using typed, directed event parameters, we can use this information. The parameter's typing guards define the input and output state-spaces. In Event-B, guards play the dual role of typing and event-enabling. So, we need to be very clear about the semantics of synchronization, and about when we expect synchronization and communication to take place. In the existing Event-B approach, there is no requirement (in the form of proof obligations) to show that an event is ever enabled. However, we believe that when components are assembled (especially pre-existing components) we require assurance that the data flow across component boundaries is compatible. There should be some common set of input and output states that will allow the events to synchronize. A similar concept was explored in work on feature composition [18]. Our solution is related to the idea of feasibility in Event-B; feasibility proof obligations for non-deterministic assignment, for instance, ensure that there is some initial value in the pre-state that allows a transition to a given post-state.

We believe that, in our approach, we should provide some proof of the feasibility of synchronization/communication. To do this we introduce pre-condition semantics for communication of data across the interface boundary where we show that, for each parameter, the range A of output parameter values is a subset of the range B of input parameter values, $A \subseteq B$. This is determined by the parameter's range as defined in the event guard.

6.2   Preconditions for Communicating Event Parameters

Design-By-Contract (DBC) [14] is an approach for composing modules using contracts. In DBC, pre-conditions and post-conditions are defined in a specification, pre-conditions should be satisfied by users of the contract and post-conditions should be satisfied by implementers of the contract. It can be seen that contracts define an interface specification for a module, and part of their use deals with ensuring that the communicating parameter values are always within acceptable bounds. In our work, the input and output parameters, and their range (a (non-strict) subset of their type) and direction information, form part of the interface specification. We wish to ensure that, for any input/output pair, the output parameter's value falls within the range of the allowable inputs. To do this, we introduce two functions, to differentiate between the ranges of the inputs $p?$ and the outputs $p!$. Given an event e and input parameter $q?$, function $rangeOfIn$ returns the range T of $q?$ as defined in the guard.

$$rangeOfIn(e, q?) = T \tag{9}$$

Also, given an event e and output parameter $q!$, function $rangeOfOut$ returns the range T of $q!$.

$$rangeOfOut(e, q!) = T \tag{10}$$

We call the pre-condition style feasibility proof obligation $\boldsymbol{FIS_{preStyle}}$. For the combined-event $e_a \parallel e_b$, we have,

$$FIS_{preStyle}(e_a(p?_a, p!_a), e_b(p?_b, p!_b))$$
$$= \tag{11}$$
$$\forall q!, q? \cdot (q! \in p! \wedge q? \in p?) \implies (rangeOfOut(e_a, q!) \subseteq rangeOfIn(e_b, q?))$$

where $q!$ represents an individual output parameter from the set of output parameters $p!$ of an event, and $q?$ represents an individual input parameter from the set of input parameters $p?$ of an event. To satisfy this proof obligation, for each pair of communicating parameters in an event, the output value must fall within the acceptable range of the input. Consider a concrete example of a combined-event $evt1 \parallel evt2$, where $evt1$ has an output parameter named $prm!$ of range $0..256$ and event $evt2$ has an input parameter $prm?$ of range $\mathbb{N}1$, then,

$$rangeOfOut(evt1, prm!) \subseteq rangeOfIn(evt2, prm?)$$
$$= 0..256 \subseteq \mathbb{N}1 \tag{12}$$
$$= \perp$$

In this case, the $FIS_{preStyle}$ proof obligation is not satisfied, since 0 does not belong to $\mathbb{N}1$. If the input range was changed to $prm? \in \mathbb{N}$, it would be satisfied.

7   An Example Illustrating the Required Tool Support

7.1   Specifying a FIFO Buffer Component

We now describe how components might be defined in a version of iUML-B [26] adapted to component (or interface) specification. Figure 1 on Page 4 shows the FIFO class diagram. The *FIFO* class diagram contains the attributes: buffer, head and tail, and three interface events, annotated with the letter i. In the model, but not shown in the diagram, the *FIFO* instance is represented by the parameter *this_FIFO*. It is automatically generated by the iUML-B tool. We now provide details of *inToBuffOK* and *retrvFromBuffOK*, two of the events shown in the diagram. The *inToBuffOK* event models successful receipt of a value and the return of TRUE as an acknowledgement. The *retrvFromBuffOK* models retrieval of a value (by a consumer) from a buffer. We do not show the *inToBuffFail* event, it handles the case of failure to receive a value, due to a full buffer, and returns a FALSE acknowledgement.

$inToBuffOK \triangleq$
    ANY $x?$, $ack!$, *this_FIFO*
    WHERE $ack \in BOOL \land x \in BYTE\_16 \land ack = TRUE \land$
        $tail(this\_FIFO) - head(this\_FIFO) < buffSize \land \dots$
    THEN $buffer(this\_FIFO) := buffer(this\_FIFO) \nleftarrow \{tail(this\_FIFO) \mapsto x\} \parallel$
        $tail(this\_FIFO) := tail(this\_FIFO) + 1 \parallel \dots$
    END

The *inToBuffOK* event shows the input parameter $x?$ of range $BYTE\_16$: the value to put in the buffer. It also has an acknowledgement, an output parameter $ack!$ of range BOOL, restricted to $ack = TRUE$. This is returned to the sender on success. The action shows the value x being written to the tail of the buffer in a statement that overrides an existing value or adds a new value. The value of *tail* is incremented in parallel.

$retrvFromBuffOK \triangleq$
    ANY y, *this_FIFO*
    WHERE $y \in BYTE\_16 \land y = buffer(this\_FIFO)(head(this\_FIFO)) \land \dots$
    THEN $head(this\_FIFO) := head(this\_FIFO) + 1$
    END

In the *retrvFromBuffOK* event, we have an output parameter $y!$. The output is modelled in the guard $y = buffer(this\_FIFO)(head(this\_FIFO))$ where y gets the value of the head of the buffer. The *head* value is incremented in the action.

7.2   Using the FIFO Component

We now introduce a *Producer* class, shown in Fig. 3, that uses two instances of the *FIFO* library component $f1$ and $f2$. Figure 4 shows how the diagram might look, with two *FIFO* instances connected to a *Producer*, and two *Consumers*. The combined-events, labelled $a \ldots e$, specify synchronizations between the *FIFO* interface, and the Producer/Consumers. Event f does not synchronize with any other event. It should be noted that there is only one machine modelling all instances of the *FIFO*, and another modelling all instances of the *Consumer*. Tool support, for the component instance diagram, can provide stubs for the synchronizing events in the *Producer* and *Consumer* machines, when the connections between an interface event and another machine are defined. The stub event for the *Producer*, called *Producer.inToBuffOK1*, would be provided with the appropriate parameters as follows,

$Producer.inToBuffOK1 \triangleq$

ANY $x!$, $ack?$, *this_Producer*, *this_FIFO*

WHERE $ack \in BOOL \wedge x \in BYTE\_16 \wedge$

$this\_Producer \in Producer \wedge this\_FIFO = f1(this\_Producer)$

END

The x and ack parameters modelling the communication, are shown, along with two additional parameters that are introduced by the iUML-B translators, these are used to model the component instances. Namely, the parameters

Fig. 3. The producer class

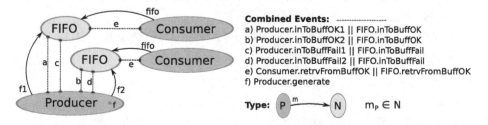

Fig. 4. A component instance diagram

this_Producer and *this_FIFO*. The developer of *Producer* should decide which value to output, and where to assign the input. A possible solution would be to model the output using the variable *value* $\in$ *BYTE_8* for output (it is possible that this would be driven by other design concerns) and use *success* $\in$ *BOOL* for modelling the acknowledgement. We could then refine the stub with these additions, note the use of the strengthened typing guard,

*Producer.inToBuffOK*1 $\triangleq$

  ANY *x!*, *ack?*, *this_Producer*, *this_FIFO*

  WHERE *ack* $\in$ *BOOL* $\land$ *x* $\in$ *BYTE_8* $\land$ *x* = *generatedA(this_Producer)* $\land$

   this_Producer $\in$ *Producer* $\land$ *this_FIFO* = *f*1(*this_Producer*)

  THEN *success(this_Producer)* := *ack* **END**

We can see here, that we model the output assignment, of the variable *generatedA* to the parameter *x*, in the guard, and we model assignment of the return value *ack* to the *success* variable in the action. This is a typical pattern in shared-event synchronization.

Let us now consider the combined-event, where we look at the underlying Event-B showing the *instance parameter*. From Fig. 3, we see that the *Producer* has two *FIFO* instances, *f*1 and *f*2. To synchronize with a library component, the user of the interface requires a separate event for each instance of the component. This is why we have two events in the *Producer* related to the event *FIFO.inToBuffOK*. In the event, *Producer.inToBuffOK*1, below, we can see *f*1(*this_Producer*) being used to identify which *FIFO* it is related to. The parameter *this_FIFO* and the guard could be generated automatically in the *Producer*, with additional tool support. This relates to the *this_FIFO* parameter in the *FIFO*, which can be generated by iUML-B tools, from the *FIFO* class diagram. Also, since the shared parameter *x* has two different ranges in the individual machines, we take the view that the stronger guard should appear in the clause since it makes the weaker guard redundant. The combined event follows,

*Producer.inToBuffOK*1 $\parallel$ *FIFO.inToBuffOK* $\triangleq$

  ANY *x*, *ack*, *this_Producer*, *this_FIFO*

  WHERE *ack* $\in$ *BOOL* $\land$ *x* $\in$ *BYTE_8* $\land$ *x* = *generatedA(this_Producer)* $\land$

   this_Producer $\in$ *Producer* $\land$ **this_FIFO = f1(this_Producer)** $\land$

   ack = *TRUE* $\land$ *tail(this_FIFO)* $-$ *head(this_FIFO)* $<$ *buffSize* $\land$ ...

  THEN *success(this_Producer)* := *ack* $\parallel$

   buffer(this_FIFO) := *buffer(this_FIFO)* $\lhd$\kern-0.5em- {*tail(this_FIFO)* $\mapsto$ *x*} $\parallel$

   tail(this_FIFO) := *tail(this_FIFO)* $+$ 1 $\parallel$ ...

  END

Now we consider the *pre-style* proof obligation of Eq. 11. In our example *Byte_16* = 0 .. 65535 and *Byte_8* = 0 .. 255, and we can discharge the proof obligation.

$$rangeOfOut(Producer.inToBuffOK1, x!)$$
$$\subseteq rangeOfIn(FIFO.inToBuffOK, x?)$$
$$= Byte\_8 \subseteq Byte\_16 \tag{13}$$
$$= 0..255 \subseteq 0..65535$$
$$= \top$$

7.3   A Composition Invariant

In our example, we may want the FIFO buffer $f1$ to hold odd numbers, and $f2$ to hold even numbers. This is a property of the composition, and should be specified in the composition invariant clause. To do this, we add an invariant stating that values in the producer's $f1$ buffers must have $mod\ 2 = 1$ and those in $f2$ buffers must have $mod\ 2 = 0$. The invariant that constrains $f1$ follows,

$$\forall p \cdot p \in dom(f1) \implies (\forall v \cdot v \in ran(buffer(f1(p)))) \implies v\ mod\ 2 = 1)$$

It states that for each producer p in the domain of the variable $f1$, and for each value v in its buffer, $v \in ran(buffer(f1(p)))$, $v\ mod\ 2 = 1$ must hold. There is a similar guard stating that $f2$'s values must be even. It would not be possible to specify this in the *Producer* machine since it does not have visibility of FIFO's *buffer* variable.

8   Related Work

A concept that is closely related to our approach is that of *Modularisation*. It is an approach for describing components and interfaces in Event-B, by Iliasov et al. [2]. It is based on the shared-variable composition approach. The authors use a pre- and post-condition syntax to specify the component interfaces and behaviour, and they introduce proof obligations to prove refinement. In contrast, shared-event composition provides an appropriate abstraction for the encapsulation that is often seen in object-oriented software components, sharing of variables is usually prohibited here. We also keep the introduction of new syntactic elements to a minimum by extending the existing class diagram techniques. In this way, an implementation of an interface is simply a refinement of that interface. A more detailed discussion of the issues can be found in [7].

Eiffel [14] is another modular approach, based on Design-by-Contract; this too, makes use of pre- and post-condition specifications. We prefer not to use pre- and post-conditions, and present a more integrated method that does not diverge so greatly from the existing iUML-B approach [26]. The CODA component model, of Butler et al. [5], describes how components can be represented on a UML-B style diagram [27]. The underlying model is used to simulate communication between components which are joined using ports and connectors. This makes use of the ProB model checker [13] and uses an oracle to compare various simulation runs. In CODA, the focus is not on reuse. Rather, it is a way of modelling message queuing over time, and it embodies the communication style found in VHDL [17] which makes it very domain specific.

Hallerstede and Hoang describe interface refinement in [9]. This makes use of the shared-variable composition approach, where external variables, and a corresponding external invariant are specified in the interface. We believe that by using the shared-event approach, we avoid having to consider the effects of sharing variables. By using interface events, as the means for interacting with components, this simplifies reasoning and proof: the encapsulation of traditional software components is closely represented by the shared-event abstraction. Banach extends the interface refinement concepts in [4], by using a CONNECTS construct to make use of interface events: continuing with the shared-variable style.

In other work on components, Kessel and Atkinson discuss reuse of software components [12] focussing on partial matches for suitability in situations where a component's intended use differs from its ultimate use. For Event-B, in the development of high-integrity systems, it will be very important to fully understand the behaviour of a component and underspecification must be judiciously applied to accommodate unforeseen variability. Other notions include location aware components, such as the distributed computation notion of components in CommUnity, which is presented by Oliveira and Wermelinger in [15]; and another concept is the component approach used in the formal modelling of agent interactions with Event-B, from [16]. Both of the latter use quite different notions of components; our components' main purpose is reuse.

9 Conclusions

In the domain of software engineering, the concept of a *component* has many different meanings. Our use of the term component is comparable with its use in the object-oriented software world [29], where a component is an element that is intended for reuse and the flow of data across the component boundary is described by its interface. In the work presented here, we propose an extension to the existing composition approach by introducing Event-B components. The existing composition approach was primarily designed to work as a top-down decomposition method. We wish to have bottom-up composition for re-use. That is not to say that we intend to dispose of the top-down approach, rather, we should have the flexibility to include, and work with, existing artefacts as and when required.

Using our diagrammatic extension we can describe a collection of communicating components. We introduce *interface events* as a concept to describe which events can be synchronized with other events. Non-interface events cannot be synchronized, but they can be non-deterministically chosen by the environment, as usual. We add input, and output specifiers, "?" and "!" to annotate the event parameters, in order to clarify the flow of information across the interface boundary. We introduce some new features to a class diagram, to create an interface class, which is annotated to show the interface events. In all other ways, class diagrams are unchanged. We introduce a new composition diagram to describe machines that are included in the composition. It is a diagrammatic representation of the composed machine construct, and is used to aid visualization of the composed machines and library components. We can describe which events synchronize and show which events are promoted to the interface of a composite machine, but we do not provide information about specific instances. To do this, we introduce a new component instance diagram describing the composition of components as class instances showing the links that describe their synchronizations. We model multiple components using the existing concept of instance parameters where there may be several instances of a particular component in the composition. All instances of a particular component are modelled in a single machine, and there may be multiple components.

Properties involving a number of components may be described in the composition invariant (CI) of the composed machine. These properties extend beyond component boundaries and should be used to describe properties that cannot be described in a single component. The guards related to the CI should go in the combined event. The feasibility of communication across the interface boundaries, for composed events, can be checked by generating additional proof obligations which ensure that, for each parameter, the output values fall completely within the range of values accepted by the corresponding input parameter. This style of feasibility proof will be particularly useful when composing pre-existing components since it is necessary to ensure that the data flow across component boundaries is compatible.

As future work, we plan to do more investigation into the use of components and compositions for team-working, and to provide the additional diagrammatic tool support. In addition, new translators will be required for generating Event-B from the diagrams. Additional tool changes are also required, to add guards to the composed machine's combined event, in order to satisfy the composition invariant.

References

1. The Rodin User's Handbook. http://handbook.event-b.org/
2. Iliasov, A., Troubitsyna, E., Laibinis, L., Romanovsky, A., Varpaaniemi, K., Ilic, D., Latvala, T.: Supporting reuse in Event B development: modularisation approach. In: Frappier, M., Glässer, U., Khurshid, S., Laleau, R., Reeves, S. (eds.) ABZ 2010. LNCS, vol. 5977, pp. 174–188. Springer, Heidelberg (2010)
3. Abrial, J.R.: Modeling in Event-B: System and Software Engineering. Cambridge University Press, Cambridge (2010)
4. Banach, R.: The landing gear case study in hybrid Event-B. In: Boniol, F., Wiels, V., Ait Ameur, Y., Schewe, K.-D. (eds.) ABZ 2014. CCIS, vol. 433, pp. 126–141. Springer, Heidelberg (2014)
5. Butler, M., Colley, J., Edmunds, A., Snook, C., Evans, N., Grant, N., Marshall, H.: Modelling and refinement in CODA. In: Refine, pp. 36–51 (2013)
6. Edmunds, A., Butler, M.: Tasking Event-B: an extension to Event-B for generating concurrent code. In: PLACES 2011, February 2011
7. Edmunds, A., Walden, M.: Modelling "operation-calls" in Event-B with shared-event composition. Technical report 1144 (2015)

8. Hallerstede, S.: Justifications for the Event-B modelling notation. In: Julliand, J., Kouchnarenko, O. (eds.) B 2007. LNCS, vol. 4355, pp. 49–63. Springer, Heidelberg (2006)
9. Hallerstede, S., Hoang, T.S.: Refinement by interface instantiation. In: Derrick, J., Fitzgerald, J., Gnesi, S., Khurshid, S., Leuschel, M., Reeves, S., Riccobene, E. (eds.) ABZ 2012. LNCS, vol. 7316, pp. 223–237. Springer, Heidelberg (2012)
10. Hoare, C.A.R.: Communicating Sequential Processes. Prentice Hall, Upper Saddle River (1985)
11. Abrial, J.R., et al.: Rodin: an open toolset for modelling and reasoning in Event-B. Softw. Tools Technol. Transf. **12**(6), 447–466 (2010)
12. Kessel, M., Atkinson, C.: Ranking software components for pragmatic reuse. In: 2015 IEEE/ACM 6th International Workshop on Emerging Trends in Software Metrics (WETSoM), pp. 63–66. IEEE (2015)
13. Leuschel, M., Butler, M.: ProB: a model checker for B. In: Proceedings of Formal Methods Europe 2003 (2003)
14. Meyer, B.: Design by contract: the Eiffel method. In: TOOLS, vol. 26, p. 446. IEEE Computer Society (1998)
15. Oliveira, C., Wermelinger, M.: The community workbench. In: Proceedings of the 26th International Conference on Software Engineering, pp. 709–710. IEEE Computer Society (2004)
16. Pereverzeva, I.: Formal development of resilient distributed systems. Ph.D. thesis, Åbo Akademi University (2015)
17. Perry, D.L.: VHDL, 2nd edn. McGraw-Hill, New York (1994)
18. Poppleton, M.R.: The composition of Event-B models. In: Börger, E., Butler, M., Bowen, J.P., Boca, P. (eds.) ABZ 2008. LNCS, vol. 5238, pp. 209–222. Springer, Heidelberg (2008)
19. Back, R., Wright, J.: Refinement Calculus: A Systematic Introduction. Springer Science & Business Media, New York (2012)
20. Romanovsky, A., Thomas, M.: Industrial Deployment of System Engineering Methods. Springer, Heidelberg (2013)
21. Said, M.Y., Butler, M., Snook, C.: Language and tool support for class and state machine refinement in UML-B. In: Cavalcanti, A., Dams, D.R. (eds.) FM 2009. LNCS, vol. 5850, pp. 579–595. Springer, Heidelberg (2009)
22. Silva, R.: Towards the composition of specifications in Event-B. In: B 2011, June 2011
23. Silva, R.: Supporting development of Event-B models. Ph.D. thesis, University of Southampton, May 2012
24. Silva, R., Butler, M.: Shared event composition/decomposition in Event-B. In: FMCO Formal Methods for Components and Objects, November 2010
25. Snook, C.: Event-B Statemachines (2011). http://wiki.event-b.org/index.php/Event-B_Statemachines
26. Snook, C.: iUML-B Statemachines. In: Proceedings of the 5th Rodin User and Developer Workshopp (2014)
27. Snook, C., Butler, M.: UML-B: formal modelling and design aided by UML. ACM Trans. Softw. Eng. Methodol. **15**, 92–122 (2006)
28. The ADVICeS Team: The ADVICeS Project. https://research.it.abo.fi/ADVICeS/
29. Wikipedia: Component-Based Software Engineering - Software Component. https://en.wikipedia.org/wiki/Component-based_software_engineering

Using B and ProB for Data Validation Projects

Dominik Hansen, David Schneider, and Michael Leuschel(✉)

Institut Für Informatik, Heinrich-Heine-Universität Düsseldorf, Universitätsstr. 1,
40225 Düsseldorf, Germany
{dominik.hansen,david.schneider,michael.leuschel}@hhu.de

Abstract. Constraint satisfaction and data validation problems can be expressed very elegantly in state-based formal methods such as B. However, is B suited for developing larger applications and are there existing tools that scale for these projects? In this paper, we present our experiences on two real-world data validation projects from different domains which are based on the B language and use ProB as the central validation tool. The first project is the validation of university timetables, and the second project is the validation of railway topologies. Based on these two projects, we present a general structure of a data validation project in B and outline common challenges along with various solutions. We also discuss possible evolutions of the B language to make it (even) more suitable for such projects.

Keywords: B method · Constraint programming · Timetabling · Scheduling · Railway

1 Introduction

Data validation[1] ensures that software operates on correct, clean data and is typically done by checking validation rules or constraints. We have previously argued that B [3] is a very expressive language to encode constraint satisfaction problems [20,24], and many data validation problems can be expressed as such. Other works have demonstrated that B is useful to express properties about data and to validate them using ProB [17], particularly in the railway domain [2,4–6,15,19].

In this paper we report on our experiences using B in combination with ProB to create tools for data validation. We have used the B language to express parts of our program's domain logic and the rules to validate data, and embedded these B models into running applications by executing the formal models with ProB without relying on code generation. It would also be possible to express these kinds of validation problems in other formal languages such as Alloy [13] and TLA+ [14]. Based on our experiences with these languages and the corresponding tools, we believe that the combination of B and ProB best meets the requirements for the data validation task. Our explicit goal is to explore

---

[1] http://www.data-validation.fr.

© Springer International Publishing Switzerland 2016
M. Butler et al. (Eds.): ABZ 2016, LNCS 9675, pp. 167–182, 2016.
DOI: 10.1007/978-3-319-33600-8_10

the applicability and scalability of this combination for projects of industrial strengths.

Based on two projects, described below, we will discuss different aspects of using B within such an application and discuss the approaches taken as well as the limitations encountered, i.e. where we had to depart from or extend the language to suit our needs.

Curriculum Validation is a project [24] in which we are creating an interactive tool to validate timetables and curricula for various faculties and courses at our university. Curriculum validation is related to timetabling [8–10,22,23]. Timetable validation differs in the sense that we are interested in the feasibility of studying an entire curriculum, spanning several semesters instead of planning out time slots for classes within one semester. The central task is to detect whether it is possible for a student to attend all classes required for a degree in the manner described by the curriculum, by a suitable choice of alternatives. In case a course contains feasibility conflicts, the tool provides assistance to detect one of the potentially many sources of the conflict by computing a unsatisfiable core of the data with respect to the validation rules; additionally we provide support in finding alternative time slots which solve such conflicts. The largest dataset provided by one of the participating faculties currently consists of 31 courses with 1343 classes and 1578 scheduled events in these classes.

Validation of Railway Topologies is the second project discussed in this article and part of a collaborative research project with Thales Transportation Systems GmbH on applying formal methods for the software development process of the Radio Block Centre (RBC). The RBC is a communication unit of the European Train Control System (ETCS) exchanging messages with trains and interlockings. One of our challenges in this context is to validate the so-called engineering rules over concrete track data. The track data is a representation of the real railway infrastructure and signalling system. Engineering rules are implementation-related rules which result from the concrete RBC implementation. This means, that the concrete RBC implementation is guaranteed to work correctly only if the concrete track data satisfy the engineering rules. For example, a simplified engineering rule requires that two signals for the same direction should not be located at the same position. The modelled engineering rules are validated on different track topologies. The biggest topology contains 1362 track segments, 457 points, 1089 balise groups and 445 signals.

Both projects rely on PROB as the tool to evaluate the models. PROB is an animator and model-checker for the B method with support for validation and proofs. PROB also is a constraint solver for the B language, which is required in an animation and model-checking scenario to efficiently find values for constants, guards and parameters of operations.

The idea of using formal method languages and tools to perform data validation has been explored in the past, e.g. by Abo and Voisin [2] or Lecomte et al. [15] among others. Our intention here is to outline the common structure and challenging aspects of data validation projects based on what we have identified in the aforementioned projects. The domains and requirements of these two projects are quite different, and all work has been done independently (i.e. by different people). Still, similar challenges were faced during the modelling process.

In the following sections we will name these challenges, discuss different language constructs of B and argue how they can be applied in modelling data validation problems. Moreover, we will outline areas where we have extended the B language to overcome some limitations we faced evaluating the models with ProB.

2   The Big Picture

Before describing the details of the data validation process we will discuss the big picture, outlining the design and architecture that emerged from both projects mentioned in the previous section.

```
!signal1, signal2.(
  signal1 : Signals & signal2 : Signals & signal1 /= signal2
  & Signal_Direction(signal1) = Signal_Direction(signal2)
   => not(Signal_TrackSegment(signal1) = Signal_TrackSegment(signal2)
         & Signal_Position(signal1) = Signal_Position(signal2)))
```

Fig. 1. Modelling of an engineering rule as a validation predicate

The general idea is to create B models that define validation predicates which are evaluated against the state of the model. The variables and constants are derived from external data we want to validate. Figure 1 shows the formalisation of the validation rule mentioned in the introduction Section where two signals should not be located at same position if they are valid for the same direction.

The projects discussed in this paper follow the general architecture shown in Fig. 2. By building data validations tools based on the B language we have identified the following concerns: The first is getting the external data from a given source into a B model which is discussed in Sect. 3. Choosing a way to represent the data is a further concern, where it is important to choose a representation and B data types suited for the validation process while keeping the import process as simple as possible; this is discussed in Sect. 4. Some validation rules rely on derived data (e.g., signals reachable from a point) which has to be computed from the imported data. In Sect. 5 we present different approaches to structure derived data in B. One purpose of the B method is to model algorithms and prove their correctness. However, are these models suitable for use by ProB to calculate results? Section 6 describes different approaches to model an algorithm in B such that it can be efficiently evaluated by ProB. Another concern is how to control the validation process from an external application.

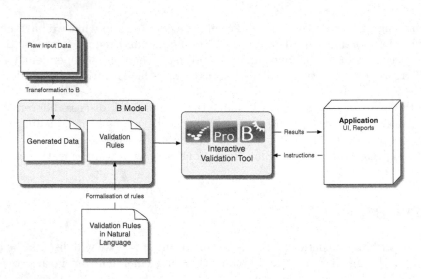

Fig. 2. Generalised architecture of PROB based data validation project

In Sect. 7 we discuss different ways to interact with the model. Finally, in Sect. 8 we briefly discuss how to reuse an existing validation model in similar projects.

3   Preparing Data for Use with a B Model

When used for data validation, our tools obviously depend on externally provided data [2,16], which has to be converted to B format, in order to be validated with PROB. Raw input data is provided in a variety of formats as used in the different domains such as Excel, CSV or XML documents.

In both projects we have opted to create tools that read and parse the externally provided data and generate a text file containing a B model of the data. The data will be accessible as a series of constants in the model.

Having an external tool keeps any knowledge about the raw data format out of the B models; but of course it raises a series of concerns. One is having to maintain an additional tool which has to generate valid B. Also the chosen data representation has to be kept in sync between the import and the validation tools.

Another concern is that, in a safety critical environment, the import tool itself has to be validated. The topology validation project takes a direct approach by avoiding putting too much knowledge into the transformation step, keeping it as simple as possible. In this approach the transformation maps the input structure of the data (XML) to B data structures and copies the values of attributes as uninterpreted strings. To ensure that all data from the input document is represented in the B model we use a back-translation (from the B model to XML) and compare the generated XML document with the source document. The back-translation is done in order to certify the translation tool and ensure that no data has been left out.

In the case of the curriculum validation tool the data is not only used for validation purposes but also to populate the application's user interface, hence we have chosen a two step approach that does not directly generate a B machine, but rather import the data into a database. The information in the database is later used to generate the actual B representation of the model at runtime. Additionally, the database is used in the application to persist changes and as the data source for the UI. Since the data is used in multiple places we map the values in the raw data to the most adequate types in the database and later to the corresponding B types.

Are There Any Alternatives? There are many alternative approaches that could be pursued to import data into a B model. E.g. instead of generating a B model with the data as constants, it would be possible to have B operations which incrementally add values to variables containing the data. These operations could be executed in various ways, e.g., using the Java API for ProB. Finally, ProB exposes external functions to B that make it possible to, e.g., load data from CSV files; these features could be extended for additional data sources (see Sect. 6.3).

4   Data Representation

Hand in hand with the decision on how to import data into a B model goes the choice of proper B data-structures to represent the data. This representation should ideally follow the structure of the source data, and additionally lend itself to be used and manipulated in B. Choosing a good representation for the problem is crucial for the complexity and readability of the model. In [11] Hayes et al. discuss some of these issues on the examples of a simple database in VDM and Z. In B, one could encode database records as nested pairs. A quaternary relation over course identifiers, semester, weekday, and starting hour could thus be represented as:

```
db = { (((course1 |-> sem2) |-> monday) |-> 14),
       (((course2 |-> sem1) |-> friday) |-> 9) }
```

In order to access the first and second element of a pair, B provides the prj_1 and prj_2 operators. However, in B accessing a certain field of a nested pair is very cumbersome, as we have to unfold the nested pair until we reach the desired field.[2] Another alternative is to use records with named fields:

```
db = { rec(course_id: course1, semester: sem2, weekday: monday,
       starting_hour: 14), rec(course_id: course2, semester: sem1,
       weekday: friday, starting_hour: 9) }
```

We can easily access a field of a record r by using the quote operator: $r'course\_id$. Compared to the encoding as nested pairs, records are more readable, especially if there are a large number of fields. Otherwise, constructing a

---

[2] In addition, the types of the arguments have to be provided for prj_1 and prj_2; e.g., $prj_2((COURSE \times SEMESTER) \times WEEKDAY, \mathbb{Z})(v)$.

record is more verbose than constructing nested pairs. Since, this part of the model is automatically generated, the verbose encoding is not an issue.

A third alternative is to create B functions for each attribute of the data record mapping a unique identifier to the corresponding attribute value. An identifier of a data record could be a unique number generated by the translator or a certain attribute of the data record. In case of our example, we could choose the attribute `course_id` as the unique identifier:

```
course_id__semester = {course1 |-> sem2, course2 |-> sem1}
course_id__weekday = {course1 |-> monday, course2 |-> friday}
course_id__start_hour = {course1 |-> 14, course2 |-> 9}
```

While this approach works well for simple tables such as in Excel or CSV documents, it would become inconvenient for nested data structures, e.g. if a value of a field is itself a set of data records such as a sub-tag of a XML document. In this case, the translation tool first has to transform the nested data structure to a relational database schema. Subsequently, the translator has to create a B function for each attribute of each table of the relational database.

One advantage of the last alternative is the handling of optional fields. Indeed, when no field value is present for a data record, we just omit the corresponding identifier from the domain of the field accessor function (i.e., we use partial functions rather than total functions). For the other two approaches, optional fields pose more of a challenge. Due to the strong and strict typing of B it is not possible to create partial records or to omit a field of a nested pair. One solution is to introduce a special NULL value for each B datatype, e.g. the empty string (" ") for the STRING type. However, we have to ensure that the NULL value is not a regular value in the source data. For other data types such as INTEGER it is more intricate (which integer to use?) and for the BOOL type impossible.

This directly leads to a further aspect of the translation. How to represent the values of the data records? They could be represented either as uninterpreted strings of data copied verbatim from the raw data input in the transformation step. Alternatively the data values could be represented using the most appropriate B data types, e.g. INTEGER for numbers, and enumerated sets for values from a set of known values. The first approach has the advantage of a very simple translation process and all the relevant knowledge is encoded in the B model. The drawback is now, however, that the data has to be translated in the B model, which typically requires extensions to the B language which are available in PROB (e.g. transforming a STRING value to an INTEGER value).

In both projects, we have chosen the record representation for the data. As already mentioned in the previous section, the timetabling tool maps the raw data to the corresponding B data types. In case of the topology validation project, all data values are represented as uninterpreted strings and the processing of these strings is part of the B model.

5   Means of Abstraction − Structuring and Auxiliary Constructs

Abstractions [1] in programs and also models control complexity, encourage reuse and make testing easier. Different parts of the B language offer different ways to abstract and structure models and programs. There are certain concepts that are applicable at the machine and operation level while others are applicable on the predicate and expression level.

Machines and Operations. On the machine level sub-problems can be structured as machines for each sub-aspect which communicate through the execution of operations. The visibility of machines and their variables and operations can be controlled using different machine composition mechanisms such as SEES, USES and INSTANCE.

On the level of a single operation the substitution language provides several expressions that are useful, either if-then-else for control flow or LET constructs to introduce scoped variables.

Expressions and Predicates. Within the mathematical language of B, constants can be used to globally save precomputed values whose computation might be expensive and should not be evaluated more than once. Figure 3 shows the calculation of the conflict relations of two different signals placed at the same position and valid for the same direction. Note, that the calculation corresponds to a SQL statement making a self join on a signal table. Moreover, constants can be used to store certain calculations in the form of lambda functions which can be used in different parts of the model. However, constants are not applicable for intermediate results which can not be precomputed globally because they depend on additional information or parameter values.

```
CONSTANTS
   ConflictRelation
PROPERTIES
   ConflictRealtion =
     UNION(r1,r2).(r1 : SignalRecords & r2 : SignalRecords
       & r1'elementID /= r2'elementID & r1'trackSegment = r2'trackSegment
       & r1'position = r2'position & r1'direction = r2'direction
       | {r1'elementID |-> r2'elementID})
```

Fig. 3. Calculating the conflict relation of two signals placed at the same position.

LET for Predicates and Expressions. In complex expressions or predicates it is often useful to introduce a shorthand for certain values or expressions, B only supports LET in the context of substitutions, nonetheless it might be useful for predicates and expressions. For example, an existential quantification (#x.(x=E & P)) can be used within predicates to achieve a result similar to a LET. PROB tries to identify existential quantifications that only have a single value and treats them specially. Within set-comprehensions an existential quantification could also be used (`{x| #y.(y=E & P)}`), but the following pattern using the domain of a set of pairs is (generally) more efficient in PROB: `dom({x,y| y=E & P})`. For expressions which denote a set of values, one can use `UNION(y).(y=E| S)`. Ideally, however, rather than using these workarounds, we would argue for adding explicit LET constructs to the B language for expressions and predicates.

DEFINITIONS. One of the available methods of decomposing larger predicates or expressions into smaller reusable components are `DEFINITIONS` (comparable to macros). The use of `DEFINITIONS` carries some issues that have to be kept in mind. Although they are textual replacements, PROB requires every definition to be syntactically correct on its own, so certain compositions patterns are not possible. Care is also needed with regard to naming conflicts, quantifications not captured in the `DEFINITION` where variables escape the scope (see, e.g., [12]). Take for example the definition `even(x) == (#y.(y:1..x & 2*y=x))`. Evaluating the predicate `even(4)` yields true. However, if we have a machine variable y whose value is 4 and evaluate `even(y)` we obtain false; the definition call was rewritten to `#y.(y:1..y & 2*y=y)`. Another issue is unintended repeated computation of arguments. Indeed, the arguments of a definition may get replaced multiple times and then also executed multiple times by PROB. Take, e.g., the definition `POW3(x)==x*x*x` and the call `POW3(f(1))`. The latter gets transformed into `f(1)*f(1)*f(1)`, resulting in repeated computations of `f(1)`. A pattern we have used to avoid this is to create a variable within each `DEFINITION`, which is assigned with the passed argument and used instead of the original parameter to avoid unintentionally causing repeated computations of the same expression:

```
DEFINITIONS
EXAMPLE(aa, bb) == #(va, vb).(
  va = aa & vb = vb & <predicate over va and vb> );
```

For the reasons described above, `DEFINITIONS`, although they are a useful method to store and structure expressions and predicates, should be used carefully, in particular for big expressions with parameters.

6   Using B to Express Computations

Sometimes data validation relies on complex concepts, which cannot be easily described as B predicates. In those cases it can be more convenient to describe these concepts using recursive rules or as fixpoints of iterative algorithms. In this section we show how this can be achieved in a natural B style, while also ensuring that the resulting algorithms can be executed efficiently.

We will discuss different approaches to model an algorithm for sorting a set of numbers into an ordered sequence. Note that the B method does not provide a built-in operator to sort a set. In the particular applications we used the techniques for more complicated constructs, such as a search on a rail way topology with various termination conditions.

6.1   Machines and Operations

First we will discuss the approach of using machines and operations to express the required functionality. Using the machine and substitution semantics of B to express computations has the clear advantage of having all tools and features of the B method at our disposal. Figure 4 shows a stateless query operation calculating the sorted sequence for a given input set.

```
out_sortedSequence <-- Sort_OP(p_set) =
  PRE p_set : POW(INTEGER) THEN
    out_sortedSequence : (
        out_sortedSequence : iseq(p_set)
        & ran(out_sortedSequence) = p_set
        & !i.(i : 1 .. size(out_sortedSequence) - 1
              => out_sortedSequence(i) < out_sortedSequence(i + 1)))
  END
```

Fig. 4. Query operation

However, ProB is not able to evaluate the operation efficiently, i.e. it does not scale for large input sets. Indeed, a naive execution of Sort_OP would calculate all possible permutations of the input set to then reject all but one, which is the sorted sequence. ProB's constraint solving can overcome this exponential complexity to some extent,[3] but for larger sequences we are a far cry from the performance of ordinary sorting algorithms. Following the refinement principles of the B method we can implement the abstract operation by a concrete sorting algorithm. Figure 5 shows a selection sort (MinSort) implementation in B. The operation Sort_OP exposes the algorithm as a single operation which can be used several times and embedded in different machines.

ProB provides various optimisations for while loops. First, an interesting point is that the variant is evaluated upon entry and gives ProB an upper-bound on the number of iterations.[4] If a certain threshold is exceeded, ProB will pre-compile the body of a while loop, by pre-computing all parts which do not depend on variables modified in the loop. Furthermore, the state of the interpreter is projected onto those variables that are modified.

In our approach, we are not interested in proving the concrete algorithm to be a correct refinement of the abstraction. However, we are interested in the correctness of the sort implementation. Therefore, we use the predicate of the abstract operation as an invariant respectively an assertion on the output of the concrete operation. Note, that in this case ProB is able to check that the predicate holds for a concrete value even for a large input set. Moreover, the termination of the sort algorithm is ensured using a loop variant which

---

[3] ProB can compute Sort_OP({3,55,22,44,1,100,20,40,55,88,10,90,200,0,5}) in 0.18 s, despite there being 15!=1,307,674,368,000 permutations.

[4] In many models, the variant actually corresponds exactly to the number of iterations.

```
out_sortedSequence <-- Sort_OP(p_set) =
  PRE p_set : POW(INTEGER) THEN
    VAR v_set, v_seq
    IN
      v_set := p_set; v_seq := [];
      WHILE v_set /= {}
      DO
        v_seq := v_seq <- min(v_set);
        v_set := v_set \ {min(v_set)}
      INVARIANT
        v_set : POW(p_set) & v_seq : iseq(p_set)
        & !i.(i : 1 .. size(v_seq)-1 => v_seq(i) < v_seq(i + 1))
      VARIANT card(v_set)
      END;
      ASSERT ran(v_seq) = p_set THEN out_sortedSequence := v_seq END
    END
  END
```

Fig. 5. Implementation of a sorting algorithm

is observed by PROB. For more complex algorithms such as different search algorithms on railway topologies we have modelled state machines instead of stateless query operations. However, the execution of the these state machines is controlled by a single operation of an additional interface machine.

A small disadvantage of using operations is that the output value of the operation can only be assigned to a variable and the operation can not be used as part of a set comprehension or quantification.

6.2 Recursive Functions

Recursive functions, which are supported by PROB [18], are a very effective way to compactly express certain kinds of algorithms. Figure 6 shows the selection sort algorithm modelled as a recursive function in B. By defining *Recursive_Sort* as an abstract constant we indicate that PROB should handle the function symbolically, i.e. PROB will not try to enumerate all elements of the function. The recursive function itself is composed of two single functions: a function defining the base case and a function defining the recursive case. Note, that the intersection of the domains of these function is empty, and hence, the union is still a function.

However, there are certain constructs that are harder to write (and read) using only the expression language of B, as it has no explicit support for let expressions and if-then-else. Nonetheless it is often easier to express a construct as a recursive function than it is to decompose the steps in order to express it as a machine. In general, the performance of a recursive function is slower compared to the operation/while approach.

Rather than using an explicit recursive call as in Fig. 6, we can also use B's transitive closure operator to compute the fixpoint of a relation. For our example,

```
ABSTRACT_CONSTANTS Recursive_Sort
PROPERTIES
  Recursive_Sort : POW(INTEGER) <-> POW(INTEGER*INTEGER)
& Recursive_Sort =
    %in.(in : POW(INTEGER) & in = {} | [])
      \/ %in.(in : POW(INTEGER) & in /= {}
                        | min(in) -> Recursive_Sort(in\{min(in)}))
```

Fig. 6. Recursive sort function

let us define the relation $step = \%(s,o).(s/=\{\} \mid (s\backslash\{min(s)\},o<-min(s)))$ which encodes one recursive step of selection sort (s is the set to sort, o is the output sequence so far). For a start set $in = \{4,5,2\}$ we can now compute $closure1(step)[\{(in,[])\}]$ resulting in $\{(\{4,5\} \mapsto [2]),(\{5\} \mapsto [2,4]),(\varnothing \mapsto [2,4,5])\}$. As we can see, the result of sorting a set in can be obtained by calling $closure1(step)[\{(in,[])\}](\{\})$.

6.3   External Functions

There are certain concepts that are not part of the B language, e.g. mathematical functions such as *sin*, *cos*, etc. Other computations are difficult or impossible to express using only predicates and expressions, while others might be too slow to evaluate purely in B. PROB offers a mechanism named **external functions** to add and expose new constructs to B. In our sorting example this might look as follows:

```
DEFINITIONS
  SORT(X) == [];
  EXTERNAL_FUNCTION_SORT == (POW(INTEGER)-->seq(INTEGER))
```

The function SORT is implemented in Prolog as part of the PROB core and exposed in B as a definition. In order to define a syntactically correct DEFINITION we use the empty sequence as a dummy value ensuring type correctness. The second definition tells PROB the type of the external function.

External functions provide the best performance for specific computations, by removing the interpretation overhead but at the same time are opaque to the user and at this point in time need to be integrated explicitly into the PROB Prolog kernel to be available in the language.

6.4   Further Language Extensions

In the topology validation project we have introduced further language constructs that provide a uniform schema to write validation rules. Wherein, the validation predicates are embedded in special RULE operations. Figure 7 shows a simplified schema of a RULES_MACHINE which contains several RULE operations and will be translated to an ordinary B machine. The result of a RULE operation can be stored by using the new `RULE_SUCCESS` or `RULE_FAIL(.)` keywords. The argument of the `RULE_FAIL(.)` keyword is the message reported in case of a rule violation. For each rule operation an ordinary variable is generated in the translated B machine containing the result of the rule evaluation (i.e. `''FAIL''`, `''NOT_CHECKED''` or `''SUCCESS''`). By using additional guards we are able to define dependencies between rules (using the new keyword `DEPENDS_ON_RULES`) or disable a rule if necessary. The model itself is non-deterministic in the sense that different rules can be executed at the same time if their guards are satisfied. Thus, we are not forced to define an explicit execution order of all rule operations and can use PROB's animation feature to conveniently execute a certain operation. To ease the writing of a rule we developed a new `FORALL` substitution which can be used to define an error message of a rule by conveniently accessing the variables of a universal quantification.

7   Interaction with the Model

There are several ways the main software can interact with the B validation model. Depending on the kind of application, one could animate or model check the B model, execute a B operation or evaluate B expressions or B predicates (assertions) on a certain state of the model.

The PROB Java API (aka. PROB 2.0[5]) provides facilities to use PROB in applications running on the JVM. Through this API it is possible to access the functionalities mentioned above and to translate B data types to and from appropriate Java types.

In case of the curriculum validation project, the tool itself is a Java application that embeds the model and PROB. We expose all features provided by the model as B operations that represent the public API of the model. These operations are evaluated, using PROB's animation facilities, with externally provided parameters to validate the different curricula. The validation operations return a list of variables that represent one possible choice of subjects to successfully finish a degree. Furthermore, we use the result computed for a feasible curricula to generate a PDF timetable for students with a recommend choice of subjects for their studies.

In the topology validation project the model is used as an independent validation tool with the goal to generate validation reports about the input data. Each engineering rule is modelled as one or more `RULE` operations containing the validation predicates (see Sect. 6.4). By using more than one `RULE` operation for an engineering rule the complexity of a natural language requirement can be decomposed into several simple and readable validation predicates. The advantage of a `RULE` operations, compared to a listing of all validation predicates as part of the `ASSERTIONS` section, is that a B operation defines a clean interface to perform the evaluation of the individual rules and to access result values and counterexamples. In order to generate a complete validation report and to validate all possible rules, we construct a trace of the model using PROB's `execute`

---

[5] https://www3.hhu.de/stups/prob/index.php/ProB_Java_API.

```
RULES_MACHINE Rules                    MACHINE Rules
SEES Features                          SEES Features
OPERATIONS                             VARIABLES rule1, rule2, rule3
 RULE rule1 = ...;                     INVARIANT
 RULE rule2 = ...;                      rule1 : {"NOT_CHECKED",
 RULE rule3 =                              "FAIL", "SUCCESS"}
  SELECT                               & rule2 : {"NOT_CHECKED",
   DEPENDS_ON_RULES(rule1, rule2)          "FAIL", "SUCCESS"}
   & Enabled(feature1) = TRUE          & rule3 : {"NOT_CHECKED",
  THEN                                    "FAIL", "SUCCESS"}
   FORALL                              INITIALISATION
    p1, p2                              rule1 := "NOT_CHECKED"
   WHERE                               || rule2 := "NOT_CHECKED"
    P(p1,p2)                           || rule3 := "NOT_CHECKED"
   EXPECT                              OPERATIONS
    Q(p1,p2)                            res,ce <-- rule1 = ...;
   THEN                                 res,ce <-- rule2 = ...;
    RULE_SUCCESS                        res,ce <-- rule3 =
   ELSE                                 SELECT
    VAR errorMsg                         rule3 = "NOT_CHECKED"
    IN                                   & rule1 = "SUCCESS"
     errorMsg := Exp(p1,p2);             & rule2 = "SUCCESS"
     RULE_FAIL(errorMsg)                 & Enabled(feature1) = TRUE
    END                                 THEN
   END                                  IF
  END                                    !(p1,p2).(P(p1,p2)
END                                         => Q(p1,p2))
                                         THEN
                                          rule3 := "SUCCESS"
                                         || res := "SUCCESS"
                                         || ce := ""
                                         ELSE
                                          ANY p1,p2
                                          WHERE
                                           P(p1,p2) & not(Q(p1,p2))
                                          THEN
                                           VAR errorMsg
                                           IN
                                            errorMsg := Exp(p1,p2);
                                            rule3 := "FAIL"
                                           || res := "FAIL"
                                           || ce := errorMsg
                                           END
                                          END
                                         END
                                        END
                                     END
```

Fig. 7. Translation of a RULES_MACHINE to an ordinary B machine

command until all operations are covered. By doing this, we eliminate the overhead which would be introduced by performing a complete model checking run on the non-deterministic model (i.e. evaluating an operation several times).

8   Configuration Management

Configuration management, i.e. how to reuse rules and infrastructure for similar or related projects which differ in very specific aspects, is very important in the context of data validation. For example, in the case of curricula validation, there are subtle differences amongst faculties in the overall structure or how the students choose classes. We have explored two different approaches, to tackle this issue.

One approach is that of a Software Product Line (SPL) [7], where the system would create, from a selection of predicates and evaluation rules a machine that composes them according to a provided configuration. A further approach would be to search for and find a data representation and formulation of the validation rules that is general enough to be applied to more than one particular instance. Such a generic model can contain variation points to control specific aspects of the validation process that differ from project to project. For example, the rule in Fig. 7 is only tested when two particular features are selected.

In both projects we have settled for a combination of both approaches, automatically generating certain parts of our models and additionally configuring the generic parts.

9   Conclusion and Future Work

In this paper we have presented two data validation projects where we have expressed the validation rules in B. Based on the experiences gathered and the similarities between the projects, we have discussed different relevant areas and presented our architecture and design decisions as well as possible alternatives.

We have identified the aspects of data validation that can be easily and elegantly expressed in B such as deriving intermediate data structures from the raw data, modelling complex algorithms while ensuring their correctness, and formalising validation predicates which are close to natural language counterparts. Otherwise, we presented the points were we had to diverge from B by either using language extensions supported by PROB or by moving certain features outside of the B models, e.g. the data import. Moreover, we described a way to interact with the formal model and to build various applications on top on PROB.

In both projects PROB satisfies the respective requirements on performance and execution time. For the curriculum validation, PROB is able to detect conflicts among courses in an appropriate time, making interactive use on top of PROB possible. In the topology validation project there are no strict timing constraints. However, our B and PROB based approach is able to compete with a pre-existing validation tool written in an imperative language.

The work on both projects has helped to push the development of ProB forward by highlighting performance bottlenecks that have since been resolved. Moreover, we added support for language constructs such as tree operators.

Due to the availability of higher-order data types, B can be used almost like a functional programming language. We have used this in particular to compute derived data. In the paper we have also shown various limitations of B, and have presented some ways to overcome them (e.g., how to encode let constructs). In the future, we would like to be able to use parts of B as a proper functional programming language. In that sense, we are considering adding polymorphic operators, as present in TLA$^+$, to provide a simpler way to structure predicates and allow the user to define new recursive operators. Moreover, we are pursuing an approach to embed parts of the mathematical B language into the Clojure programming language using native syntax and evaluating it with ProB, an approach comparable to aRby for Alloy [21].

Acknowledgements. We would like to thank Luis-Fernando Mejia from Alstom for pushing (and funding) B and ProB into new directions. We have also grateful to ClearSy and Systerel for exercising ProB, allowing us to discover new aspects of the B language. We are grateful to Thales for funding a collaborative research project and to the Thales team, in particular Nader Nayeri, for all their help and insights. Finally we would like to thank Frank Meier, Tobias Witt, Philip Höfges and the planning teams at the Faculty of Arts and Humanities and the Faculty of Business Administration and Economics at Heinrich Heine University for their contributions to the curriculum validation project.

References

1. Abelson, H., Sussman, G.J.: Structure and Interpretation of Computer Programs, 2nd edn. MIT Press, Cambridge (1996)
2. Abo, R., Voisin, L.: Formal implementation of data validation for railway safety-related systems with OVADO. In: Counsell, S., Núñez, M. (eds.) SEFM 2013. LNCS, vol. 8368, pp. 221–236. Springer, Heidelberg (2014)
3. Abrial, J.-R.: The B-Book. Cambridge University Press, Cambridge (1996)
4. Ayed, R.B., Collart-Dutilleul, S., Bon, P., Idani, A., Ledru, Y.: B formal validation of ERTMS/ETCS railway operating rules. In: Ait Ameur, Y., Schewe, K.-D. (eds.) ABZ 2014. LNCS, vol. 8477, pp. 124–129. Springer, Heidelberg (2014)
5. Badeau, F., Amelot, A.: Using B as a high level programming language in an industrial project: roissy VAL. In: Treharne, H., King, S., C. Henson, M., Schneider, S. (eds.) ZB 2005. LNCS, vol. 3455, pp. 334–354. Springer, Heidelberg (2005)
6. Badeau, F., Doche-Petit, M.: Formal data validation with Event-B. In: Proceedings of DS-Event-B 2012, Kyoto. CoRR, abs/1210.7039 (2012)
7. Clements, P., Northrop, L.M.: Software Product Lines: Practices and Patterns. Addison-Wesley Longman Publishing Co. Inc, Boston (2001)
8. Corne, D., Ross, P., Fang, H.-L.: Evolving timetables. In: Practical Handbook of Genetic Algorithms: Applications, vol. 1, pp. 219–276 (1995)
9. Deris, S., Omatu, S., Ohta, H.: Timetable planning using the constraint-based reasoning. Comput. Oper. Res. **27**(9), 819–840 (2000)

10. Gotlieb, C.C.: The construction of class-teacher time-tables. In: IFIP Congress, pp. 73–77 (1962). http://dblp.uni-trier.de/rec/bib/conf/ifip/Gotlieb62, http://dblp.org

11. Hayes, I.J., Jones, C.B., Nicholls, J.E.: Understanding the differences between VDM and Z. ACM SIGSOFT Softw. Eng. Notes 19(3), 75–81 (1994)

12. Herman, D., Wand, M.: A theory of hygienic macros. In: Drossopoulou, S. (ed.) ESOP 2008. LNCS, vol. 4960, pp. 48–62. Springer, Heidelberg (2008)

13. Jackson, D.: Software Abstractions: Logic, Language, and Analysis. MIT Press, Cambridge (2012)

14. Lamport, L.: Specifying Systems: The TLA+ Language and Tools for Hardware and Software Engineers. Addison-Wesley, Boston (2002)

15. Lecomte, T., Burdy, L., Leuschel, M.: Formally checking large data sets in the railways. In: Proceedings of DS-Event-B 2012, Kyoto. CoRR, abs/1210.6815 (2012)

16. Leuschel, M., Bendisposto, J., Dobrikov, I., Krings, S., Plagge, D.: From animationto data validation: the ProB constraint solver 10 years on. In: Formal Methods Applied to Complex Systems, pp. 427–446 (2014)

17. Leuschel, M., Butler, M.: ProB: a model checker for B. In: Araki, K., Gnesi, S., Mandrioli, D. (eds.) FME 2003. LNCS, vol. 2805, pp. 855–874. Springer, Heidelberg (2003)

18. Leuschel, M., Cansell, D., Butler, M.: Validating and animating higher-order recursive functions in B. In: Abrial, J.-R., Glässer, U. (eds.) Rigorous Methods for Software Construction and Analysis. LNCS, vol. 5115, pp. 78–92. Springer, Heidelberg (2009)

19. Leuschel, M., Falampin, J., Fritz, F., Plagge, D.: Automated property verification for large scale B models. In: Cavalcanti, A., Dams, D.R. (eds.) FM 2009. LNCS, vol. 5850, pp. 708–723. Springer, Heidelberg (2009)

20. Leuschel, M., Schneider, D.: Towards B as a high-level constraint modelling language. In: Ait Ameur, Y., Schewe, K.-D. (eds.) ABZ 2014. LNCS, vol. 8477, pp. 101–116. Springer, Heidelberg (2014)

21. Milicevic, A., Efrati, I., Jackson, D.: αRby—An embedding of Alloy in Ruby. In: Ait Ameur, Y., Schewe, K.-D. (eds.) ABZ 2014. LNCS, vol. 8477, pp. 56–71. Springer, Heidelberg (2014)

22. Rudová, H., Murray, K.: University course timetabling with soft constraints. In: Burke, E.K., De Causmaecker, P. (eds.) PATAT 2002. LNCS, vol. 2740, pp. 310–328. Springer, Heidelberg (2003)

23. Schimmelpfeng, K., Helber, S.: Application of a real-world university-course timetabling model solved by integer programming. OR Spectr. 29(4), 783–803 (2006)

24. Schneider, D., Leuschel, M., Witt, T.: Model-based problem solving for university timetable validation and improvement. In: Bjørner, N., de Boer, F. (eds.) FM 2015. LNCS, vol. 9109, pp. 487–495. Springer, Heidelberg (2015)

Generating Event-B Specifications
from Algorithm Descriptions

Joy Clark[1], Jens Bendisposto[1(✉)], Stefan Hallerstede[2], Dominik Hansen[1],
and Michael Leuschel[1]

[1] Institut Für Informatik, Heinrich Heine University, Düsseldorf, Germany
Joy.Clark@hhu.de, {bendisposto,hansen,leuschel}@cs.uni-duesseldorf.de
[2] Department of Engineering, Aarhus University, Aarhus, Denmark
sha@eng.au.dk

Abstract. We present a high-level algorithm description language which is translated to Event-B specifications for simulation, model checking and proof. Rather than trying to recover the program structure from a lower-level Event-B specification, we start with a high-level description of the algorithm. Our goals are more tractable code generation and more convenient modelling, while keeping the power of the Event-B method in terms of proof and refinement. We present various examples of algorithm descriptions and show that our translation ensures that they can be completely proven within Rodin while achieving a high-level of automatic proof.

Keywords: Event-B · Code generation · Sequential algorithm

1 Introduction

Event-B [1] was introduced to enable modelling of systems, while the main goal of classical B [2,3] is software construction. Whereas classical B uses software abstractions in terms of the procedural concept of operations, Event-B models state-based transition systems in terms of events that describe state transitions. As such, Event-B has a more flexible refinement concept: parameters of events can be changed, new events can be introduced, events can be split or merged. On the other hand, Event-B lacks structured programming constructs such as if-then-else, sequential composition or while loops. In practice, this means that, e.g., an if-then-else statement has to be encoded by two events: one event for the then branch and one for the else branch. Similarly, sequential composition or while loops require the introduction of auxiliary variables. They have to be encoded by introducing explicit program counter variables along with at least one event per program location. This can lead to a large number of events and auxiliary variables. In [1, Chapter 15] Abrial presents an approach, along with a series of rules, to derive structured programming constructs from an Event-B model specified as a collection of events. One motivation of this approach is code generation for Event-B models. There are, however, various drawbacks to the approach from [1] for developing larger algorithms or software artefacts:

© Springer International Publishing Switzerland 2016
M. Butler et al. (Eds.): ABZ 2016, LNCS 9675, pp. 183–197, 2016.
DOI: 10.1007/978-3-319-33600-8_11

1. Developing an Event-B model with many events and explicit program counters can be cumbersome.
2. The rules to recover the program structure are not deterministic: it can be difficult to determine whether an if-then-else statement or a while statement should be generated (both rules from [1] can apply at the same time).
3. There are issues with scaling the approach, i.e., trying to recover the program structure from an Event-B model with hundreds or thousands of events can become intractable.[1]
4. Even modestly-sized Event-B models of structured programs can become quite intricate. This has various reasons discussed in [5]. For instance, there is no concept of local variables and interference of events needs to be dealt with although in the model they may only occur sequentially; also the user has to keep track of properties like termination to be proved.

UML-B [6] is a UML-based graphical notation built on top of Event-B. UML-B state machines maintain program counters corresponding to the hierarchical states connected by event transitions. Although this approach produces more comprehensible models, it is not suited for general purpose software development. It generates state machine encodings instead of structured programs.

The work of Mery [7] presents guidelines on developing algorithms in Event-B in the style of [1]. The Event-B flow approach [8], on the other hand, is based on adding control-flow annotations which are discharged using theorem proving. There are various code generation approaches [9,10] for Event-B, which require the user to provide the algorithmic structure present in a refined Event-B model. On the side of classical B, there are plenty of tools and works [11,12]. In this setting the work of the automatic refiner BART [13] is very successful in practice.

This study was motivated by an industrial project, trying to generate correct-by-construction code for sequential algorithms on a generic railway topology. The work is inspired by the PlusCal language for TLA$^+$ [14], an algorithm description language which is translated to TLA$^+$ to be used with the TLC [15] model checker, or the TLAPS [16] interactive theorem proving system. While Plus-Cal and the translation to TLA$^+$ were invented before TLAPS, the algorithm description language and translation presented in this paper explicitly focus on the provability of the generated Event-B specifications in Rodin.

Another related approach is that of the Dafny [17] language and verifier. There are quite a few similarities; indeed the tutorial in [18] uses the same binary search example we use later in Sect. 5. One obvious difference is that we use Event-B as intermediate language, along with the Rodin provers to discharge proof obligations. The Dafny verifier translates programs to the Boogie intermediate language [19] and uses mainly Z3 for proof (Z3 can also be used within Rodin, see [20]). Another difference is that Event-B provides us with set theory and higher-order relations and functions as data structures. This makes

---

[1] Private communication with Thierry Lecomte related to [4] and experiences of the FORCOMENT project on deducing the Ariane flight sequencer from an Event-B model.

it well suited for modelling more abstract versions of algorithms, very akin to pseudo-code.

2   A Domain-Specific Language for Structured Programs

The ProB Java API[2] provides a domain specific language (DSL) for creating and manipulating Event-B machines and contexts programmatically. The manipulation was developed as an internal DSL using the programming language Groovy, which is a dynamic language that runs on the Java virtual machine. The DSL generates a Java object model of the specification which can then be traversed and manipulated programmatically. This approach makes it possible for end users to experiment with the tool and add features without having to wait for new releases and to use any library that runs on the Java virtual machine.

In this paper we present an extension to the DSL which can be used to define algorithms. Figure 1 shows an example program written using our extension on the left side. Because it is an embedded DSL, we cannot extend the parser to recognize Event-B formulas so we must input these as strings by enclosing them in quotation marks. However, because of the programmatic nature of the DSL, an external parser with a concrete syntax can be easily integrated to generate the same model. The same example with a concrete syntax can be seen on the right side of Fig. 1. We use this program to illustrate the translation to Event-B.

```
procedure(name: "mult") {
  argument "x", "NAT"
  argument "y", "NAT"
  result    "product", "NAT"
  precondition "x >= 0 & y >= 0"
  postcondition "product = x * y"
  implementation {
    var "x0", "x0 : NAT", "x0 := x"
    var "y0", "y0 : NAT", "y0 := y"
    var "p", "p : NAT", "p := 0"
    algorithm {
      While("x0 > 0",
        invariant: "p + x0*y0 = x*y") {
        If("x0 mod 2 /= 0") {
          Then("p := p + y0")
        }
        Assign("x0,y0 := x0/2,y0*2")
      }
      Assert("p = x*y")
      Return("p")
}}}
```

procedure $mult(\mathbb{N}\ x, \mathbb{N}\ y) = \mathbb{N}\ product$
precondition $x \geq 0 \wedge y \geq 0$
postcondition $product = x * y$
implementation
   var $x0$ **type** $x0 \in \mathbb{N}$
    init $x0 := x$
   var $y0$ **type** $y0 \in \mathbb{N}$
    init $y0 := y$
   var p **type** $p \in \mathbb{N}$ **init** $p := 0$
   algorithm
    while $(x0 > 0)$:
     invariant
      $p + (x0 * y0) = x * y$
     if $(x0 \bmod 2 \neq 0)$:
      $p := p + y0$
     $x0, y0 := x0/2, y0 * 2$
    assert $p = x * y$
    return p

Fig. 1. Example program in our notation

---

[2] Its documentation is available online http://www.prob2.de.

Procedure:
procedure proc(T_1 $x_1, \ldots, T_n$ x_m)=U_1 $y_1, \ldots, U_n$ y_n
    precondition $p(x_1, \ldots, x_m)$
    postcondition $q(x_1, \ldots, x_m, y_1, \ldots, y_n)$
    implementation
       I
      algorithm
        S
        return $v_1, \ldots, v_n$

Assignment:
$v_1, \ldots, v_k := e_1, \ldots, e_k$

Assertion:
assert c

Procedure call:
$r_1, \ldots, r_n := \mathbf{proc}(a_1, \ldots, a_m)$

Sequential composition:
S_1
S_2

if-then-else Statement:
if (c):
    S_1
else
    S_2

while Statement:
while (c):
    invariant i
    S

Fig. 2. Syntax of the structured program notation

The syntax of the program notation can be seen in Fig. 2. The top-level construct is the *procedure* "proc" that has value parameters T_1 $x_1, \ldots, T_n$ x_m and result parameters U_1 $y_1, \ldots, U_n$ y_n, where Z z denotes a parameter named z of type Z. All parameter names must be distinct. The procedure has a precondition $p(x_1, \ldots, x_m)$ and a postcondition $q(x_1, \ldots, x_m, y_1, \ldots, y_n)$ that specify the behaviour of the procedure. The *implementation* of the procedure begins with a declaration (and initialisation) of *local variables*. The local variable names must be distinct from the parameter names. Each local variable declaration has the shape "**var** l **type** $l \in T_l$ **init** $l := e$", where e is an expression that may refer to procedure parameters. The initialisation of all local variables is simultaneous. The body of the procedure, the *algorithm*, must implement the specified behaviour (in terms of p and q). It consists of a structured program composed from the usual constructs (sequential composition, assertion, if-then-else statement and while statement) followed by a return statement **return** $v_1, \ldots, v_n$, where $v_1, \ldots, v_n$ are parameters or local variables with the matching types of the corresponding return parameters $y_1, \ldots, y_n$. We currently prohibit recursive procedure calls since we have not yet formalized how to prove that the recursion would terminate. Each while statement specifies the loop *invariant*. A defined procedure can then be called with variables $a_1, \ldots, a_m$ and $r_1, \ldots, r_n$ whose types correspond to $x_1, \ldots, x_m$ and $y_1, \ldots, y_n$ from the procedure. The formulas which can be used in the program description must be in the Event-B syntax.

To model a procedure, e.g. mult in Fig. 1, three Event-B components are generated as shown in Fig. 3. We will now sketch the translation and correctness argument for the approach. Details on the translation and dedicated proof support for Event-B are provided in later sections. Abrial describes the correctness statement for a sequential program with a Hoare-triple,

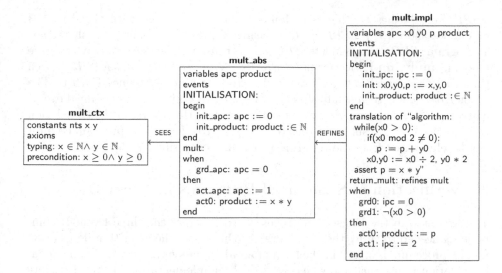

Fig. 3. How algorithms are translated

$$\{P\}\ PROG\ \{Q\}$$

where P and Q describe the precondition and postcondition of the program $PROG$ and then (manually) decomposes $PROG$ into a collection of events that is shown by refinement to implement the specification $SPEC$: $\{P\} \cdot \{Q\}$. We follow a similar approach using a single refinement step but in an automated way. We want to prove by refinement that the behaviour described by P and Q holds for a given implementation $PROG$. The arguments of the procedure are passed by-value and thus cannot be modified, so they are represented as constants in the context **mult_ctx** corresponding to the approach proposed by Abrial [1]. The precondition must only reference the arguments, so this is defined as an axiom in the context.

We define an abstract machine, **mult_abs** containing a specification event $ESPEC$ (called here mult) of the following shape, specifying the intended behaviour of our algorithm as a single event:

Event $\ ESPEC \ \widehat{=}$
  when
    grd_apc: $apc = 0$
  then
    act_apc: $apc := 1$
    act0: $y_1, \ldots, y_n : |\ Q'$
  end

where Q' is Q with $y_1, \ldots, y_n$ substituted by $y'_1, \ldots, y'_n$. If possible, we transform the nondeterministic action *act0* into a deterministic action. Because the precondition is stated in the context it always holds. In particular, it holds before the execution of $ESPEC$ when apc = 0. Furthermore, the action of

ESPEC establishes Q, that is Q holds when apc = 1. So with this Event-B model we show $\{P\}$ *ESPEC* $\{Q\}$. Using Event-B refinement we show that a program *PROG* implements *SPEC* by translating *PROG* into a concrete machine **mult_impl** with events $EPROG_1$, ..., $EPROG_n$, *ERETURN*, such that $EPROG_1$ to $EPROG_n$ all refine skip and *ERETURN* refines *ESPEC*. The event *ERETURN* then establishes Q while the precondition P is assumed everywhere. The event recomposition rules of [1] are just the reverse of the translation process and thus are sufficient to show $\{P\}$ *PROG* $\{Q\}$. The program counter can be eliminated from the resulting program as described in [5].

3   Verification of Structured Programs in Event-B

The first step in the translation process is to generate the three Event-B components needed to verify the procedure. At this stage, the algorithm description for the procedure is stored in the form of an abstract syntax tree (AST) as meta-information in the machine which models the implementation of the procedure (e.g. **mult_impl**). The translation then proceeds by extracting this AST from the machine and generating a control flow graph (CFG) [21] that describes the structure of the algorithm. The CFG is the basis for the definition of a program counter that ensures that the events generated for the algorithm get scheduled in the correct order. This process can be seen in Fig. 4.

The nodes of the CFG are the statements of the AST with the exception of assertions. They are translated in a later step using the CFG as discussed in Sect. 4. The edges show the order in which the statements are to be executed. An *unconditional edge* from a node A to a node B denotes that B is always executed after A. A *conditional edge* labelled with a predicate c from node A to node B denotes that B may be executed after A when condition c holds. In the diagram of Fig. 5 a rectangle denotes a node of the CFG associated with a statement from the AST. An unconditional edge is denoted as an unlabelled line, and a conditional edge is denoted as a labelled line whose label is the condition for the edge.

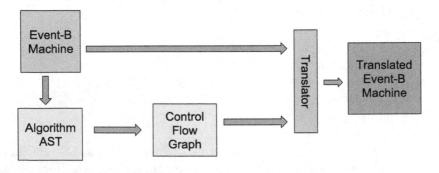

Fig. 4. Translation process

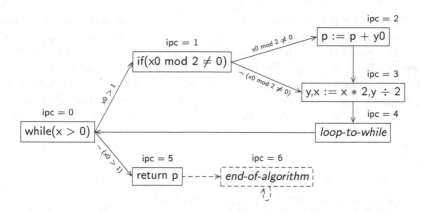

Fig. 5. Control flow graph for multiplication algorithm

The CFG for the multiplication example (Fig. 1) is shown in Fig. 5. In this graph assignments and return statements are nodes with one unconditional edge leading to the next statement in the control flow[3]. For an if statement, conditional edges are generated labelled with the condition and its negation, leading to the graphs for the then-block and else-block of the if statement respectively. The final nodes of these graphs are connected to the statement immediately following the if-statement in the control flow. If either block is empty, the corresponding edge is connected directly to the statement. For a while-statement a conditional edge labelled with the while-condition is generated leading to the CFG for the loop body. All of the outgoing nodes from this graph lead to a special statement *loop-to-while* whose outgoing edge leads back to the loop beginning. A conditional edge is also generated for the negation of the while-condition which leads to the statement immediately following the while-statement in the control flow. A node named *end-of-algorithm* is appended to the node with the final return statement. This node is for analysis with the model checker, where termination is identified with an infinite sequence of skip statements similarly to TLA+[14].

The program counter for the statements is generated by traversing the CFG. The program counter is increased for each node in the graph. We then generate an event for every edge in the CFG with the following form:

Event *unique_event_name* $\widehat{=}$ for a procedure, the program counter is "ipc"
 when
   grd1: $ipc = pcInfo[edge.from]$ value of pc for *edge.from*
   grd2: *edge.condition*   if the edge is conditional
 then
   translation of the action of *edge.from* if any
   actN: $ipc := pcInfo[edge.to]$ value of ipc for *edge.to*
 end

---

[3] The same holds for procedure calls whose treatment we omit in this article.

The translated if and while statements add only the action `actN` to modify the program counter since they affect the control flow but do not otherwise change the state of the machine. Assignments are added as actions to the translated event. The translated return statement refines $ESPEC$ from the abstract machine. The returned variables $v_0, \ldots, v_n$ correspond to the result parameters $y_0, \ldots, y_n$ from the procedure. When the return statement is translated, we generate an action for each result, setting its value to the corresponding variable:

`act0:` $y_0 := v_0$

...

`actn:` $y_n := v_n$

The special node *end-of-algorithm* is translated to

Event $end\_algorithm \;\widehat{=}$
    when
        `grd0:` $pc = last\_pc$
    then
        skip
    end

This event is similar to the *final* event in the sequential algorithms developed by Abrial [1, p. 451], with a guard which evaluates to true only after the program has completed execution. It recognises the end of the algorithm.

Merging Nodes in the Control Flow Graph. The if and while statements affect the guards of the generated events, but they do not change the state of the machine. To reduce the number of events that are generated by the translator, we can merge consecutive conditional edges. We can also combine a conditional edge with the statement which follows it if the following statement is an assignment, return statement, or procedure call. The translation of the multiplication example with this transformation is shown in Fig. 6.

Procedure Calls. Procedure calls of a procedure *proc* with a precondition P, postcondition Q, value parameters $x_1, \ldots, x_m$ and result parameters $y_1, \ldots, y_n$ can be translated in the following way. Let $r_1, \ldots, r_n := proc(v_1, \ldots, v_m)$ be the procedure call. Then we treat the call as if it was the statement

assert P_1
$r_1, \ldots, r_n : \mid Q'_1$

where P_1 and Q_1 are the precondition and postcondition with the value parameters $x_1, \ldots, x_m$ substituted by the variables $v_1, \ldots, v_m$ and the result parameters $y'_1, \ldots, y'_n$ replaced by the variables $r'_1, \ldots, r'_n$. For instance, the procedure call $res := \mathbf{mult}(a, b)$ would be translated as

assert $a \geq 0 \wedge b \geq 0$
$res := a * b$

mult_impl

```
variables x0 y0 p product ipc
invariants
    gluing1: ipc<3 ⇒ apc=0
    gluing2: ipc≥3 ⇒ apc=1
    while0_inv: pc=0 ⇒ p+x0*y0=x*y
    assert0: ipc=0 ∧ ¬(x0>0) ⇒ p=x*y

INITIALISATION:
begin
    init: x0,y0,p := x,y,0
    init_product: product :∈ ℕ
    init_ipc: ipc := 0
end
```

enter_while0_if0_then:
when
    grd0: ipc=0
    grd1: x0 > 0
    grd2: x0 mod 2≠0
then
    act0: p := p+y0
    act1: ipc := 1
end
 	
assign1:
when
    grd0: ipc=1
then
    act0: x0,y0 := x0÷2,y0*2
    act1: ipc := 2
end
	loop_to_while0:
	when
	    grd0: ipc=2
	then
	    act0: ipc := 0
	end
	
	enter_while0_if0_else:
	when
	    grd0: ipc=0
	    grd1: x0 > 0
	    grd1: ¬(x0 mod 2≠0)
	then
	    act1: ipc := 1
	end
		exit_while0:
		refines mult
		when
		    grd0: ipc=0
		    grd1: ¬(x0>0)
		then
		    act0: product := p
		    act1: ipc := 3
		end
		
		end_of_algorithm:
		when
		    grd0: ipc=3
		then
		    skip
		end

Fig. 6. Event-B translation of the multiplication algorithm

Given that $a \geq 0 \wedge b \geq 0$, the procedure will yield $res := a * b$. Note that the nondeterministic shape of $product := x*y$ is $product : | \; product' = x*y$. The lists of variables $r_1, \ldots, r_n$ and $v_1, \ldots, v_m$ do not need to be disjoint. The procedure call $a := \mathbf{mult}(a, b)$ yields the nondeterministic assignment $a : | \; a' = a * b$, and thus

assert $a \geq 0 \wedge b \geq 0$
$a := a * b$

The renaming of return parameters into their primed counterparts breaks the analogy of the approach with that using the specification "$\{ P \} \cdot \{ Q \}$". There the two variable lists must be disjoint. Otherwise for postcondition $x = y + 1$, say, one could make the call with postcondition $r = r + 1$, after which anything could be asserted, of course.

Parameters of procedure calls can only be variables. This avoids complicated parameter passing semantics such as call-by-name, or using copying schemes of parameters into local variables in order to attain call-by-value semantics.

4 Verifying Generated Models by Backwards Propagation

An Event-B model can be verified via formal proof with the Rodin tool, which generates proof obligations (POs) with the form

$$HYP \Rightarrow GOAL \tag{1}$$

where HYP is a set of hypotheses, and $GOAL$ is the predicate to be proven. Rodin also has provers which attempt to resolve these POs automatically.

By proving simulation (SIM) POs for every event in the refinement, we prove that its behaviour corresponds to the abstract machine. For the translated return statement, the postcondition Q of the procedure is the before-after predicate for the abstract action, so the SIM PO for this event with guard G and action $v : |R$ is:

$$A \wedge I \wedge G \wedge R \Rightarrow Q \tag{2}$$

Note, that G and R may refer to more variables than Q because the state of the abstraction may be enriched by the introduction of local variables in the procedure body. The other translated events refine *skip* so the SIM PO is

$$A \wedge I \wedge G \wedge R \Rightarrow y = y' \tag{3}$$

We define two gluing invariants of the form $ipc < last\_pc \Rightarrow apc = 0$ and $ipc \geq last\_pc \Rightarrow apc = 1$ for the disappearing program counter variable apc. This permits us to use case distinction on the value of the program counter: either the concrete value corresponds to the abstract initial program counter value or the abstract final program counter value. The return event sets the program counter to $last\_pc$ and refines the abstract event, all others must refine skip.

In order to help prove the SIM POs, we add information to the model with axioms A in a referenced context or by adding invariants I to the implementation. One of the ways we can add information to our model is by declaring assertions within our algorithm body. **Assertions** are predicates that must hold during execution at the location of the program where they are declared. To translate an assertion, we find the first statement following it in the control flow and the event that has been generated for it. An assertion A that precedes this statement is encoded as the invariant $G \Rightarrow A$, where G is the conjunction of the guards of the translated event. To prove that an invariant $i(v_1, ..., v_n)$ holds for a machine, Rodin generates INV POs for each event with guards G and action $v_1, ..., v_n : | R$:

$$A \wedge I \wedge G \wedge R \Rightarrow i(v'_1, ..., v'_n) \tag{4}$$

Here $i(v'_1, ..., v'_n)$ is the invariant with the variables $v_1, ..., v_n$ replaced by $v'_1, ..., v'_n$. Since we have the invariant $G \Rightarrow A$ and guards G as hypotheses, the assertion A will appear as a hypothesis in the POs for the event. G always contains some guard $ipc = X$ where X is the value of the program counter at the program location of the statement. Proving $G \Rightarrow A$ proves that A holds after the execution of the statements that precede the assertion in the algorithm because they set the program counter to the value X. Loop invariants L are translated as

$ipc = X \Rightarrow L$ where X is the value of the program counter of the while loop: the loop invariant holds both when entering the while loop and upon exiting it. The translation helps mastering the details of placing the assertions and invariants. Dealing with this manually in Event-B is possible, but rather cumbersome and error prone. A particular difficulty arises when a model needs to be changed and program counters and guards be adapted across the Event-B model.

In our example, the final assertion $p = x * y$ is translated to the invariant $ipc = 0 \wedge \neg(x0 > 0) \Rightarrow p = x*y$. The loop invariant for the machine is encoded as $ipc = 0 \Rightarrow (p + (x0 * y0) = x * y)$. We can add extra assertions so that the provers have enough information to resolve the POs automatically. However, knowing a statement S and the assertions A which follow it, we can derive the assertions N that must precede it. This is called *backwards assertion propagation*. It relies on the weakest precondition semantics of the programming notation that is mapped to Event-B models (see [2,22]).

The new assertion N is calculated from S and A in the following way:

$$A[S] = N \ ,$$

where the left-hand side denotes a substitution. Many of the POs derived from these assertions added to the model can be automatically discharged.

We iterate this process backwards following the structure of the algorithm. Hence assertions are propagated backwards through sequences of statements:

$$A[S_n] = N_n$$

$$\vdots$$

$$N_2[S_1] = N_1$$

If the statement is an assertion, this assertion is added to the assertions which are being propagated. If the statement list is empty, the original assertions are returned.

We can propagate an assertion A over a deterministic assignment of the form $v_1, ..., v_n := e_1, ..., e_n$ where $v_1, ..., v_n$ are variables of the machine and $e_1, ..., e_n$ are Event-B expressions. We propagate the assertion over this assignment by replacing all of the variables v_i which are used in the assertion with the corresponding expression e_i on the right-hand side of the assignment, the new assertion N is obtained through $A[v_1, ..., v_n := e_1, ..., e_n] = N$.

We have not yet implemented assertion propagation for a non-deterministic assignment $x :\mid P$, but in the future we can rewrite the assertions A using (see e.g. [2, Chapter 4])

$$\forall x' \cdot P \Rightarrow A[x := x'] \ .$$

This could be useful when propagating over procedure calls. If the input and output variable lists of a call as described in the preceding section would be disjoint, then we could use

$$\forall x \cdot P[x' := x] \Rightarrow A \ ,$$

avoiding the renaming of variables in the assertions.

Procedure calls are represented by an assertion followed by an assignment, so propagating over procedure calls is covered by the techniques discussed above.

In our example, we need to prove that the loop invariant $p + x0 * y0 = x * y$ holds after the execution of the last assignment $x0, y0 := x0 \div 2, y0 * 2$ in the while loop. The assertion propagation then yields

$$(p + x0 * y0 = x * y)[x0, y0 := x0 \div 2, y0 * 2] = p + (x0 \div 2) * (y0 * 2) = x * y$$

We now need to propagate this assertion over the if statement $\mathbf{if}(x0 \bmod 2 \neq 0)$: $p := p + y0$. We begin by propagating the assertion over the then and else block in the if statement. For an if statement $\mathbf{if}$ $(c) : S_1$ $\mathbf{else}$ S_2 this is done according to the formula $(c \Rightarrow A[S_1]) \wedge (\neg c \Rightarrow A[S_2])$ [2, Chapter 4].

In this example, the generated assertions are therefore

$$(p + (x0 \div 2) * (y0 * 2) = x * y)[\mathbf{if}(x0 \bmod 2 \neq 0): p := p + y0] =$$
$$x0 \bmod 2 \neq 0 \Rightarrow (p + y0) + (x0 \div 2) * (y0 * 2) = x * y \wedge$$
$$\neg(x0 \bmod 2 \neq 0) \Rightarrow p + (x0 \div 2) * (y0 * 2) = x * y)$$

We have now propagated the loop invariant over the body of the whole loop. Assertion propagation over the while only relies on the weakest liberal precondition. We have to deal with two cases: the condition of the while loop holds when the body is entered or the condition does not hold and the while loop is exited. For a while loop with syntax $\mathbf{while}$ $(c) : \mathbf{invariant}$ i S, this is translated as $(c \Rightarrow (i)[S]) \wedge (\neg c \Rightarrow A)$. In our example, this gives us

$$(p = x * y)[\mathbf{while}(x0 > 0): \ldots] =$$
$$x0 > 0 \wedge x0 \bmod 2 \neq 0 \Rightarrow (p + y0) + (x0 \div 2) * (y0 * 2) = x * y \wedge$$
$$x0 > 0 \wedge \neg(x0 \bmod 2 \neq 0) \Rightarrow p + (x0 \div 2) * (y0 * 2) = x * y \wedge$$
$$\neg(x0 > 0) \Rightarrow p = x * y$$

These generated assertions are translated in the same way as user declared assertions. The purpose of assertion propagation is to achieve better proof support but it also helps to attain an increased number of proof obligations that can be proven automatically.

5   Practical Matters: Use of the Tool and Method

We evaluated our approach by applying it to five case studies[4]. The multiplication algorithm was introduced in Fig. 1. We also modelled an algorithm for calculating the greatest-common divisor (GCD), a factorial algorithm, an algorithm to perform model checking, and a binary-search algorithm. Both the model checking algorithm and the binary search algorithm are algorithms which are used in real world applications. Using the Rodin tool with the Atelier-B provers, we were able to achieve a high level of automated proof for all of our case studies as seen in Table 1. By merging the generated events where possible, we were able to decrease the number of POs generated for a model. But as a rule this did not increase the number of POs which could be automatically proved.

---

[4] Our models and the standalone translation tool can be found at https://github.com/joyclark/eventb_gen.

Table 1. Automatic Proof Resolution for Naive and Merged Translation

Algorithm	Automatic Proof		Manual Proof		Total POs		Percent	
	Naive	Merged	Naive	Merged	Naive	Merged	Naive	Merged
Multiplication	149	83	1	1	150	84	99.33 %	98.81 %
GCD	367	168	9	11	376	179	97.61 %	93.85 %
Factorial	528	271	4	4	532	275	99.25 %	98.55 %
Model-Checking	772	428	7	5	779	433	99.10 %	98.85 %
Binary-Search	1019	462	25	29	1044	491	97.61 %	94.09 %

All of the algorithms were modelled using the procedure abstraction. The model checking algorithm was modelled using four separate procedures to decompose the algorithm into smaller components that could be proven separately. We believe this approach will scale to larger algorithms.

The speed for modelling and verifying algorithms depends greatly on the person who is doing the modelling and their familiarity with the tool chain. In these case studies, the modeller had some familiarity with the tool, but very little practical experience in performing proof with it. With these limitations, all of the case studies were modelled and completely proven within two days. We consider this to be a reasonable time frame in which to prove the correctness of an algorithm.

6 Conclusion

In this paper we have presented a method for defining algorithm descriptions and translating them to Event-B models so that they can be verified. Our approach focuses on verifying an algorithm using formal proof, however, the translation also supports verification by model checking. Within our algorithm descriptions, we can define assertions and loop invariants which can be proven correct for the model. A key technique in our approach is *backwards propagation*. Indeed, using the algorithm, we can propagate the assertions and loop invariants through the model to generate new assertions in order to assist the user and the tool in proving the model automatically. We evaluated our approach by applying it to five case studies and found that it was possible to model and verify all of the algorithms with a high level of automated proof.

The procedure abstraction we implemented allows us to use refinement to verify an algorithm in a single refinement step. This was possible for our case studies, but since this may not scale, in the future we will consider how to perform the translation with multiple refinement steps. However, this abstraction has allowed us to break an algorithm into smaller sections which can be independently verified, so we believe that it will be possible to scale our approach for larger algorithms.

Permitting several refinement steps will also make it possible to use the technique of termination verification used in Event-B. Instead of introducing loops (and nested loops) in one step, one could first introduce an abstraction of the loop containing a nondeterministic assignment as body. This would produce an event that is either anticipated or convergent depending on whether a variant is specified or not. Termination proofs would then proceed the way they do in Event-B.

References

1. Abrial, J.-R.: Modeling in Event-B: System and Software Engineering, 1st edn. Cambridge University Press, New York (2010)
2. Abrial, J.-R.: The B-Book. Cambridge University Press, New York (1996)
3. Schneider, S.: The B-Method: An Introduction. Palgrave Macmillan, Basingstoke (2001)
4. Lecomte, T.: Ten years disseminating the B method. In: Attiogbe, C., Mery, D. (eds.) Proceedings of TFM-B 2010, pp. 65–72. APCB, June 2010
5. Hallerstede, S., Leuschel, M.: Experiments in program verification using Event-B. Formal Aspects Comput. **24**(1), 97–125 (2012)
6. Snook, C.F., Butler, M.J.: UML-B: formal modeling and design aided by UML. ACM Trans. Softw. Eng. Methodol. **15**(1), 92–122 (2006)
7. Méry, D.: Refinement-based guidelines for algorithmic systems. Int. J. Softw. Inf. **3**(2–3), 197–239 (2009)
8. Iliasov, A.: Use case scenarios as verification conditions: event-B/Flow approach. In: Troubitsyna, E.A. (ed.) SERENE 2011. LNCS, vol. 6968, pp. 9–23. Springer, Heidelberg (2011)
9. Edmunds, A., Butler, M., Maamria, I., Silva, R., Lovell, C.: Event-B code generation: type extension with theories. In: Derrick, J., Fitzgerald, J., Gnesi, S., Khurshid, S., Leuschel, M., Reeves, S., Riccobene, E. (eds.) ABZ 2012. LNCS, vol. 7316, pp. 365–368. Springer, Heidelberg (2012)
10. Edmunds, A.: Templates for Event-B code generation. In: Ait Ameur, Y., Schewe, K.-D. (eds.) ABZ 2014. LNCS, vol. 8477, pp. 284–289. Springer, Heidelberg (2014)
11. Petit, D., Poirriez, V., Mariano, G.: The B method and the component-based approach. Trans. SDPS **8**(1), 65–76 (2004)
12. Bert, D., Boulmé, S., Potet, M.-L., Requet, A., Voisin, L.: Adaptable translator of B specifications to embedded C programs. In: Araki, K., Gnesi, S., Mandrioli, D. (eds.) FME 2003. LNCS, vol. 2805, pp. 94–113. Springer, Heidelberg (2003)
13. Requet, A.: BART: a tool for automatic refinement. In: Börger, E., Butler, M., Bowen, J.P., Boca, P. (eds.) ABZ 2008. LNCS, vol. 5238, p. 345. Springer, Heidelberg (2008)
14. Lamport, L.: Specifying Systems, The TLA+ Language and Tools for Hardware and Software Engineers. Addison-Wesley, Boston (2002)
15. Yu, Y., Manolios, P., Lamport, L.: Model checking TLA+ specifications. In: Pierre, L., Kropf, T. (eds.) CHARME 1999. LNCS, vol. 1703, pp. 54–66. Springer, Heidelberg (1999)
16. Chaudhuri, K., Doligez, D., Lamport, L., Merz, S.: Verifying safety properties with the TLA + proof system. In: Giesl, J., Hähnle, R. (eds.) IJCAR 2010. LNCS, vol. 6173, pp. 142–148. Springer, Heidelberg (2010)
17. Rustan, K., Leino, M.: Developing verified programs with dafny. In: Proceedings ICSE 2013, pp. 1488–1490 (2013)

18. Koenig, J., Rustan, K., Leino, M.: Getting started with dafny: a guide. In: Nipkow, T., Grumberg, O., Hauptmann, B. (eds.) Software Safety and Security - Tools for Analysis and Verification. NATO Science for Peace and Security Series - D: Information and Communication Security, vol. 33, pp. 152–181. IOS Press (2012)
19. Leino, K.R.M., Rümmer, P.: A polymorphic intermediate verification language: design and logical encoding. In: Esparza, J., Majumdar, R. (eds.) TACAS 2010. LNCS, vol. 6015, pp. 312–327. Springer, Heidelberg (2010)
20. Déharbe, D., Fontaine, P., Guyot, Y., Voisin, L.: SMT solvers for rodin. In: Derrick, J., Fitzgerald, J., Gnesi, S., Khurshid, S., Leuschel, M., Reeves, S., Riccobene, E. (eds.) ABZ 2012. LNCS, vol. 7316, pp. 194–207. Springer, Heidelberg (2012)
21. Allen, F.E.: Control flow analysis. SIGPLAN Not. **5**(7), 1–19 (1970)
22. Hallerstede, S.: On the purpose of Event-B proof obligations. Formal Asp. Comput **23**(1), 133–150 (2011)

Formal Proofs of Termination Detection for Local Computations by Refinement-Based Compositions

Maha Boussabbeh[1,2(✉)], Mohamed Tounsi[2], Mohamed Mosbah[1],
and Ahmed Hadj Kacem[2]

[1] LaBRI Laboratory, University of Bordeaux, Talence, France
{maha.bousabbah,mohamed.mosbah}@labri.fr
[2] ReDCAD Laboratory, University of Sfax, Sfax, Tunisia
{ahmed.hadjkacem,mohamed.tounsi}@fsegs.rnu.tn

Abstract. In this paper, we propose a formal framework enhancing the termination detection property of distributed algorithms and reusing their specifications as well as their proofs. By relying on refinement and composition, we show that an algorithm specified with local termination detection, can be reused in order to compute the same algorithm with global termination detection. The main idea relies upon the development of distributed algorithms following a top/down approach and the integration of additional computation steps developed in a pre-defined module. This module is specified in a generic and scalable way in order to be composed with particular developments. Once the composition link is proven, the global termination emerges automatically.

Keywords: Distributed algorithms · Local computations · Termination detection · SSP algorithm · Composition · Event-B method

1 Introduction

1.1 Overview

It is widely agreed that implementing distributed systems poses major problems and remains a real challenge. Distributed termination detection is one of the most important problems in distributed computing [11]. It is closely related to many other problems such as determining a causally consistent global state [10], detecting deadlocks [12], etc. Contrary to sequential algorithms, the termination of distributed computing is neither simple nor clear: what does termination mean in distributed algorithms? Can processors be aware of global termination? Is it essential for a processor to distinguish between its termination and the termination of the entire computation? Different termination detection modes have been defined [8] to relate the local state of processors with the global state of the network. The two modes we are interested in are the following: *Local Termination Detection* (LTD), i.e., each processor is able to determine only its own

© Springer International Publishing Switzerland 2016
M. Butler et al. (Eds.): ABZ 2016, LNCS 9675, pp. 198–212, 2016.
DOI: 10.1007/978-3-319-33600-8_12

termination condition; and *Global Termination Detection* (GTD), i.e., at least one processor knows when the entire computation has finished on the network. P. Castéran et al. [7] have proved that it is quite interesting for a processor to detect that the algorithm has globally terminated. However, if we don't have a global perspective and we are only interested in local interactions, it will not be evident to know if the distributed algorithm has finished.

In this paper, we propose a general framework, based on Event-B, for transforming correct algorithms with LTD into algorithms with GTD. We rely on the high level abstraction of local computations [20], and we focus on formal proofs of termination, using a refinement-based composition. The prime objective is to provide a proof-based development which can be reused for building and ensuring global termination. Another objective is to show the effectiveness of combining a *correct-by-construction* [19] approach with a compositional reasoning for preserving properties and reusing proofs.

1.2 Related Works

Formal specifications are often beneficial, and provide a real help for expressing correctness with respect to safety properties in the study of distributed computing. This paper is not an exception in this respect. Formal approaches have been proposed to deal with the correctness of such algorithms in different contexts: solving gathered problems [13], revisiting snapshot algorithms [3], etc. Numerous studies related to the termination detection problem have been done. However, no clear idea, about the way which would be better for transforming distributed algorithms from LTD to GTD mode and reusing their proofs, has been come out from these works. To the best of our knowledge, no reusable formal approach has been published yet, clarifying how far it can save efforts of designers and how far it can be reused in particular developments.

Related works have been proposed to suggest solutions for algorithms detecting only the LTD mode. Two major algorithms are used to cope with this problem and build the *GTD* mode: the Dijkstra-Scholten algorithm [14] and the algorithm by Szymanski, Shi and Prywes (the SSP algorithm for short) [22]. E. Godard et al. [16] proposed to compose two graph relabelling systems, one encoding a given algorithm and another encoding a termination detection algorithm such as the Dijkstra-Scholten algorithm or the SSP algorithm. They proved that the resulting relabelling system transforms the first algorithm from LTD mode to GTD mode. However, this transformation modifies the algorithm and makes the computation steps of the nodes more complex.

V. Filou et al. [15] proposed to compose formal specifications of distributed algorithms with formal specifications of the Dijkstra-Scholten algorithm: let A be a distributed algorithm specified with LTD mode. Based on the Event-B method, authors proved that a node being in a terminal state of A can execute the Dijkstra algorithm in order to detect the instant where every other node has computed its final value. The specification as well as the proofs of the first algorithm (algorithm A) are reused when detecting the global termination. However, a composition of the Dijkstra algorithm cannot be proposed as a general approach for dealing with algorithms encoded with LTD mode: this algorithm is based on the election process. Several conditions were found to allow election algorithms: P. Castéran et al. [7] and J. Chalopin et al. [9] characterized families of graphs that admit this algorithm.

1.3   Contribution

In a previous work [5], we proposed a proof based development for transforming a spanning tree algorithm with LTD mode into the same algorithm with GTD mode. We relied on Event-B refinement and we specified a combination of the SSP and the spanning tree algorithm following a top down approach. In this paper, we generalize our approach and we propose a formal framework enhancing the termination detection property of distributed algorithms without altering their specifications. Our framework is based on formal specifications and proofs of termination, encapsulated in a separate module according to the *SSP* algorithm. This module is developed in a generic and scalable way in order to be composed with particular developments.

Let A be a distributed algorithm where processors do not detect the global termination of A. Our main goal is to compose A with SSP and produce a correct algorithm which (i) reuses specifications as well as proofs associated to the algorithm A and (ii) enables processors to detect the global termination of the computation. We investigate necessary requirements for such a composition and give users guidelines for reusing the *SSP* algorithm in several developments with correctness.

The refinement of models is the key element, allowing preservation of correctness proofs. Moreover, using pre-defined modules, developed in a high level abstraction, makes our approach reusable. Based on the *correct-by-construction* approach and on the modularization [18] technique, we achieve our aim and show, with examples, what users gain with the proposed approach. In this paper, we illustrate our approach by the *3-colouring* of a ring specified with the *LC1* synchronisation. The main objective of this simple example is to demonstrate the use of our work during models development. Our approach is also applied to complex case studies such as the Mazurkiewicz [21] algorithm, specified with the *LC2* synchronization. The illustration of this algorithm gives us new results which do not appear in this paper, but they will be presented in a future publication.

1.4   Organization of the Paper

The paper is organized as follows: Sect. 2 recalls basic concepts of local computations and Event-B method. Section 3 presents the *SSP* algorithm and introduces our approach where we describe the composition process of *SSP* with Event-B developments. Section 4 details formal specifications and proofs of this composition. Section 5 illustrates our approach by an example. Finally, a short discussion, conclusion and ongoing work round the paper up.

2   Preliminaries

2.1   Local Computations Model

In this section, we illustrate, in an intuitive way, the notions of local computations, and particularly those of graph relabelling systems by showing how some algorithms on networks of processors may be encoded within this framework [20]. As usual, such a network is represented by a graph whose vertices (nodes) stand for processors and edges for (bidirectional) links between processors. Each vertex represents an entity that is capable of performing computation steps, sending and receiving messages. We consider anonymous networks with asynchronous message passing, i.e., each computation may take an unpredictable, but finite, amount of time. At every time, each vertex and each edge are in some particular state which will be encoded by a vertex or an edge label. According to its own state and to the states of its neighbours, each vertex may decide to do an elementary computation step. After this step, the state of this vertex, its neighbours and the corresponding edges may be changed according to some specific computation rules. Moreover, it is supposed that once a node reaches a final state it remains in such a state until the end of the algorithm.

The graph relabelling systems meet the following requirements: (i): they do not change the underlying graph, but they change only the labelling of their components (edges and/or vertices). The final labelling is the result. (ii) they are local, that is, each rewriting changes only a connected subgraph of a fixed size in the underlying graph. A sub-graph contains a subset of the vertices and edges in a graph G. (iii) they are locally generated, that is, the applicability condition of the rewriting depends only on the local context of the relabelled sub-graph.

The distributed aspect comes from the fact that several rewriting steps can be performed simultaneously on "far enough" subgraphs, giving the same result as a sequential realization of them, in any order. A large family of classical distributed algorithms encoded by graph rewriting systems is given in [20].

2.2   Event-B

The Event-B [1] modelling language defines mathematical structures as contexts and formal model of the system as machines. The context is defined by abstract sets, constants, and axioms which describe properties of constants. An Event-B machine describes a reactive system, using a set of invariant properties and a finite list of events modifying state variables. Recently the Event B language and its tool support Rodin [2] have been extended with the possibility to define a *module interface* i.e., logical unit containing *callable* operations. The important characteristic of these modules is that they can be developed separately and, when needed, incorporated and instantiated in the main system development. According to A. Iliasov et al. [18], a *module interface* is a separate Event-B component specifying a set of services. It encapsulates external variables, constants, invariants, and a collection of operations characterised by their *pre-/post-conditions*.

When a *module interface M* is *imported* into an Event-B machine (via the clause *USES*), an instance is created. Several instances can be created for the same module. To avoid name clashes, each instance is added with a prefix *pre* chosen by the user. Consequently, all the names *imported* from the module appear with the corresponding prefix. The importing machine can invoke the operations by means of events and read the external variables of *M*. As presented below, an interface operation *Op* is characterized by its *pre-* and *post-conditions* [18]. The *pre-conditions* contain a list of predicates applied on parameters *par* and on external variables v to define the states when an operation may be invoked. The primed variables v' and res', defined in the operation *post-condition* (*post*), stand for the final variable values after the operation execution. If some primed variables are not mentioned, the corresponding variables are unchanged by the operation.

The execution of a called operation is abstractly modelled by an Event-B event, named *calling_Op*. Internal parameters (*in_par*) are evaluated and passed to the operation *Op*. In addition to calculating a result (*result*), an operation call can also update the external variables (v). A set of proof obligations are generated to guarantee that the state of the module is protected by the operations.

```
OPERATION  Op  ≙
any    par
pre    pre(par, v)              Event   calling_Op  ≙
return  result           ⊑      any    in_par
post   post(par, v', res')  =   where   guard(in_par, pre_v)
end                             then    resultOp := pre_Op(in_par)
                                end
```

3   SSP Composition with *Correct-by-Construction* Developments

3.1   The SSP Algorithm

We consider a distributed algorithm which terminates when all nodes reach their local termination conditions. The *SSP* algorithm [22] detects an instant in which the entire computation is achieved in the network. Let G be a graph. Each node $n0$ is associated with a predicate $P(n0)$ and an integer $a(n0)$. $P(n0)$ depends on the local termination of $n0$. Once a node $n0$ detects its local termination, the value of $P(n0)$ can be transformed from $FALSE$ to $TRUE$. $a(n0)$ is introduced to specify the fact that all nodes being in a distance equal to $a(n0)$ have locally terminated. Initially $P(n0)$ is $FALSE$ and $a(n0)$ is equal to -1. Transformations of the value of $a(n0)$ are defined by the following rules.

- *if $P(n0) = FALSE$ Then $a(n0) = -1$,*
- *if $P(n0) = TRUE$ Then $a(n0) = 1 + min\{a(n_k)|k \geq 0 \text{ and } k \leq d\}$.*

Let $n0$ be a node and let $\{n_1, ..., n_d\}$ be the set of nodes adjacent to $n0$; d stands for the number of edges the node $n0$ has to other nodes, called the degree of $n0$. The new value of $a(n0)$ depends on values associated with $n0$ and its neighbours.

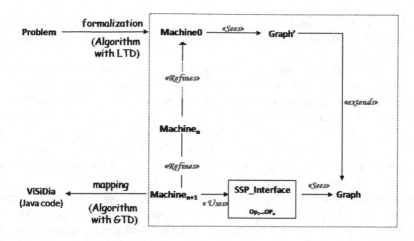

Fig. 1. SSP composition with Event-B development

3.2 Proposed Approach

We summarize in this section the idea of the SSP composition with a distributed algorithm expressed in Event-B. Our aim is to prove that for any algorithm satisfying LTD, expressed in event-B and proved independently, the algorithm obtained by the composition with SSP satisfies the same specification with GTD. Note that an Event-B development is based on a *correct-by-construction* approach [19] which supports an incremental process controlled by the refinement of models. In previous works [15, 23], authors proposed Event-B patterns containing proof-based guidelines and describing how a distributed algorithm can be correctly designed: the development can start with a very abstract model, then, by successive refinements, we obtain a concrete one that expresses the local behaviour (state) of processors in the network. As presented in Fig. 1, we consider a problem, which can be formalized through a distributed algorithm with *LTD*, developed by a chain of Event-B machines. We suppose that the size of the graph is known by all the nodes. Let A be the distributed algorithm. The formalization process is based on the refinement of models, using the RODIN [2] platform.

– The chain of refinement $Machine_0...Machine_n$ expresses the *Problem* in an incremental development.
– The $Machine_n$ defines local interactions between nodes according to the computation process of the algorithm A.
– The refinement of $Machine_n$ by $Machine_{n+1}$ produces a set of events corresponding to some additional computation steps to detect the global termination of the algorithm A. These computations are specified according to the *SSP* rules.
– The *SSP_Interface* is a predefined logical entity that we have developed and proved independently. More precisely, it is an Event-B Interface containing

callable operations, corresponding to the execution process of the *SSP* rules. Moreover, it contains general proofs of termination detection for local computations.

- The $Machine_{n+1}$ *uses* the *SSP_Interface*, means that it can have access to all the proofs discharged in the SSP_Interfce and execute its operations.
- The *Graph* is an Event-B context that we have developed to specify basic properties of a network which represents the application field of distributed algorithms. Basically, a network is defined as a connected, simple and undirected graph. The *Graph'* context *extends Graph*, means that it can use all the specifications defined in the *Graph* context and introduce other static properties, describing the particularity of the algorithm *A*.
- A translation of Event-B specifications into a java code can be generated by relying on *B2Visidia* tool [24]. Thus, a solution can be mapped from the $Machine_{n+1}$ into *ViSiDiA* [4], i.e., *ViSiDiA* is a platform for simulating, visualizing and testing local computations. Note that this item is not studied in this paper.

4   Formal Descriptions

4.1   Network Specification: The *Graph* Context

The *Graph* context describes basic properties of the network on which distributed algorithms are running. Formally, a network can be straightforwardly modelled as a connected, undirected and simple graph where nodes denote processors and edges denote point-to-point communication links. An undirected graph means that there is no distinction between the two nodes associated with each edge (see axm4). A graph is simple if it has at most one edge between any two nodes (see axm2 and axm3) and no edge starts and ends at the same node (see axm5). A graph (directed or not) is connected if, for each pair of nodes, a path joining these two nodes exists(see axm6). According to D. Cansell et al. [6], a connected graph g over a set of finite nodes ND (see axm1) can be presented as follows:

$$
\begin{array}{l}
axm1 : finite(ND) \\
axm2 : g \subseteq ND \times ND \\
axm3 : dom(g) = ND \\
axm4 : g = g^{-1} \\
axm5 : ND \lhd id \cap g = \varnothing \\
axm6 : \forall s \cdot s \subseteq ND \wedge s \neq \varnothing \wedge g[s] \subseteq s \Rightarrow ND \subseteq s
\end{array}
$$

4.2   The *SSP_Interface*

The *SSP_Interface* is an Event-B component, specifying a set of additional computation steps that can be associated to the nodes in order to build a global termination detection. We assume that the local termination of the algorithm that we want to compose is stable, i.e., a terminated processor will not again be woken up in the course of further computation. We mean by termination, the termination of the algorithm that we want to compose. This component encapsulates formal specifications of the *SSP* algorithm into callable operations, modifying a set of external variables. These variables are defined as follows:

```
Interface SSP_Interface
SEES Graph
Variables
Local_TD
counter
Global_TD
...
OPERATION Update_Termination(...)
...
OPERATION To_Global_Termination(...)
...
OPERATION Global_Termination(...)
...
OPERATION Diffusion(...)
```

- $Local\_TD$ characterizes the local termination detection of a node:
  $Local\_TD \in ND \rightarrow BOOL$. Let n be a node. $Local\_TD(n) = True$ means that n has computed its final value. Initially, there is no node detecting the local termination.
- $counter$ leads a node to detect the global termination:
  $counter \in ND \rightarrow \mathbb{P}(\mathbb{N} \times \mathbb{Z})$. If n is a node, $(i \mapsto j) \in counter(n)$ means that at the computation step i, all the nodes, being in a j distance from n, detected locally the termination. Initially $counter(n) = \{0 \mapsto -1\}$. After i computation steps, $counter(n) = \{0 \mapsto -1, 1 \mapsto 0, ..., i \mapsto j\}$. We store the entire history of counters in order to investigate invariants of the model and simplify proofs during the development. In this paper, the *last counter* of n means the value of j calculated at the last computation step i. Formally, it can be specified as follows: last counter(n)= max(ran(counter(n))).
- $Global\_TD$ characterizes the global termination state of a node:
  $Global\_TD \in ND \rightarrow BOOL$. $Global\_TD(n) = True$ means that n is aware of the termination of the algorithm. Initially, there is no node detecting the global termination.

***Update Termination* Operation.** A node being in a terminal state, updates its $Local\_TD$ and acts on its $counter$. Let $counter_i(n)$ be the *last counter* computed by a node n. The new value of the $counter$ of n is defined at the computation step $(i + 1)$. This computation is specified by the *Update_Termination* operation. The pre-conditions (Pre) stand for the requirements of the operation execution. The terminal state of a node depends naturally on the algorithm that we want to compose. Thus the condition of the updating action is defined while calling this operation (see the example in Sect. 5). Note that the execution of this operation should be blocked once a node updates its $Local\_TD$ from $FALSE$ to $TRUE$. This can be achieved by strengthening its pre-conditions by $(Local\_TD(n) = FALSE)$ and $(counter_i(n) = -1)$. The primed variables $Local\_TD'(n)$ and $counter'(n)$ defined in the post-conditions stand for the new values of $Local\_TD(n)$ and $counter(n)$ after the operation execution. In our case, we are interested in updating external variables without returning particular results.

pre-conditions: $Local\_TD(n) = FALSE$ and $counter_i(n) = -1$
post-conditions: $Local\_TD'(n) = TRUE$ and $counter'_{i+1}(n) = 0$

OPERATION $Update\_Termination$ $\hat{=}$
any n, i
pre $pre1 : Local\_TD(n) = FALSE$
             $pre2 : i = max(dom(counter(n)))$
             $pre3 : max(ran(counter(n))) = -1$
return $result1$
post $post1 : Local\_TD'(n) = TRUE$
             $post2 : counter'(n) = counter(n) \cup \{(i+1) \mapsto 0\}$
             $post3 : result1' = TRUE$

To Global Termination Operation. Once a node n updates its $Local\_TD$ state, it computes a new value of the $counter$. Let i be the last computation step in which n has modified its $counter$. The new value of the $counter$ of n is defined at the computation step $(i+1)$, and depends on the *last counter* computed by the neighbours of n. Let $Ng(n)$ be the set of nodes adjacent to n. $Ng(n) = \{ng_1, ..., ng_d\}$. Let $\{counter_j(ng_1), ..., counter_k(ng_d)\}$ be the set of the *last counter* computed by each node in $Ng(n)$. Let C be the set of the *last counter* computed by each node in $(Ng(n) \cup \{n\})$. $C = \{counter_j(ng_1), ..., counter_k(ng_d), counter_i(n)\}$. $counter_{i+1}(n) = 1 + min(C)$. This computation step is specified by the operation $To\_Global\_Termination$. The execution of this operation should be blocked if the *counter* of the node n reaches the size of the graph (S), or if one of the nodes in $(Ng(n) \cup n)$ detects the global termination.

pre-conditions: $Local\_TD(n) = TRUE$ and $counter_i(n) < S$
                n has no neighbour detecting the global termination
post-conditions: $counter'_{i+1}(n) = 1 + min(C)$

OPERATION $To\_Global\_Termination$ $\hat{=}$
any n, i, Ng, C
pre $pre1 : Local\_TD(n) = TRUE$
             $pre2 : max(ran(counter(n))) < card(ND)$
             $pre3 : i = max(dom(counter(n)))$
             $pre4 : Ng = \{vi \cdot vi \mapsto n \in g|vi\}$
             $pre5 : C = \{a1, n1 \cdot n1 \in Ng \cup \{n\} \wedge a1 = max(ran(counter(n1)))|a1\}$
             $pre6 : \forall vi \cdot vi \in Ng \cup \{n\} \Rightarrow Global\_TD(vi) = FALSE$
return $result2$
post $post1 : counter'(n) = counter(n) \cup \{(i+1) \mapsto (1 + min(C)\}$
             $post2 : result2' = TRUE$

Global Termination Operation. Within a finite number of steps, the *counter* of a node n can reach the size of the graph S, i.e., all the nodes being in a distance $\leq S$ from n have locally terminated. Thus, the node n can detect the fact that the entire computation is achieved in the network. Hence, it can update its $Global\_TD$ state and diffuse this information to its neighbours to make them aware of this global termination. The updating action of $Global\_TD$ state is specified by the operation $Global\_Termination$. The execution of this operation is blocked once the node updates its $Global\_TD$ from $FALSE$ to $TRUE$. This can be achieved by strengthening its pre-condition by $(Global\_TD(n) = FALSE)$.

pre-conditions: $counter_i(n) = S$ and $Global\_TD(n) = FALSE$
post-conditions: $Global\_TD'(n) = TRUE$

```
┌─────────────────────────────────────────────────────────┐
│ OPERATION  Global_Termination  ≙                        │
│ any    n                                                │
│ pre    pre1 : max(ran(counter(n))) = card(ND)           │
│        pre2 : Global_TD(n) = FALSE                      │
│ return   result3                                        │
│ post    post1 : Global_TD'(n) = TRUE                    │
│         post2 : result3' = TRUE                         │
└─────────────────────────────────────────────────────────┘
```

***Diffusion* Operation.** Once a node n is aware of the termination of the entire computation in the network, it can transmit this information to its neighbours by updating the value of their *Global_TD* states. Each neighbour can in turn transmit the same information. This computation step is specified by the operation *Diffusion*. The execution of this operation is blocked when all the neighbours of the node n detect the global termination. This can be achieved by strengthening its pre-condition via a predicate, showing that n has at least one neighbour that is not detecting the global termination.

pre-conditions: $Global\_TD(n) = TRUE$

n has at least one neighbour v where $Global\_TD(v) = FALSE$

post-conditions: n updates the $Global\_TD'$ values of its neighbours.

```
┌───────────────────────────────────────────────────────────────────┐
│ OPERATION   Diffusion  ≙                                          │
│ any    n, Ng                                                      │
│ pre    pre1 : Global_TD(n) = TRUE                                │
│        pre2 : Ng = {v · v ∈ g[{n}]|v}                            │
│        pre3 : ∃v · v ∈ Ng ∧ Global_TD(v) = FALSE                │
│ return   result4                                                 │
│ post    post1 : Global_TD' = Global_TD ⊲ {v · v ∈ g[{n}]|v ↦ TRUE}│
│         post2 : result4' = TRUE                                  │
└───────────────────────────────────────────────────────────────────┘
```

4.3   Formal Proofs

The intention behind our approach is to compose distributed algorithms with the *SSP* specifications in order to build the global termination detection. Such a composition is achieved via calling the previous operations. In this section, we prove that the execution of these operations ensures the correctness of the resulting algorithm. More precisely, we investigate the invariants of the model and prove the following properties.

(P1): the resulting algorithm preserves the *LTD* property of the initial algorithm (Theorem 1). This property can be easily proved. In fact, we can prove that the *counter* computed by a node n increases during the different computation steps (Invariant 1). Thus, once a node computes a positive *counter*, the new value of this *counter* remains positive.

Invariant 1. $\forall n, i, a, i', a' \cdot i' < i \wedge (i \mapsto a) \in counter(n) \wedge i' \mapsto a' \in counter(n) \Rightarrow a' < a$

Invariant 2. $\forall n \cdot Global\_TD(n) = TRUE \Rightarrow (max(ran(counter(n))) \geq 0)$

Moreover, we can prove that the last counter $((max(ran(counter(n))))$, computed by a node n detecting the global termination, is a positive value (Invariant 2). Furthermore, once a node detects locally the termination, it sets the new value of its *counter*, through the execution of the *Update_Termination* operation, to zero. Thus, we can prove that a node n detects the local termination $(Local\_TD(n) = TRUE)$ if and only if the value of the last *counter* calculated by this node is positive (Invariant 3).

Invariant 3. $\forall n \cdot Local\_TD(n) = TRUE \Leftrightarrow (max(ran(counter(n))) \geq 0)$

Theorem 1. $\forall n \cdot Global\_TD(n) = TRUE \Rightarrow Local\_TD(n) = TRUE$

(P2): If a node n updates its global termination state $(Global\_TD(n) = True)$, every node v on the network has locally terminated (Theorem 2): once a node updates its *Local_TD* state, it increments the value of its *counter* by executing the operation *To_Global_Termination*. The new value of the *counter* depends on the values associated to the neighbours. We can prove that the difference between the maximum and the minimum of the *last counter* computed by two neighbours does not exceed 1 (Invariant 4).

Invariant 4. $\forall n, v, a, b \cdot v \in g[\{n\}] \wedge a = max(ran(counter(n))) \wedge b = max(ran(counter(v))) \Rightarrow (max(\{a,b\}) - min(\{a,b\}) \leq 1)$

Let *chains* be the set of possible chains in the graph, i.e., connected edges, and $Nodes(ch)$ be the set of the nodes concerned in a chain *ch*.

```
axm7 : chains = {x1, x2, t, nodes · x1 ∈ ND ∧ x2 ∈ ND ∧ x1 ≠ x2 ∧ nodes ⊆ ND
          ∧{x1, x2} ⊆ nodes ∧ t ∈ nodes \ {x2} ↣ nodes \ {x1} ∧ t ⊆ g ∧ (t ≠ t⁻¹)|t}
axm8 : Nodes ∈ chains → ℙ(ND)
axm9 : ∀ch · ch ∈ chains ⇒ Nodes(ch) = ran(ch) ∪ dom(ch)
```

Let $E = \{n \cdot n \in Nodes(ch)|max(ran(counter(n)))\}$ be the set of the *last counter* computed by $Nodes(ch)$, and $card(ch)$ be the size of the chain *ch*. Note $(Max - Min)$ is the difference between the maximum and the minimum of E. We prove, by induction on the size of *ch*, that $(Max - Min \leq card(ch) - 1)$ (Invariant 5). Furthermore, the *last counter*, computed by the first node detecting the global termination, reaches the size of the graph S $(S = card(ND))$. Thus, the maximum of E is equal to S. Moreover, $card(ch) \leq S$. Hence, $Max - Min \leq S$. Therefore, $Min \geq 0$. Consequently, we prove that if a node detects the global termination, all the other nodes have locally terminated (Theorem 2).

Invariant 5. $\forall ch, N \cdot ch \in chains \wedge N = Nodes(ch) \Rightarrow max(\{n \cdot n \in N|max(ran(counter(n)))\}) - min(\{n \cdot n \in N|max(ran(counter(n)))\}) \leq card(N) - 1$

Theorem 2. $(\forall n \cdot Global\_TD(n) = TRUE) \Rightarrow (\forall v \cdot v \in ND \Rightarrow Local\_TD(v) = TRUE)$

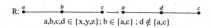

$$a,b,c,d \in \{x,y,z\}; b \in \{a,c\}; d \notin \{a,c\}$$

Fig. 2. 3-Colouring of a ring

5   Example: 3-Colouring of a Ring

Consider a ring with at least 3 nodes. The 3-colouring problem consists in assigning a color to each node from a set of three ones. Two neighbours have different colors. Let $\{x, y, z\}$ be the set of colors. The corresponding relabelling system is defined by considering the rule R (Fig. 2). A new context should be added to extend the *Graph* context in order to specify formal properties of a ring. Building a correct model may start with a very abstract machine and then, by successive refinements, we obtain a concrete one in which we specify the relabelling rule. We refine the last level by introducing a new machine in which we clarify the local termination of the nodes, and we use the *SSP_Interface* in order to build a global termination. According to the computation steps of this algorithm, we affirm that a node n reaches its final state when n and its neighbours get different colors. Hence, the following invariant should be added: assume that *Col* is a function introduced to characterize the color of the nodes.

Invariant 6. $\forall s \cdot s \in ND \Rightarrow (Col(s) \notin \{Col(g(s))\} \Leftrightarrow SSP\_Terminaison(s) = TRUE)$

The calling of the previous operations is similar to the previous example, except the execution of the *Update_Terminaton* operation which should be strengthened by a new guard (grd4), specifying when a node can update its termination state.

> **EVENT**   *Calling_Op1*
> **any** n, i
> **where**
> $grd1 : SSP\_Local\_TD(n) = FALSE$
> $grd2 : i = max(dom(SSP\_counter(n)))$
> $grd3 : max(ran(SSP\_counter(n))) = -1$
> **grd4** $: (Col(n) \notin \{Col(g(n))\}$
> **then** $act1 : res\_OP1 := SSP\_Update\_Termination(n \mapsto i)$

5.1   What We Gain

It seems that we have to do more work in order to compose an algorithm with the proposed specifications: we have to develop the algorithm in a progressive way controlled by the refinement of models. Then we add a new machine to specify a suitable implementation of the predefined operations. But we do have the following advantages:

– We don't need to prove the computation steps of the *SSP* algorithm. This is because we have already done this, when developing the *SSP_Interface*.

Table 1. Proof Statistics

Model	Total	Automatic		Interactive	
SSP_Interface	91	39	43 %	52	57 %
SpTree *M3* (with the *SSP_interface*)	44	31	70 %	13	30 %
SpTree *M3* (without the *SSP_interface*)	112	36	32 %	76	68 %
3-colouring *M3* (with the *SSP_interface*)	45	36	80 %	9	20 %
3-colouring *M3* (without the SSP_interface)	117	41	35 %	76	65 %

Thus, we have saved efforts of users on discharging proofs (see Table 1). The additional proof obligations, generated while introducing a new machine using the *SSP_Interface*, are not very complex to discharge: 70 % and 80 % of them are respectively proved automatically for the two examples (the spanning tree as well as the 3-colouring of a ring). Note that we have tested building and proving the global termination of the two algorithms without using the proposed SSP_Interface. The new generated machines for the spanning tree and the 3-colouring algorithm produce respectively only 32 % and 35 % of proof obligations discharged automatically.

- We reuse all the proofs associated to the first algorithm (the Spanning tree and the 3-colouring of a ring in our cases). The incremental proof-based process of refinement provides a way to preserve the correctness of the algorithm and to validate the integration of new requirements.
- We can reuse the proposed specifications in other case studies. The SSP_Interface is defined in a high level abstraction in order to build the global termination detection of distributed algorithms.

6 Discussion, Conclusion and Future Work

In this paper, we have proposed a proof-based framework for composing distributed computing with the *SSP* algorithm in order to build a global termination detection. The main characteristic of our approach is that it transforms algorithms from LTD to GTD mode, enables reuse in development, and saves efforts on proving distributed computing: by relying on the *SSP* algorithm, we specified a generic module containing formal specifications and proofs for the global termination detection. This module can be composed with simple and complex cases studies with different synchronizations. During the development, a list of proof obligations is generated by the Rodin [2] platform to ensure the safety of the development. We believe that this work has a number of benefits. In a nutshell, we say that composing a *correct-by-construction* development with the *SSP* algorithm enhances the termination detection property of distributed computing. Moreover, specifying *SSP* in a pre-defined module greatly simplifies the reuse of specifications and proofs.

It is worth noticing that if we look carefully at what the global termination means in distributed algorithms, we have to distinguish between the termination of the computing and the transmitting messages between processors. In this work, the global termination, we are interested in, is to detect the instant when all processors have computed their final values, i.e., no processor can modify its state. We don't detect the instant when there is no message in transit in the network. Moreover, our approach might be improved if we used the diameter of the graph instead of its size. But we made this choice for the sake of simplicity. As a future work, it would be interesting to take into account these limitations and study the case that local termination of processors is not stable. Moreover, added to safety properties, we think that it would be more relevant to ensure liveness properties [17]: when all nodes have locally terminated, the algorithm will eventually detect global termination. Furthermore, we aim to study and detail the last item presented in Sect. 3.2. Thus, we can implement *Java* codes of the proposed framework and simulate algorithms into the *Visidia* [4] platform. Starting with previous studies [24], we can translate our formal specifications into Java codes and propose a certified tool for transforming automatically distributed algorithms from LTD to GTD.

References

1. Abrial, J.: Modeling in Event-B - System and Software Engineering. Cambridge University Press, New York (2010)
2. Abrial, J.R., Butler, M., Hallerstede, S., Hoang, T.S., Mehta, F., Voisin, L.: Rodin: an open toolset for modelling and reasoning in event-b. Int. J. Softw. Tools Technol. Transf. **12**(6), 447–466 (2010)
3. Andriamiarina, M.B., Méry, D., Singh, N.K.: Revisiting snapshot algorithms by refinement-based techniques. Comput. Sci. Inf. Syst. **11**(1), 251–270 (2014)
4. Bauderon, M., Mosbah, M.: A unified framework for designing, implementing and visualizing distributed algorithms. Electr. Notes Theor. Comput. Sci. **72**(3), 13–24 (2003). http://dx.doi.org/10.1016/S1571-0661(04)80608-X
5. Boussabbeh, M., Tounsi, M., Hadjkacem, A., Mosbah, M.: Towards a general framework for ensuring and reusing proofs of termination detection in distributed computing. In: 24rd Euromicro International Conference on Parallel, Distributed, and Network-Based Processing, PDP 2016, Heraklion Crete, Greece, 17th-19th February 2016 (2016)
6. Cansell, D., Méry, D.: The event-B modelling method: concepts and case studies. In: Bjørner, D., Henson, M.C. (eds.) Logics of Specification Languages. Monographs in Theoretical Computer Science, pp. 47–152. Springer, Berlin (2008)
7. Castéran, P., Filou, V.: Tasks, types and tactics for local computation systems. Stud. Inform. Univ. **9**(1), 39–86 (2011)
8. Chalopin, J., Godard, E., Métivier, Y.: Local terminations and distributed computability in anonymous networks. In: Taubenfeld, G. (ed.) DISC 2008. LNCS, vol. 5218, pp. 47–62. Springer, Heidelberg (2008)
9. Chalopin, J., Godard, E., Métivier, Y.: Election in partially anonymous networks with arbitrary knowledge in message passing systems. Distrib. Comput. **25**(4), 297–311 (2012). http://dx.doi.org/10.1007/s00446-012-0163-y

10. Chandy, K.M., Lamport, L.: Distributed snapshots: determining global states of distributed systems. ACM Trans. Comput. Syst. **3**(1), 63–75 (1985). http://doi.acm.org/10.1145/214451.214456
11. Chandy, K.M., Misra, J.: Parallel program design - a foundation. Addison-Wesley, UK (1989)
12. Chandy, K.M., Misra, J., Haas, L.M.: Distributed deadlock detection. ACM Trans. Comput. Syst. **1**(2), 144–156 (1983). http://doi.acm.org/10.1145/357360.357365
13. Courtieu, P., Rieg, L., Tixeuil, S., Urbain, X.: A certified universal gathering algorithm for oblivious mobile robots. CoRR abs/1506.01603 (2015)
14. Dijkstra, E.W., Scholten, C.S.: Termination detection for diffusing computations. Inf. Process. Lett. **11**(1), 1–4 (1980). http://dx.doi.org/10.1016/0020-0190(80)90021-6
15. Filou, V., Mosbah, M., Tounsi, M.: Towards proved distributed algorithms through refinement, composition and local computations. In: 2013 Workshops on Enabling Technologies: Infrastructure for Collaborative Enterprises, Hammamet, Tunisia, 17–20 June 2013, pp. 353–358 (2013). http://dx.doi.org/10.1109/WETICE.2013.67
16. Godard, E., Métivier, Y., Mosbah, M., Sellami, A.: Termination detection of distributed algorithms by graph relabelling systems. In: Corradini, A., Ehrig, H., Kreowski, H.-J., Rozenberg, G. (eds.) ICGT 2002. LNCS, vol. 2505, pp. 106–119. Springer, Heidelberg (2002)
17. Hoang, T.S., Abrial, J.-R.: Reasoning about liveness properties in event-b. In: Qin, S., Qiu, Z. (eds.) ICFEM 2011. LNCS, vol. 6991, pp. 456–471. Springer, Heidelberg (2011)
18. Iliasov, A., Troubitsyna, E., Laibinis, L., Romanovsky, A., Varpaaniemi, K., Ilic, D., Latvala, T.: Supporting reuse in event B development: modularisation approach. In: Frappier, M., Glässer, U., Khurshid, S., Laleau, R., Reeves, S. (eds.) ABZ 2010. LNCS, vol. 5977, pp. 174–188. Springer, Heidelberg (2010)
19. Leavens, G.T., Abrial, J., Batory, D.S., Butler, M.J., Coglio, A., Fisler, K., Hehner, E.C.R., Jones, C.B., Miller, D., Jones, S.L.P., Sitaraman, M., Smith, D.R., Stump, A.: Roadmap for enhanced languages and methods to aid verification. In: Proceedings of 5th International Conference of Generative Programming and Component Engineering GPCE 2006, Portland, Oregon, USA, 22–26 October 2006, pp. 221–236 (2006). http://doi.acm.org/10.1145/1173706.1173740
20. Litovsky, I., Métivier, Y., Sopena, E.: Graph relabelling systems and distributed algorithms. In: Handbook of Graph Grammars and Computing by Graph Transformation, pp. 1–56. World Scientific Publishing Co., Inc., River Edge (1999)
21. Mazurkiewicz, A.W.: Distributed enumeration. Inf. Process. Lett. **61**(5), 233–239 (1997)
22. Szymanski, B.K., Shi, Y., Prywes, N.S.: Terminating iterative solution of simultaneous equations in distributed message passing systems. In: Proceedings of the Fourth Annual ACM Symposium on Principles of Distributed Computing, Minaki, Ontario, Canada, 5–7 August 1985, pp. 287–292 (1985). http://doi.acm.org/10.1145/323596.323623
23. Tounsi, M., Mosbah, M., Méry, D.: Proving distributed algorithms by combining refinement and local computations. ECEASST 35 (2010) http://journal.ub.tu-berlin.de/eceasst/article/view/442
24. Tounsi, M., Mosbah, M., Méry, D.: From event-b specifications to programs for distributed algorithms. In: 2013 Workshops on Enabling Technologies: Infrastructure for Collaborative Enterprises, Hammamet, Tunisia, 17–20 June 2013. pp. 104–109 (2013). http://dx.doi.org/10.1109/WETICE.2013.44

How to Select the Suitable Formal Method for an Industrial Application: A Survey

Felix Kossak and Atif Mashkoor[✉]

Software Competence Center Hagenberg GmbH,
Hagenberg, Austria
{felix.kossak,atif.mashkoor}@scch.at

Abstract. The share of formal methods is still marginal in contemporary systems and software engineering. One of the reasons is the absence of systematic guidelines and evaluation criteria that help software practitioners choose the right formal method for the problem at hand. In this paper, we present a comprehensive set of criteria, based on a systematic literature review and decade-long personal experience in industrial projects, for evaluating and comparing different formal methods. We argue that besides technical grounds (e.g., modeling capabilities and supported development phases), formal methods should also be evaluated from social and industrial perspectives. At the end of the paper, we present an evaluation of "ABZ" methods based on the stipulated criteria.

1 Introduction

Despite many years of advocacy, numerous success stories in safety-critical systems and the availability of various *easy to use* methods and tools, the application of formal techniques is still sparse in mainstream software development. Several factors can be held accountable for this result. One of them is that no proper guidelines are available at the disposal of software practitioners to enable them to navigate through the intricate process of choosing the formal method suitable for their problem domain.

Different formal methods are generally suitable for different kinds of software projects, domains, and social and economic settings. For instance, the development of safety-critical systems will require elaborate evidence for compliance with safety requirements and standards, while in other projects, budget and time restrictions will not allow for expansive verification efforts. As another example, it makes a difference whether mostly mathematicians or specially trained engineers are involved in a project, and will also be available for maintenance later on,

The research presented in this paper is supported by the Austrian Ministry for Transport, Innovation and Technology, the Federal Ministry of Science, Research and Economy, and the Province of Upper Austria in the frame of the COMET center SCCH. The writing of the paper is partially supported by the Austrian Science Fund project: *Behavioral Theory and Logics for Distributed Adaptive Systems* (FWF-P26452-N15).

M. Butler et al. (Eds.): ABZ 2016, LNCS 9675, pp. 213–228, 2016.
DOI: 10.1007/978-3-319-33600-8_13

or whether the methods used must be suitable for ordinary software developers. Several studies have already been published where individual formal methods are compared. However, as we will detail in Sect. 2, many of these studies are either outdated or concentrate on limited aspects (e.g., technical criteria of predominantly academic interest or a particular domain of application). None has presented general guidelines/evaluation criteria which may help software practitioners in choosing the right formal method for their problem at hand.

In this paper, we present a comprehensive list of criteria for the comparison of formal methods with respect to general industrial interest, drawn from a structured literature review as well as personal experience in industrial and academic projects, for example, hemodialysis machines [42], an aircraft landing gear system [34], machine control systems [43], transportation systems [45], platooning systems [44], and business process modeling [35]. In contrast to many other publications, we include a wide range of criteria which we deem crucial for a wider adoption of formal methods in the industry.

The main goal of this study is to provide guidelines to software practitioners to help them choose a particular formal method, or maybe a small set of methods, for a particular software (or software-hardware co-development) project. Thereby the focus is laid on industrial projects, including large-scale projects. The motivation behind this goal is to provide necessary means to help propagate the use of formal methods in day-to-day systems and software engineering.

The prime research question of our work is: What criteria are useful in order to select a particular formal method for a particular setting? Additionally, we demonstrate the use of the criteria with several selected formal methods in tabular form. A much more detailed description of the criteria and our evaluation of different methods is available in [36].

This paper is structured as follows: First we present our research approach and the list of literature reviewed (Sect. 2). Then in Sect. 3, we present a structured list of criteria for selecting a suitable formal method for an industrial application. In Sect. 4, we compare particular methods by means of the previously described criteria. The paper is concluded in Sect. 5.

2   Approach and Literature Reviewed

2.1   The Research Approach

In this paper, we answer the following research questions:

1. What criteria are useful in order to select a particular formal method for a particular setting?
2. Why are the criteria important for the evaluation of a particular method?
3. How do various state-based methods fare with respect to these criteria?

Our research approach is based on a structured literature review complemented by our own experiences with several formal methods. We limited the literature research to an Internet search with the following search strings:

- "formal methods" AND "evaluation criteria"
- "formal methods" AND "comparison"
- "formal methods" AND "state of the art"
- "formal methods" AND "literature review"

We stopped after seven pages of search results, after which relevance dropped markedly. We further included literature which we were already aware of.

The literature research showed that several comparisons between different classical formal methods were conducted in the 1990s and around 2000. Recently, more comparisons were made in special settings, typically in the context of university courses. We noted a recent surge in formal method-related tools which can be integrated in traditional development platforms. These are typically static checkers or model checking tools that only partially cover specification and model-based verification against custom safety properties. Most existing studies compare only a few methods, often only two or three. Evaluation criteria vary widely, revealing different possible viewpoints.

Evaluations of formal methods from the 1990s must certainly be considered outdated, for much has changed since, in particular with respect to tool support, the amount of practical experience, and how widespread a method is used. This does not leave much material for a concrete evaluation of particular methods. Still, older publications can yield interesting contributions to the *criteria* by which formal methods should be evaluated (sometimes presented as wishes). Often the focus is on a purely academic viewpoint in this respect, but not always.

2.2 Literature Reviewed

We now present the literature (in order of relevance) which we found relevant for the current study (excluding sources on a single method).

Information on concrete evaluations within the industry appears to be scarce, though we assume that such evaluations happen. A notable exception is a recent paper by Chris Newcombe, "Why Amazon Chose TLA+" [50], though it largely only describes experience with TLA+ [38] and, to a lesser extent, Alloy [29] and Microsoft VCC [17]. The criteria are drawn from the very demanding domain of cloud infrastructure services, where key demands include a high level of distribution, high performance, and high availability.

A position paper by Sifakis [54] also discusses industry-centric evaluation criteria and provided useful input for us. Sifakis discusses, amongst others, the crucial point of usability and human factors in general.

The papers from Ardis et al. [4] and Knight et al. [33] also provide frameworks for the evaluation of formal specification languages. They first present criteria and then evaluate several formal languages. The latter also present the perspectives of developers, engineers, and computer scientists on these languages.

Woodcock et al. [58] contribute an overview of historical experiences with formal methods, in particular from industrial projects. We could extract several important criteria from this paper, in particular with regard to tool support.

McGibbon [46] discusses different evaluation criteria from a government viewpoint, including more detailed requirements for tools.

Several evaluation criteria can be extracted from the seminal papers by Clarke and Wing [15] and Bowen and Hinchey [13]. The former present the state of the art and future directions of formal methods and the latter present some guidelines to help propagate the use of formal methods in industry.

Liu et al. [40] list a number of evaluation criteria and compare a great number of methods. Amongst others, the authors bring in the additional criterion of applicability in re-engineering, in particular in reverse engineering and restructuring. Although they are primarily concerned with support for re-engineering, this paper is also of general interest; it includes interesting characterizations of many different methods and their state at the time, though unfortunately much of this information is now (potentially) outdated.

Banach, in "Model Based Refinement and the Tools of Tomorrow" [5], compares B [2], Event-B [3], Z [55], and the ASM method [10] from a mathematical/technical point of view.

Also a book by Gannon, Zelkowitz, and Purtilo, entitled *Software Specification: A Comparison of Formal Methods* [23], focuses on mathematical issues; it discusses only VDM [30] as a formal method in a closer sense, together with temporal logic in general as well as "risk assessment."

Also Kaur et al. in "Analysis of Three Formal Methods - Z, B and VDM" [32], stress mathematical and modeling issues, but they also mention, e.g., tool support, code generation, and testing.

In *Software Specification Methods* (ed. by Frappier and Habrias) [21], many different methods are introduced through a case study. In the last chapter, some of the methods are qualitatively compared. The criteria include some which we chose not to adopt here, including graphical representation (lack of relevance for state-based methods), object-orientated concepts (design-centric, see further below), use of variables (too detailed), and event inhibition (too detailed).

In "A practical comparison of Alloy and Spin," Zave [59] compares Alloy and Spin/Promela [28], two methods of general interest. However, we did not find any new criteria there.

In a master's thesis, Rainer-Harbach [53] compares several different proving tools for software verification, but not any comprehensive method which could support other project phases and aspects.

ter Beek et al. [9] wrote a paper on "Formal Methods for Service Composition," dealing with a very narrow field of application. They compare only automata as a basis for model checking, Petri Nets, and process algebras.

A paper by Dondossola [18] specializes on the application domain of safety-critical knowledge-based components and on the method TRIO [25]. Towards the end, it also offers a comparison of different formal methods, including VDM and Z; however, the criteria used there are only very coarsely described, so we could not extract much extra information useful for our purposes.

A technical report by Barjaktarovic [8] names in particular industrial requirements for formal methods throughout the text; most of those requirements are also found in other sources, but this paper provides a good confirmation.

From an article by Pandey and Batra [52], we obtained useful assessments of Z and VDM, in particular.

3 Criteria for Evaluating Formal Methods

Now we will present a structured list of criteria which we deem relevant for assessing and comparing formal methods for their usefulness in concrete industrial projects, depending on the concrete settings of a project. We first give an overview of the criteria we found and deemed relevant before describing each of them in more detail and explaining their significance. Please note that the classification of certain criteria under a particular category may be cross-cutting and overlapping to some degree. This is by choice as it makes each category an independent unit of analysis that can also be taken into consideration in isolation for concentration on a particular class of criteria. Please see [36] for a detailed discussion on the criteria.

3.1 Overview

We found five categories of criteria relevant for industrial projects:

1. **Modeling Criteria:** What possibilities and scope for modeling and refinement does the method offer?
   - Support for composition/de-composition
   - Support for abstraction/refinement and what notion of refinement is employed
   - Support for parallelism/concurrency/distribution
   - Support for non-determinism
   - The possibility to express global system properties of correctness
   - Support for the modeling of time and performance properties
   - Expressibility of various special (domain-specific) concepts (e.g., differential equations or user interface aspects)
   - The possibility to express rich concepts *easily*
2. **Supported Development Phases:** Which phases of a software (and/or hardware) development project can be supported (and how)?
   - Specification
   - Validation
   - Verification
   - Bug diagnosis
   - Architecture and design
   - Coding/code generation
   - Testing
   - Maintenance
   - Reverse engineering

3. **Technical Criteria:** What tools are available, and how do the method and the tools interact with other development requirements from a technical point of view?
   - Overall tool support
   - Commercial support for tools
   - Traceability of requirements and during refinement/code transformation
   - Support for change management (how much stability of the initial specification is presupposed? What about maintenance of the finished product?)
   - Effect of the method on development time (for specification, validation, verification, etc.)
   - Efficiency of generated code (can the generated code be used as it is? how much manual tweaking is necessary?)
   - Efficiency of code generation (how fast does code generation work? what does a small change in the model mean for subsequent code re-generation?)
   - Interoperability with other methods and/or other tools
   - Integration of methodology and tools with the usual development methods and tools (IDEs)
4. **Human/Social Criteria:** How easily can people with different backgrounds and expertise handle the method and its results? How can people collaborate when using the method?
   - Learning curve (how fast can one learn the method from scratch, and what prior expertise is required?)
   - General understandability (is the model understandable for non-experts? can the model be made accessible via visualization/animation?)
   - Available documentation (including case studies)
   - Support for collaboration
5. **Industrial Applicability:** How well can the method be used in potentially large and complex industrial projects, and what industrial experience is there so far?
   - Support for industrial deployment
   - Scalability
   - Amount of (industrial) experience so far
   - Success rate in industrial application
   - Is specialized staff required, and if so, to what extent?
   - Standardization
   - Availability and licensing of method and tools

3.2   Modeling Criteria

Modeling criteria concern the scope of systems and requirements which can be modeled. *Composition* is important for large models, including their verification. *Refinement* is even more important for constructing increasingly large models and can support validation, design, and coding.

Support for modeling *parallelism, concurrency, and distribution* is essential for a wide range of real-life applications. *Support for non-determinism* is very useful for keeping models abstract. *The possibility to express global properties of system correctness* is necessary to be able to prove respective safety and liveness requirements such as temporal constraints (termination, deadlock freeness, fairness). *Support for modeling time* must regard sparse and dense models for time separately (see [40]). *Performance properties* refer to the complexity of algorithms, both with respect to time and to memory. Many domains of application require that *special concepts* be easily expressed in a modeling language. One important example is hybrid systems.

We can generally expect a desire in industry to "be able to capture rich concepts without tedious workarounds" [50]. In a related note, [15] demand support for sufficient data structures and algorithms.

Additionally, [21,32] have suggested support for the object-oriented concept as an evaluation criterion. However, we think that this criterion is too implementation-centric (or at least design-centric) for specifications.

3.3   Supported Development Phases

[15] state that it should be possible to amortize the cost of a formal method or tool over many uses; this means it should be possible to use a model throughout as many development phases as possible.

A special phase which is not regularly considered is that of *reverse engineering* – extracting the high-level functionality and a respective specification from a (typically ill-documented) legacy system. We owe attention to this additional project phase to [40].

Bug diagnosis is an issue which deserves special mention beside verification, because finding that some property does not hold does not mean that one can then easily identify the source of error. [50] points out the importance of this issue; [8] states even that "Industry is mostly interested in tools that find bugs rather than tools that prove correctness."

3.4   Technical Criteria

Technical criteria concern tool support and how the method and the available tools interact with other aspects of system development. Besides the range of *overall tool support*, the availability of professional support for those tools is important, for which reason companies typically prefer *commercial support*.

An important issue stressed by many industrial sources is the *traceability of requirements* throughout the development process. *Support for change management* addresses the fact that the waterfall model is actually unrealistic. The *effect of the method on overall development time* is crucial for the industry.

Regarding code generation, we can consider the efficiency of the generated code as well as the efficiency of code generation. The *efficiency of the generated code* is the quality of the code that has been generated by an automatic tool from a more abstract model: runtime behavior, use of memory, or the amount of manual fixing which is required after generation. The *efficiency of code generation* concerns the speed (and use of resources) with which code is generated. This is important for "playing" with the model and testing different designs.

The demand for *interoperability with other methods and/or other tools* arises from the insight that different methods and tools are differently suitable for different tasks and project phases. Moreover, such a possibility will greatly enhance reuse. A related criterion is *Integration of methodology and tools with the usual development methods and tools* to facilitate the transition between different project phases and requirements tracking, amongst others.

3.5   Human/Social Criteria

In industrial settings, specially trained people will not be available for every development task. The easier a method is applicable for normal engineers and developers, the easier it can be adopted by the industry. Moreover, certain products of the method should be accessible to people outside the development team, including domain experts, managers, or even lawyers (cf. [37]).

The *learning curve* of a method concerns the speed with which an average modeler (specifier, designer or developer) can learn the method from scratch and obtain useful results in practice. *General understandability* is important because formal models often need to be understood by various stakeholders. The importance of *documentation*, including reference handbooks or tutorials, is self-evident. *Support for collaboration* is easily forgotten when academics develop a new method, but it is an important issue in larger real-life projects.

3.6   Industrial Applicability

There are still further criteria particularly concerning the capability of employing a formal method in a typical industrial setting. Industrial application very often means large and complex systems, as well as certain economic and legal constraints.

The criterion of *support for industrial deployment* is designed to capture the availability of outside help. *Scalability* is the ability of the method to be well applicable to arbitrarily large and complex projects. Certainly the actual *amount of industrial experience* which has been gathered with a method is very interesting for decision makers who ponder newly introducing formal methods. Also the *success rate* would be interesting, but would be extremely difficult to assess objectively. A cliché that formal methods would require *specially trained, "expensive" personnel* is actually well-founded. There are considerable differences between particular methods in this respect.

Standardization can be very helpful for the industry: it enhances the probability of long-term availability of commercial tools and facilitates training as well as exchangeability of results.

Related is the availability and *licensing* of the method and related tools. Most of the widely used methods and their tools are open source, but open-source software requires a large and stable community to maintain and further develop. Moreover, the availability of commercial support and training is essential for more widespread uptake in the industry.

4   Comparison of Methods

4.1   Comparison

We now compare the different "ABZ" methods through simplified tables, see Tables 1, 2, 3, 4 and 5.

Table 1. Modeling criteria

	Alloy	ASMs	B	Event-B	TLA+	VDM	Z
(De-)Compos.	Y	Med.	Med.	Med.	Y	Y	Med.
Refinement	Med.	Good	Med.	Med.	Good	Good	Med.
Parall./concur.	Med.	Good	Part.	N	Good	Y	N
Nondeterminism	Impl.	Y	Y	Y	Y	(VDM++)	Y?
Global propert.	Y	Med.	Med.	Y	Y	N?	(N)
Time/perform.	(N)	N	N	N	Y	Y	N
Spec. concepts	-	(Hybr.)	N	(Hybr.)	-	Few	-
Rich conc. easy	(N)	Y	Med.	Med.	Y	(N)	?

Table 2. Supported development phases

	Alloy	ASMs	B	Event-B	TLA+	VDM	Z
Verification	(Good)	Med	V.Good	V.Good	Good	Y	Y
Bug diagnosis	Med.	Y	Y	Y	Med.	Y	-
Archit./design	Med.	Med.	(Y)	-	(Y)	(Y)	Good
Coding	Poor	Man.	Y	Poor	(N)	Y	N?
Testing	Med.	Y	Y	Poor	Y	Y	Good
Maintenance	-	Poor	-	-	N	-	-
Reverse engin.	(Y)	Y	Y	-	-	N	Good

In the tables, "Y" means "yes/supported" (quality unknown), "N" means "not supported." A dash "-" means that we could not find (sufficient) information. A "?" means that we have inconsistent or even contradicting information. "(Y)" means restricted support, "(N)" means little support, "(Good)" means "Good" with some proviso, etc.; parentheses may also indicate that special versions or prototypes support this feature, but not the standard version.

Table 3. Technical criteria

	Alloy	ASMs	B	Event-B	TLA+	VDM	Z
Tool support	Y	Med.	Good	Good	(Good)	Med.	Y
Comm. support	N	N	Y	Part.	N	Y	Part.
Time effort	-	Adapt.	(Long)	(Long)	(Short)	-	(Long)
Efficient code	-	n/a	Med.	n/a	n/a	-	n/a
Efficient code gen.	-	n/a	Y	n/a	n/a	-	n/a
Traceability	Poor	Good	(Y)	Good	-	-	Med.
Interoperability	N	(N)	Part.	Part.	Part.	N	N
Integration/IDE	-	(N)	N	(N)	-	-	(N)

Table 4. Human/social criteria

	Alloy	ASMs	B	Event-B	TLA+	VDM	Z
Learning curve	Med.	Good	Med.	Med.	Good	(Good)	Bad
Understandability	Med.	Good	Med.	Med.	Good	Bad?	Bad
Documentation	Good	(Good)	Good	Good	Good	Good	Good
Collaboration	-	N	-	Y	-	-	-

Table 5. Industrial applicability

	Alloy	ASMs	B	Event-B	TLA+	VDM	Z
Deployment sup.	N	(N)	Y	Y	N	Y	Y
Scalability	Bad	Med.	(Good)	Med.	-	Y	(Good)
Experience	(Much)	Med	Much	(Much)	Much	Much	Much
Special staff	Y	(N)	Y	Y	N	(N)	Y
Standardization	N	N	N	N	N	Y	Y
Licensing	OS	OS	Cm	OS	OS	Cm/OS	OS

"Med." abbreviates medium quality, "Part." partial support, "Man." manual, i.e., no tool support, "Impl." abbreviates only implicit support, and "Adapt." adaptable. "Cm." abbreviates "commercial" (licensing), "OS" open source. "n/a" means "not applicable." The entry "Hybr." denotes the possibility to model hybrid (discrete-continuous) systems.

A couple of criteria have been omitted due to either uniform support or lack of information: in Table 2, *specification*, *validation* and *performance checking*; and in Table 3, *change management*.

4.2   Justification

A detailed justification of the entries in the above tables is given in [36]. Here we only give an extract of the potentially contentious points regarding the modeling criteria.

Alloy features an explicit *composition* operation [22]. Temporal composition of functions is possible with operators `merge` and `override`. According to [50], *refinement* is not very flexible. Regarding *concurrency*, according to [50], the method is not suitable for large complex systems. *Non-determinism* can only be implicitly modeled [59]. For examples of the modeling of *global system properties*, see e.g., [14,16,31]. Alloy has no direct *notion of time* [24]; however, [1,19,57] have shown how to express timing properties. [41] selected Alloy for its "expressive power," amongst others, but according to [50], the *expression of rich concepts* is *not* easy.

For **ASMs**, *(de-)composition* is judged as medium by [5]; however, from a practical point of view, we consider it to be quite flexible. *Refinement* is good, according to [5] as well as by our own experience. The ASM method supports n-to-m refinement and both data refinement and procedural refinement [10]. ASMs are well suited for modeling *parallel and concurrent systems* [20]. *Non-determinism* is supported by the "choose" operator and via abstract rules and derived functions. *Global properties* can be expressed via the state space, but there is no explicit support. There is no explicit *notion of time*. Regarding *special concepts*, Banach and others have used ASMs for modeling continuous systems [7]. Regarding *easy expression of rich concepts*, the simple notation can be easily adapted and expanded, but tool support may always be limited.

In **B**, *(de-)composition* is possible by including other machines. According to Banach [5], "The [...] INCLUDES, USES, SEES mechanisms are certainly composition mechanisms, but they just act at the top level." B supports only 1-to-1 *refinement* (cf. [5]), and also as per our own experience, the support for refinement in B can be rated as 'medium." B supports *parallelism*, except for code generation, but not *concurrency* [32,40]. B supports *non-determinism* through non-deterministic choice of values as well as by operators "ANY" and "CHOICE". Regarding *system properties*, it is possible to express typical safety properties through invariants, but B has no explicit means for modeling *time* or *temporal properties* (see also [40]). *Reliability* properties can be expressed via invariants. Regarding the *easy expression of rich concepts*, B provides a rich language for set theory and relational theory. However, expressing certain concepts such as data structures can often be awkward and unintuitive.

For **Event-B**, [5] assesses *(de-)composition* as "good"; however, we find the decomposition/recomposition facilities not straightforward. Event-B only supports 1-to-1 *refinement*. Event-B does not explicitly support *parallelism and concurrency*; however, both parallel (cf. [27]) and concurrent (cf. [11]) programs can be defined using decomposition and refinement. Event-B supports *non-determinism* by allowing for non-deterministic choice of values for variables and through event parameters. *Global system properties* can effectively be specified using invariants. Event-B has no explicit means for *modeling time or temporal* *properties.* Regarding *special concepts*, there exist proposals regarding hybrid and continuous systems, e.g., in [6]. Regarding the *easy expression of rich concepts*, our comments on B apply here as well.

TLA+ supports *composition* through different mechanisms such as logical connectives of implication, conjunction, and quantification [38, 47]. *Refinement* is assessed as "good" by [50, p. 28]; according to [47, p. 445], "A distinctive feature of TLA is its attention to refinement and composition." Support for modeling *parallel, concurrent, and distributed systems* is good, as confirmed by [50, p. 36]. *Non-determinism* is also supported [38, Section 6.6]. TLA+ does not formally distinguish between specifications and *system properties*: both are written as logical formulas and concepts such as refinement and composition [47]. It uses set theocratic constructs to define safety properties and temporal logic to define liveness properties. *Modeling of time* is explicitly supported [51, p. 69], enabling modeling and checking of *performance properties*. According to [50, pp. 27, 36], *rich concepts can be easily expressed* in TLA+.

In **VDM**, *Composition* is possible according to [40], but [32, 46] deny it. Classical VDM models can be structured into data types and modules, while VDM++ models can be structured into classes. *Refinement* is achieved through data reification and operation decomposition. According to [40, 46], VDM does not support *parallelism*, but [39] describes the use of VDM for *distributed*, embedded, *real-time* systems. [52] state that "VDM emphasizes on the feature of concurrency control" (cf. [56]). Support for *non-determinism* is only given in VDM++ [32]. [46] states that VDM has no explicit *notion of time*, but [39] describes timing analysis for identifying performance bottlenecks. [49, 56] also deal with real-time systems. Regarding *special modeling concepts*, [46, 52] note that VDM has explicit *exception handling*. Support for e.g., performance and reliability modeling has been introduced more recently.

In **Z**, *(De-)Composition* is achieved through "schemas" [5, 12] or by means of "promotion" [5]. [5] notes that the schema calculus is not monotonic with respect to refinement. Also taking [32] into account, we assess composition as medium. Regarding *refinement*, [5] notes that "spurious traces, not corresponding to real world behaviour, can be generated." Refinement cannot completely go down to the code level [12]. Z does not directly support *concurrency* [40, 46, 52]. Regarding *non-determinism*, [12, 21] state it is given but [32] denies it. Z does not support non-determinism explicitly, but e.g., several after-state valuations for a single pre-state binding are possible (cf. [48]). *Expression of global system properties* is not straightforward (cf. [26]). An explicit *concept of time* is obviously not given [40, 46].

4.3   Project-Specific Assessment

We have furthermore tried to condense the available information into a much simplified table that can be found in [36] for fast management decisions. Here, we just sketch the structure of the table as follows:

– **Project setting:** Is the product safety-critical? How severe is time pressure? Is the project conducted in an agile setting? Does the method allow to quickly start using it without prior experience – at least with *initial* help by experts? Or will continued support by experts be required?

- **Company:** Do we deal with a big company or with a small or medium-sized company? Can the company afford a transition phase for introducing formal methods?
- **Goal of using formal methods:** Do we want to improve product quality, or process quality? Reduce specification errors? Improve requirements definitions, documentation, understanding of the design? Explore a model before implementation? Obtain a sound foundation for maintenance and/or testing? Meet safety requirements?

5 Conclusion

The main contribution of this work is to consolidate and further develop a system of criteria for assessing particular formal methods especially with respect to their potential usefulness in industrial projects.

Most of the criteria were assembled from a structured literature review, supplemented by our own experience, whereby we tried to put a special focus on sources close to industry. We came up with five categories into which to sort the criteria, which focus on different aspects to enable more focused assessments. Thereby also a certain amount of redundancy was retained so as to enable assessments based on one or two categories of interest only. We exemplarily evaluated the "ABZ" methods on the stipulated criteria.

We hope that our work will contribute to better acceptance of formal methods in industry, as practitioners and managers should now find it easier to assess the possible impacts of introducing such methods in real-life projects and to select the best suitable methods for their needs.

References

1. Abdunabi, R., Sun, W., Ray, I.: Enforcing spatio-temporal access control in mobile applications. Computing **96**(4), 313–353 (2014)
2. Abrial, J.R.: The B-Book: Assigning Programs to Meanings. Cambridge University Press, Cambridge (1996)
3. Abrial, J.R.: Modeling in Event-B System and Software Design. Cambridge University Press, Cambridge (2010)
4. Ardis, M.A., Chaves, J.A., Jagadeesan, L.J., Mataga, P., Puchol, C., Staskauskas, M.G., Von Olnhausen, J.: A framework for evaluating specification methods for reactive systems. IEEE Trans. Softw. Eng. **22**(6), 378–389 (1996)
5. Banach, R.: Model based refinement and the tools of tomorrow. In: Börger, E., Butler, M., Bowen, J.P., Boca, P. (eds.) ABZ 2008. LNCS, vol. 5238, pp. 42–56. Springer, Heidelberg (2008)
6. Banach, R., Zhu, H., Su, W., Huang, R.: Formalising the continuous/discrete modeling step. In: Proceedings Refine 2011. EPTCS, vol. 55, pp. 121–138 (2011)
7. Banach, R., Zhu, H., Su, W., Wu, X.: A continuous ASM modelling approach to pacemaker sensing. ACM Trans. Softw. Eng. Methodol. **24**(1), 2 (2014)

8. Barjaktarovic, M.: The state-of-the-art in formal methods. Technical report/Wilkes University and WetStone Technologies (1998). http://www.cs.utexas.edu/users/csed/formal-methods/docs/StateFM.pdf
9. ter Beek, M.H., Bucchiarone, A., Gnesi, S.: Formal methods for service composition. Ann. Math. Comput. Teleinformatics 1(5), 1–10 (2007)
10. Börger, E., Stärk, R.: Abstract State Machines: A Method for High-Level System Design and Analysis. Springer, Berlin (2003)
11. Boström, P., Degerlund, F., Sere, K., Waldén, M.: Derivation of concurrent programs by stepwise scheduling of Event-B models. Formal Aspects Comput. 26(2), 281–303 (2014)
12. Bowen, J.P.: Z: a formal specification notation. In: Frappier, M., Habrias, H. (eds.) Software Specification Methods: An Overview Using a Case Study, pp. 3–19. Springer, London (2001)
13. Bowen, J.P., Hinchey, M.G.: Ten commandments of formal methods. Computer 28(4), 56–63 (1995)
14. Brunel, J., Rioux, L., Paul, S., Faucogney, A., Vallée, F.: Formal safety and security assessment of an avionic architecture with Alloy. In: Third International Workshop on Engineering Safety and Security Systems (ESSS 2014), pp. 8–19 (2014)
15. Clarke, E.M., Wing, J.M.: Formal methods: state of the art and future directions. ACM Comput. Surv. 28(4), 626–643 (1996)
16. Cochran, D., Kiniry, J.R.: Formal model-based validation for tally systems. In: Heather, J., Schneider, S., Teague, V. (eds.) Vote-ID 2013. LNCS, vol. 7985, pp. 41–60. Springer, Heidelberg (2013)
17. Cohen, E., Dahlweid, M., Hillebrand, M., Leinenbach, D., Moskal, M., Santen, T., Schulte, W., Tobies, S.: VCC: a practical system for verifying concurrent C. In: Berghofer, S., Nipkow, T., Urban, C., Wenzel, M. (eds.) TPHOLs 2009. LNCS, vol. 5674, pp. 23–42. Springer, Heidelberg (2009)
18. Dondossola, G.: Formal methods in the development of safety critical knowledge-based components. In: Proceedings of the KR 1998 European Workshop on Validation and Verification of Knowledge-Based Systems, pp. 232–237 (1998)
19. Dwivedi, A.K., Rath, S.K.: Model to specify real time system using Z and Alloy languages: a comparative approach. In: International Conference on Software Engineering and Mobile Application Modelling and Development (ICSEMA 2012), pp. 1–6 (2012)
20. Ferrarotti, F., Schewe, K., Tec, L., Wang, Q.: A new thesis concerning synchronised parallel computing - simplified parallel ASM thesis. CoRR abs/1504.06203 (2015)
21. Frappier, M., Habrias, H. (eds.): Software Specification Methods. ISTE, London (2006)
22. Frias, M.F., Pombo, C.G.L., Aguirre, N.M.: An equational calculus for alloy. In: Davies, J., Schulte, W., Barnett, M. (eds.) ICFEM 2004. LNCS, vol. 3308, pp. 162–175. Springer, Heidelberg (2004)
23. Gannon, J.D., Zelkowitz, M.V., Purtilo, J.M.: Software Specification: A Comparison of Formal Methods. Greenwood Publishing, Westpoint (1994)
24. Georg, G., Bieman, J., France, R.B.: Using Alloy and UML/OCL to specify runtime configuration management: a case study. In: Workshop of the pUML-Group Held Together with the UML 2001 on Practical UML-Based Rigorous Development Methods - Countering or Integrating the eXtremists, pp. 128–141, GI (2001)
25. Ghezzi, C., Mandrioli, D., Morzenti, A.: TRIO: a logic language for executable specifications of real-time systems. J. Syst. Softw. 12(2), 107–123 (1990)
26. Haughton, H.P.: Using Z to model and analyse safety and liveness properties of communication protocols. Inf. Softw. Technol. 33(8), 575–580 (1991)

27. Hoang, T.S., Abrial, J.-R.: Event-B decomposition for parallel programs. In: Frappier, M., Glässer, U., Khurshid, S., Laleau, R., Reeves, S. (eds.) ABZ 2010. LNCS, vol. 5977, pp. 319–333. Springer, Heidelberg (2010)

28. Holzmann, G.J.: The SPIN Model Checker: Primer and Reference Manual. Addison-Wesley, Reading (2004)

29. Jackson, D.: Software Abstractions: Logic, Language, and Analysis. MIT Press, Cambridge (2006)

30. Jones, C.B.: Systematic Software Development Using VDM, 2nd edn. Prentice-Hall Inc., Upper Saddle River (1990)

31. Kang, E., Jackson, D.: Formal modeling and analysis of a flash filesystem in Alloy. In: Börger, E., Butler, M., Bowen, J.P., Boca, P. (eds.) ABZ 2008. LNCS, vol. 5238, pp. 294–308. Springer, Heidelberg (2008)

32. Kaur, A., Gulati, S., Singh, S.: Analysis of three formal methods - Z, B and VDM. Int. J. Eng. Res. Technol. (IJERT) 1(4), 1–4 (2012)

33. Knight, J.C., DeJong, C.L., Gibble, M.S., Nakano, L.G.: Why are formal methods not used more widely? In: The Fourth NASA Langley Formal Methods Workshop (LFM 1997) (1997)

34. Kossak, F.: Landing gear system: an ASM-based solution for the ABZ case study. In: Boniol, F., Wiels, V., Ait Ameur, Y., Schewe, K.-D. (eds.) ABZ 2014. CCIS, vol. 433, pp. 142–147. Springer, Heidelberg (2014)

35. Kossak, F., Illibauer, C., Geist, V., Kubovy, J., Natschläger, C., Ziebermayr, T., Kopetzky, T., Freudenthaler, B., Schewe, K.D.: A Rigorous Semantics for BPMN 2.0 Process Diagrams. Springer, Heidelberg (2015)

36. Kossak, F., Mashkoor, A.: How to Evaluate the Suitability of a Formal Method for Industrial Deployment? A Survey. Technical report SCCH-TR-1603, Software Competence Center Hagenberg GmbH, Hagenberg, Austria (2016). http://www.scch.at/en/rse-news/fm_comparison

37. Kossak, F., Mashkoor, A., Geist, V., Illibauer, C.: Improving the understandability of formal specifications: an experience report. In: Salinesi, C., van de Weerd, I. (eds.) REFSQ 2014. LNCS, vol. 8396, pp. 184–199. Springer, Heidelberg (2014)

38. Lamport, L.: Specifying Systems: The TLA+ Language and Tools for Hardware and Software Engineers. Addison-Wesley, Boston (2002)

39. Larsen, P.G., Wolff, S.: Development process of distributed embedded systems using VDM, Overture – Open-source Tools for Formal Modelling TR-2010-02 (2010)

40. Liu, X., Yand, H., Zedan, H.: Formal methods for the re-engineering of computing systems. In: Proceedings of the 21st Computer Software and Applications Conference (COMPSAC 1997), pp. 409–414 (1997)

41. Maoz, S., Ringert, J.O., Rumpe, B.: Semantically configurable consistency analysis for class and object diagrams. In: Whittle, J., Clark, T., Kühne, T. (eds.) MODELS 2011. LNCS, vol. 6981, pp. 153–167. Springer, Heidelberg (2011)

42. Mashkoor, A., Biro, M.: Towards the trustworthy development of active medical devices: a hemodialysis case study. IEEE Embed. Syst. Lett. 8(1), 14–17 (2016)

43. Mashkoor, A., Hasan, O., Beer, W.: Using probabilistic analysis for the certification of machine control systems. In: Cuzzocrea, A., Kittl, C., Simos, D.E., Weippl, E., Xu, L. (eds.) CD-ARES Workshops 2013. LNCS, vol. 8128, pp. 305–320. Springer, Heidelberg (2013)

44. Mashkoor, A., Jacquot, J.P.: Stepwise validation of formal specifications. In: 18th Asia-Pacific Software Engineering Conference (APSEC 2011), pp. 57–64. IEEE, Ho Chi Minh City, Vietnam (2011)

45. Mashkoor, A., Jacquot, J.P.: Utilizing Event-B for domain engineering: a critical analysis. Requirements Eng. **16**(3), 191–207 (2011)
46. McGibbon, T.: An analysis of two formal methods: VDM and Z. Technical report, DoD Data and Analysis Center for Software (DACS) (1997). https://www.csiac.org/sites/default/files/An%20Analysis%20of%20Two%20Formal%20Methods%20-%20VDM%20and%20Z%20-%20SOAR.pdf
47. Merz, S.: The specification language TLA+. In: Bjørner, D., Henson, M. (eds.) Logics of Specification Languages. Monographs in Theoretical Computer Science, pp. 401–451. Springer, Heidelberg (2008)
48. Mirian-HosseinAbadi, S.H., Mousavi, M.R.: Making nondeterminism explicit in Z. In: Proceedings of the Iranian Computer Society Annual Conference (CSICC 2002), Tehran, Iran (2002)
49. Mukherjee, P., Bousquet, F., Delabre, J., Paynter, S., Larsen, P.G.: Exploring timing properties using VDM++ on an industrial application. In: Proceedings of the Second VDM Workshop (2000)
50. Newcombe, C.: Why Amazon chose TLA+. In: Ait Ameur, Y., Schewe, K.-D. (eds.) ABZ 2014. LNCS, vol. 8477, pp. 25–39. Springer, Heidelberg (2014)
51. Newcombe, C., Rath, T., Zhang, F., Munteanu, B., Brooker, M., Deardeuff, M.: How amazon web services uses formal methods. Commun. ACM **58**(4), 66–73 (2015)
52. Pandey, S., Batra, M.: Formal methods in requirements phase of SDLC. Int. J. Comput. Appl. **70**(13), 7–14 (2013)
53. Rainer-Harbach, M.: Methods and tools for the formal verification of software. An analysis and comparison. Diplomarbeit, Fakultät für Informatik, Technische Universität Wien. https://www.ads.tuwien.ac.at/publications/bib/pdf/rainer-harbach_11.pdf
54. Sifakis, J.: Formal methods and their evaluation. Position Paper Presented at FEM-SYS in Munich (1997). http://www-verimag.imag.fr/~sifakis/RECH/FEMSYS/paper.ps
55. Spivey, J.M.: The Z Notation: A Reference Manual. Prentice-Hall Inc., Upper Saddle River (1989)
56. Verhoef, M., Larsen, P.G., Hooman, J.: Modeling and validating distributed embedded real-time systems with VDM++. In: Misra, J., Nipkow, T., Sekerinski, E. (eds.) FM 2006. LNCS, vol. 4085, pp. 147–162. Springer, Heidelberg (2006)
57. Wang, T., Ji, D.: Active attacking multicast key management protocol using Alloy. In: Derrick, J., Fitzgerald, J., Gnesi, S., Khurshid, S., Leuschel, M., Reeves, S., Riccobene, E. (eds.) ABZ 2012. LNCS, vol. 7316, pp. 164–177. Springer, Heidelberg (2012)
58. Woodcock, J., Larsen, P.G., Bicarregui, J., Fitzgerald, J.: Formal methods: practice and experience. ACM Comput. Surv. **41**(4), 19 (2009)
59. Zave, P.: A practical comparison of Alloy and Spin. Formal Aspects Comput. **2015**(2), 239–253 (2015)

Short Articles
(Work in Progress)

Unified Syntax for Abstract State Machines

Paolo Arcaini[2]([✉]), Silvia Bonfanti[3], Marcel Dausend[1], Angelo Gargantini[3],
Atif Mashkoor[4], Alexander Raschke[1], Elvinia Riccobene[5], Patrizia Scandurra[3],
and Michael Stegmaier[1]

[1] Ulm University, Ulm, Germany
{marcel.dausend,alexander.raschke,michael-1.stegmaier}@uni-ulm.de
[2] Faculty of Mathematics and Physics,
Charles University in Prague, Prague, Czech Republic
arcaini@d3s.mff.cuni.cz
[3] Università degli Studi di Bergamo, Bergamo, Italy
{silvia.bonfanti,angelo.gargantini,patrizia.scandurra}@unibg.it
[4] Software Competence Center Hagenberg, Hagenberg im Mühlkreis, Austria
atif.mashkoor@scch.at
[5] Università degli Studi di Milano, Milano, Italy
elvinia.riccobene@unimi.it

Abstract. The paper presents our efforts in defining UASM, a unified syntax for Abstract State Machines (ASMs), based on the syntaxes of two of the main ASM frameworks, CoreASM and ASMETA, which have been adapted to accept UASM as input syntax of all their validation and verification tools.

1 Introduction and Goals of the Project

Abstract State Machines (ASMs) are a flexible, yet mathematically well-founded method and language for rigorous system engineering. The formalism can be seen as "pseudocode over abstract data" [2]. Although this pseudocode notation is formally defined, in practice many ways exist to encode algebraic concepts and many abbreviations can be used to improve model conciseness and readability.

Among the different frameworks for the ASM method (like AsmL, ASM Workbench, ASMGofer, KIV), two of the main ones are ASMETA [1] and Core-ASM [4]. These platforms provide industrial strength tools to specify, verify, simulate, and test ASM models. However, they implement different dialects of the pseudocode notation and support slightly different extensions of the original definition. For example, AsmetaL (the textual notation of the ASMETA framework [5]) provides the concept of module that is not present in CoreASM, while CoreASM allows the use of abstract rules, a feature that is not present in

The research reported in this paper has been partly supported by the Charles University research funds PRVOUK, and by the Austrian Ministry for Transport, Innovation and Technology, the Federal Ministry of Science, Research and Economy, and the Province of Upper Austria in the frame of the COMET center SCCH.

M. Butler et al. (Eds.): ABZ 2016, LNCS 9675, pp. 231–236, 2016.
DOI: 10.1007/978-3-319-33600-8_14

AsmetaL. There are some differences on the way they represent signature: Core-ASM is not typed, and so it permits an agile modeling style, while AsmetaL is strongly typed. Moreover, the two syntaxes are sometimes slightly different in representing the same concept: for example, the keyword for a sequential block is `seq ...endseq` in AsmetaL and `seqblock ...endseqblock` in CoreASM.

Therefore, while the availability of multiple support platforms is obviously an advantage, it may also be confusing for new adopters of the method. Moreover, designers cannot share models among the tools (unless a translator or adapter is defined) and thus can not easily take advantage of each tool's strengths.

To overcome these limitations, the idea of a common syntax definition "Unified ASM" (UASM), driven by the community, open to any actors, has grown in the last two years. This paper presents the activities performed so far. The challenges of this project are both to preserve various useful extensions of the different tools and support a variety of application scenarios. On the one hand, the UASM language should be usable for communication with customers and non-experts, and, on the other hand, precise enough to allow automatic analysis (like type checking, property verification, etc.). Moreover, we will try to identify unifying solutions for those ASM aspects for which the two frameworks made different design decisions. Two examples are the syntax and semantics for state initialization and the definition of basic data types.

2   Insights into the UASM Grammar

As mentioned in the previous section, the applications of ASMs are manifold. Due to this, we decided to keep the new common grammar as flexible as possible. We tried to include as many useful constructs from the contributing languages as possible, but naturally, some design decisions had to be made.

In order to allow for a more legible specification, UASM offers textual notations for basic constructs. Instead of keywords for mathematical constructs, we also allow Unicode characters (e.g., $\in$ instead of `in`, $\forall$ instead of `forall`).

UASM does not require type annotations. If no type information is given, the types are checked dynamically during runtime (e.g., $*$ (multiplication) can only be applied on two numbers). Whenever type information is required at a later date, it can be added on demand. If a type information is given, the type correctness is checked when the specification is parsed. Currently, only a few basic (boolean, numbers, chars, strings) and set-based types (set, list, bag, map) are defined. For the future, we plan to integrate a notation for algebraic data types in order to allow for arbitrary complex (recursive) data types.

The module concept of ASMETA was adopted to allow for a better modularization of large specifications.

UASM also provides definition for some aspects that have been left open up to now, e.g., the new keyword `exec` followed by a rule name defines the rule to be executed by the initial agent. Usually, this rule introduces new agents and their programs and initializes the abstract state. Alternatively, initial values for all locations or only specific locations of a function can be described as part of its definition (see example below). The defined constructs allow in CoreASM as well in ASMETA to write and execute multi-agent specifications. Despite of that, it might be useful in the future to define special constructs for creating, removing, or assigning programs to agents.

In the following, we introduce an excerpt of the UASM language definition[1]. We focus on the definition of functions and their initialization. First, this part is a substantial supplement to the existing syntax and semantic definition of the underlying literature; second, this part reflects some design decisions originating from different existing realizations of ASM languages.

The aforementioned decision that types are optional strongly influences the UASM language. This is reflected, for example, in the definition of function parameters that allows identifiers as well as domains or a combination of both.

```
ParamameterDef ::= '(' (Id 'in' Domain | Id | Domain)
    (',' (Id 'in' Domain | Id | Domain))* ')'
```

For the initialization of a function, we support a fixed value for all its locations, or specific location values by using maps and terms. The following example illustrates these different concepts by the initialization of the controlled function *gateStatus* of a rail road crossing that is defined according to the following definition.

```
ControlledFunction ::= ( .. | 'controlled' 'function') IdFunction
    ParamameterDef? ('->' Domain)? ('initially' 'from'? Term )?
```

controlled function *gateStatus*(gate **in** GATES) **initially from**
(1)   {gate1 → *open*, gate2 → *closed*}
(2)   **if** *isTrainApproaching*(gate) **then** *closed* else *open*
            where *isTrainApproaching*(g) = ∃ s **in** SENSORS **with**
                g ∈ *observedGates*(s) **and** *trainOnTrack*(s)

The above example (1) illustrates how maps can be used to assign different initial values to specific locations, i.e., gate status for specific gates is different. A more flexible approach is the dynamic initialization of the state based on derived functions. In this case, the initialization is done lazily, i.e., before a function is accessed (read), it is checked whether this particular location has been previously initialized or updated. If not, the given derived function is evaluated returning the initial value. Under the assumption that a railroad crossing control module should take control at a random time, i.e., under different circumstances, we use this dynamic initialization for the status of the rail road gates (2). The initial values of *gateStatus* are computed on demand by the derived function *isTrainApproaching*, whose result is based on current sensor data.

UASM allows the declaration and definition of static functions. For instance, a function **sum** that takes two integers and returns the sum, can be defined as:

static function *sum*(a **in** INTEGER, b **in** INTEGER) **always** a + b

---

[1] The syntax of our language definition is conform to the W3C EBNF notation.

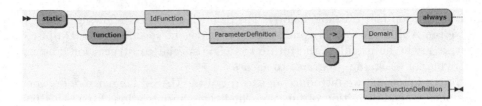

Fig. 1. Static function definition

Several tools can be used to build a visualization of the grammar rules by means of syntax diagrams (or railroad diagrams)[2]. For instance, the definition of static functions is shown in Fig. 1.

3   Re-engineering Existing Tools

In order to allow the CoreASM and the ASMETA frameworks to accept the new common ASM syntax, we had to do some re-engineering, as described as follows.

Reference Parser and Integration into CoreASM. CoreASM is an open-source project defining an ASM language implementing tools that focus on high-level design and experimental validation of ASM specifications.

The CoreASM tool architecture defines a highly modular system based on a minimal kernel. This architecture enables to seamlessly integrate additional language constructs as well as tools and yields in manifold extensions and domain specific applications.

CoreASM's major strengths are in the creation, refinement, and debugging of specifications. For example, starting with an abstract and untyped or only partially typed specification that can already be executed, refining this specification, and performing comprehensive debugging and testing.

As our goal is to provide an easy to read and understandable definition of the UASM syntax. The current grammar definition is not optimized for automatic processing, yet. It contains ambiguities resulting from i.e., optional end-constructs and operator precedences which are not reflected by the grammar. Hence, we are going to derive a grammar definition for UASM that facilitates automatic processing like using parser generators.

Other than usual bottom-up or top-down parsers, JParsec-Framework can deal with our grammar definition as it can handle left recursion and it resolves ambiguities by applying strategies that make parsing deterministic. Therefore, we implemented a reference parser for UASM using the JParsec-Framework by merely transcribing our grammar into the JParsec syntax. This parser is already publicly available[3]. Because JParsec is a parser combinator, the reference implementation can be easily extended. We also integrated it into CoreASM without

---

[2] We use the web service http://www.bottlecaps.de/rr/ui.

[3] https://github.com/uasm/uasm-reference-parser.

any limitations to the application of existing tools, e.g., the interpreter and the debugger [3].

Integration into ASMETA. The ASMETA framework [1] is based on the *ASM Metamodel* (AsmM) [5], an abstract syntax description (defined with the Eclipse Modeling Framework (EMF)) of a language for ASMs. From AsmM, a concrete textual syntax (AsmetaL), a parser, Java APIs, etc., have been developed for model editing, storage, and manipulation. On the top of these, more complex tools for validation, verification, and testing have been developed. They all manipulate AsmM models (i.e., instances of AsmM). So, in order to use ASMETA tools on UASM specifications, we must map UASM specifications to AsmM models.

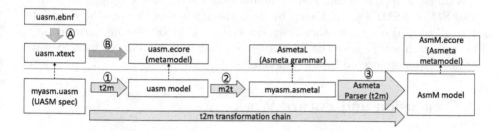

Fig. 2. UASM and ASMETA integration

We followed the approach shown in Fig. 2. We derived an Xtext grammar starting from the UASM EBNF grammar (step A); from the Xtext grammar, a metamodel and a parser have been obtained automatically (step B). Transforming a UASM specification to an AsmM model (bottom of Fig. 2) consists in parsing the specification with the Xtext parser (step 1), producing an AsmetaL specification from the ecore objects produced by Xtext (step 2), and finally obtaining an AsmM model with the AsmetaL parser (step 3).

4   Validation of the Approach

After the integration of UASM in the existing frameworks (see Sect. 3), we have devised two validation activities that one has to apply to check whether a tool correctly supports the new syntax. These activities will be initially applied to CoreASM and ASMETA, but, in the future, to any tool willing to support UASM (to get a sort of UASM compliance certification).

As a first validation activity, we plan to create a repository of syntactically correct and non-correct UASM specifications. They should be representative of the kind of models that can be written in UASM, i.e., they should cover all grammar elements. We will then check that a UASM compliant tool correctly accepts/rejects the specifications.

As a second validation activity, we want to check that the semantics is the same in any framework. We will establish a way for accepting a sequence of inputs and saving the machine behavior as sequences of update sets (and outputs). Any UASM tool must be able to produce the behavioral sequences in that format. We will have a way to compare if two behaviors are identical. We will save, together with the benchmarks, also their expected behaviors and we will then check if CoreASM and ASMETA (or any other tool) correctly capture the intended semantics. This approach can only validate deterministic single-agent ASMs that, given an input sequence, produce only one possible output sequence. As future work, we plan to devise ways to validate our tools also using nondeterministic and/or multi-agents ASMs. In that case, we could record the output in terms of trees representing all possible evolutions of the system. However, the approach could not scale or even be not applicable in case of infinite-state models. A different approach could be to simulate the model with a framework (either CoreASM or ASMETA) and, step by step, check whether the produced update set is allowed also in the other framework (in a kind of runtime monitoring). This approach would have the advantage of being scalable also to infinite-state systems, although it could miss some faults.

5 Conclusion and Future Work

We have presented our efforts in defining UASM, a unified syntax for ASMs, based on the syntaxes of the main two ASM frameworks, CoreASM and ASMETA, which have been adapted to accept UASM as input syntax of all their validation and verification tools.

As further future work, to check that the two frameworks interpret the UASM models in the same way, we plan to apply some validation activities based on comparison of simulation traces. Moreover, we also plan to extend UASM with constructs not part of the two starting syntaxes, but that are part of other ASM syntaxes (e.g., classes of AsmL).

References

1. Arcaini, P., Gargantini, A., Riccobene, E., Scandurra, P.: A model-driven process for engineering a toolset for a formal method. Softw. Pract. Experience **41**, 155–166 (2011)
2. Börger, E., Stärk, R.F.: Abstract State Machines. A Method for High-Level System Design and Analysis. Springer, Heidelberg (2003)
3. Dausend, M., Stegmaier, M., Raschke, A.: Debugging abstract state machine specifications: an extension of coreASM. In: Proceedings of the Posters and Tool Demos Session, iFM 2012 and ABZ 2012. pp. 21–25 (2012)
4. Farahbod, R., Gervasi, V., Glässer,.U.: CoreASM: an extensible ASM execution engine. Fundamenta Informaticae **77**(1–2), 71–104 (2007)
5. Gargantini, A., Riccobene, E., Scandurra, P.: A metamodel-based language and a simulation engine for abstract state machines. J. UCS **14**(12), 1949–1983 (2008)

A Relational Encoding for a Clash-Free Subset of ASMs

Gerhard Schellhorn$^{(\boxtimes)}$, Gidon Ernst, Jörg Pfähler, and Wolfgang Reif

Institute for Software and Systems Engineering, University of Augsburg,
Augsburg, Germany
{schellhorn,ernst,pfaehler,reif}@isse.de

Abstract. This paper defines a static check for clash-freedom of ASM rules, including sequential and parallel composition, nondeterministic choice, and recursion. The check computes a formula that, if provable, makes a relational encoding of ASM rules possible, which is an important prerequisite for efficient deduction. The check is general enough to cover all sequential rules as well as many typical uses of parallel composition.

Keywords: Abstract State Machines · Synchronous parallelism · Clashes

1 Introduction

ASM rules are very expressive. Compared to other state-based formalisms they do not just give a transition relation as a formula $\varphi(\underline{x}, \underline{x}')$ in terms of the prestate $\underline{x}$ and the post state $\underline{x}'$ (like e.g. Z, TLA or Event-B do). The additional concepts like function updates, parallel and sequential composition, nondeterministic choice, and defined rules with recursion give ASMs a lot of additional expressiveness that allows refinement from very abstract models down to ASMs which can easily be seen to be equivalent to real programs. For formalisms based on transition relations translating to real programs is hard, typically only the reverse is done: encoding programs to transition relations with the help of program counters. On the flip side a relational encoding for ASM rules is difficult. As a consequence, we are not aware of any deduction approach with tool support for arbitrary ASM rules. Most verification tools, such as e.g. KIV [3], have allowed the purely sequential fragment only with parallel assignments restricted to different function symbols. Others have avoided sequential composition and recursion, and used assignment for functions with arity zero only. With these restrictions however we are in essence back to transition systems.

As soon as parallel rules R are allowed, it becomes hard to define a relation $\mathsf{rel}(R)(\underline{f}, \underline{f}')$, which characterizes the effect of R in terms of the dynamic functions $\underline{f}$ it assigns. Consider the simple parallel rule $f(t_1) := u_1$ **par** $f(t_2) := u_2$. If we define $\mathsf{rel}(f(t_i) := u_i)(f, f') \equiv f'(t_i) = u_i$ and use conjunction for **par** then the relation will not ensure that f is unchanged for arguments other than t_1 and t_2. The formula will also miss the clash for the case $t_1 = t_2$ but $u_1 \neq u_2$,

M. Butler et al. (Eds.): ABZ 2016, LNCS 9675, pp. 237–243, 2016.
DOI: 10.1007/978-3-319-33600-8_15

which results in undefined behavior. Clashes are the main obstacle for a relational encoding and in most applications rules with clashes are undesirable anyway.

The contribution of this paper is a predicate $\mathsf{con}(R)$ that statically computes a first-order formula from ASM rules R. If provable, then all executions of rule R are guaranteed to be clash-free[1] and a relational encoding is possible.

The predicate $\mathsf{con}(R)$ is related to the one used in the logic for ASMs defined by Stärk and Nanchen [7] (also given in [2]). While we use the syntax and semantics of ASMs given there, our predicate $\mathsf{con}(R)$ differs in several aspects. It does not use modal constructs ($[R]\,\varphi$) but statically computes a formula even for recursive rules, where the definitions in [7] would lead to an infinite computation (note that the completeness theorem that permits to eliminate modal constructs is for *hierarchical* ASMs only, where recursion is forbidden). Our computation stops at calls and therefore allows one to check each (sub-)rule separately. Different from [7] our $\mathsf{con}(R)$ does not imply that executing R terminates (via $\mathsf{def}(R)$) — termination must be shown using well-founded orders otherwise. We support nondeterministic choice, replaced by choice functions in [7] (making rules and verification conditions at least harder to read).

The new approach in [4] extends [7] to nondeterminism, but does not consider recursion at all. The rules of our relational encoding rel have some similarity to the ones for $\mathsf{upd}(R, X)$ in [4], in particular higher-order functions are used in both. However, our consistency check is purely first-order.

The price we pay for having a computable $\mathsf{con}(R)$ for all rules is that our predicate only approximates clash-freedom. There are clash-free rules which our predicate rejects. The scheme is however strong enough to trivially return true for all rules of the sequential fragment, as well as for some typical parallel rules. In general a theorem prover (or a decision procedure, when the data structures used by the rule are decidable) is needed to prove the computed $\mathsf{con}(R)$, and an SMT solver should suffice for many practical cases to establish clash-freedom.

2   Syntax and Semantics of ASM Rules

We assume the reader to be familiar with first order logic and the syntax of ASM rules R and their semantics as given in e.g. [7]. We only repeat a few basic notations. Given an algebra $\mathfrak{A}$ and a valuation ξ, term t is evaluated to $t_\xi^{\mathfrak{A}}$ and formula φ by $\mathfrak{A}, \xi \models \varphi$. An ASM rule R modifies an algebra $\mathfrak{A}$ to $\mathfrak{A}'$. The basic assignment is $f(\underline{t}) := u$ for a *dynamic* function f. The set of assigned functions in a rule is denoted $mod(R)$. The main rule does not use any free variables, but subrules within **choose** x **with** $\varphi(x)$ **in** R or **forall** x **with** $\varphi(x)$ **in** R may use free variables denoted as $free(R)$. Given an algebra $\mathfrak{A}$ executing a rule R computes $\mathfrak{A}'$ in two steps. First, a set of updates U is computed recursively over the structure of R as $[\![R]\!]_\xi^{\mathfrak{A}} \triangleright U$. An update is of the form $(f, \underline{a}, b)$. Applying it on

---

[1] We regard potential clashes that occur only under some specific non-deterministic choices to be even worse than guaranteed clashes in every run. Even simulating runs of the ASM may fail to detect them. We also regard computing the same update twice as undesirable, and our approach will return $\mathsf{con}(R) = \text{false}$ in both cases.

an algebra $\mathfrak{A}$ modifies function $f^{\mathfrak{A}}$ at arguments $\underline{a}$ to be b. A set of updates U is consistent, if it does not contain two updates $(f, \underline{a}, b_1)$ and $(f, \underline{a}, b_2)$ with $b_1 \neq b_2$. In this case the whole set U can be applied to give $\mathfrak{A}' := \mathfrak{A} \oplus U$. If a rule always computes consistent sets of updates, it is called clash-free. A rule is *defined* if it computes an update set at all as recursive rules may fail to terminate. In the following we assume that all dynamic functions are unary, or have no arguments ("program variables" typically named z). Declarations of a subrule named ρ have the form $\rho(x; z).R$ where variable x is one value parameter and z is a program variable. None of these restrictions is essential, they just allow us to save notation for sequences of arguments. The body R of ρ is restricted to have $mod(R) = z$, all updated locations must be explicitly given. A call of ρ is of the form $\rho(t; f(u))$ where u may contain static function symbols only (to avoid problems with lazy evaluation), semantically $[\![\rho(t; f(u))]\!]_{\xi}^{\mathfrak{A}} \rhd U$ iff $[\![R_z^{f(u)}]\!]_{\xi\{x \mapsto t_{\xi}^{\mathfrak{A}}\}}^{\mathfrak{A}} \rhd U$ where $R_z^{f(u)}$ replaces all occurrences of z with $f(u)$, so $f(u)$ is read and updated instead of z.

3   Syntactic Consistency

Syntactic consistency uses the function $\mathsf{asg}(R, f)$ that computes a formula over a dedicated variable f_{arg} and $free(R)$. The values $\xi(f_{\mathrm{arg}})$ for which it hold give an overapproximation of the arguments where f is assigned.

$$\mathsf{asg}(g(u) := t, f) \equiv \mathsf{asg}(\rho(t; g(u)), f) \equiv \begin{cases} f_{\mathrm{arg}} = u, & f = g \\ \mathrm{false}, & \mathrm{otherwise} \end{cases}$$

$$\mathsf{asg}(R_1 \ \mathbf{seq} \ R_2, f) \equiv \begin{cases} \mathsf{asg}(R_1, f) \vee \mathsf{asg}(R_2, f), & mod(R_1) \cap dep(R_2, f) = \emptyset \\ \mathrm{true}, & f \in mod(R_1 \ \mathbf{seq} \ R_2) \\ \mathrm{false}, & \mathrm{otherwise} \end{cases} \quad (\star)$$

$$\mathsf{asg}(R_1 \ \mathbf{par} \ R_2, f) \equiv \mathsf{asg}(R_1, f) \vee \mathsf{asg}(R_2, f)$$

$$\mathsf{asg}(\mathbf{if} \ \varphi \ \mathbf{then} \ R_1 \ \mathbf{else} \ R_2, f) \equiv (\varphi \wedge \mathsf{asg}(R_1, f)) \vee (\neg \varphi \wedge \mathsf{asg}(R_2, f))$$

$$\mathsf{asg}(\mathbf{choose} \ x \ \mathbf{with} \ \varphi(x) \ \mathbf{in} \ R, f) \equiv \mathsf{asg}(\mathbf{forall} \ldots, f) \equiv \exists \ x. \ \varphi(x) \wedge \mathsf{asg}(R, f)$$

For assignments to $f(u)$, we keep $f_{\mathrm{arg}} = u$. The case for sequential composition considers whether R_1 assigns to some g that controls the argument of f as $f(g(u))$ in R_2 ($g \in dep(R_2, f)$). In this case, possible values for f_{arg} are unconstrained if f is modified at all. Conditionals strengthen the check of the branches by the assumption from the test. In a **forall** or **choose** rule, f could be affected by any execution of the body for an x that satisfies the condition φ. Note that we do not impose any constraint on φ, infinitely many choices for x are possible.

The set of dynamic function symbols $dep(R, f)$ that the final value of f after execution of R depends on, can be defined recursively as follows:

$$dep(g(u) := t, f) = dep(\rho(t; g(u)), f) := \begin{cases} \{h : h \ \mathrm{occurs} \ \mathrm{in} \ t \ \mathrm{or} \ \mathrm{in} \ u\}, & f = g \\ \emptyset, & \mathrm{otherwise} \end{cases}$$

$$dep(R_1 \ \mathbf{seq} \ R_2, f) := dep(R_1, f) \cup dep(R_2, f) \cup \bigcup_{g \in dep(R_2, f)} dep(R_1, g)$$

$$dep(R_1 \ \mathbf{par} \ R_2, f) := dep(R_1, f) \cup dep(R_2, f)$$

$dep(\textbf{if } \varphi \textbf{ then } R_1 \textbf{ else } R_2, f) := \{h : h \text{ occurs in } \varphi\} \cup dep(R_1, f) \cup dep(R_2, f)$

$dep(\textbf{choose } x \textbf{ with } \varphi(x) \textbf{ in } R, f)$

$\quad = dep(\textbf{forall } x \textbf{ with } \varphi(x) \textbf{ in } R, f) := \{h : h \text{ occurs in } \varphi(x)\} \cup dep(R, f)$

For assignments, dependencies come from the argument terms and the right hand sides. Sequential composition chains the dependencies transitively. For **if**, **choose**, and **forall**, the dynamic functions h occurring in the respective test φ potentially have an influence on the final value of f as well.

Syntactic consistency $con(R)$ of a rule R is defined over the structure of rules:

$con(g(u) := t) \equiv con(\rho(t; f(u))) \equiv \text{true}$

$con(R_1 \textbf{ seq } R_2) \equiv con(R_1) \wedge con\left(R_2{}^{f'}_{mod(R_1)}\right)$ where the $\underline{f'}$ are globally fresh

$con(R_1 \textbf{ par } R_2) \equiv con(R_1) \wedge con(R_2) \wedge \bigwedge_f \neg \exists\, y.\ \mathsf{asg}(R_1, f)(y) \wedge \mathsf{asg}(R_2, f)(y)$

$con(\textbf{if } \varphi \textbf{ then } R_1 \textbf{ else } R_2) \equiv (\varphi \wedge con(R_1)) \vee (\neg\, \varphi \wedge con(R_2))$

$con(\textbf{choose } x \textbf{ with } \varphi(x) \textbf{ in } R) \equiv \forall\, x.\ \varphi(x) \rightarrow con(R)$

$con(\textbf{forall } x \textbf{ with } \varphi(x) \textbf{ in } R) \equiv \forall\, x.\ (\varphi(x) \rightarrow con(R)) \wedge$

$\bigwedge_f \neg \exists\, x_1, x_2, y.\ x_1 \neq x_2 \wedge \varphi(x_1) \wedge \mathsf{asg}(R_x^{x_1}, f)(y) \wedge \varphi(x_2) \wedge \mathsf{asg}(R_x^{x_2}, f)(y)$

Assignments and calls do not impose any additional constraints. as they do not provoke clashes when viewed in isolation (provided that the body of procedures ρ is checked separately). In a sequential composition, consistency of R_2 must be checked for possibly modified values of dynamic functions, expressed by fresh symbols $\underline{f'}$ that are unconstrained. An example, where we lose precision is

$R^* \quad \equiv \quad g(u_1) := u_2 \textbf{ seq } f(g(u_3)) := u_4,$ \qquad where the u_i are static terms.

For the second assignment, we get $dep(f(g(u_3)) := u_4, f) = \{g\}$. Therefore case $(\star)$ applies and $\mathsf{asg}(R^*, f) = \text{true}$. Informally, we do not know statically whether the first assignment affects the argument of f, i.e., whether $g(u_1)$ aliases $g(u_3)$. Note that R^* is still clash-free ($con(R^*) = \text{true}$).

Parallel execution of R_1 and R_2 conservatively excludes assignments to the same location, where $\mathsf{asg}(R, f)(y)$ renames f_{arg} to a fresh variable y in $\mathsf{asg}(R, f)$. To continue the example, putting R^* in parallel with any assignment to f will make con false for the combined rule. Note that this combination of sequential and parallel composition, and the fact that we assume that the argument $f(u)$ is *always* assigned in a recursive call are the only two sources for imprecision (when assigning the same value to a location twice is regarderd as a clash).

Nondeterministic choice hides the bound variable x and adds the assumption $\varphi(x)$ about the choice for that x to the consistency check of the body. For **forall** we additionally exclude conflicts between two pairwise parallel executions of the body where two fresh distinct representants x_1 and x_2 of the index x both cause assignments to $f(y)$ for the same y that replaces f_{arg} locally in the body.

Lemma 1. *Given that R yields update set U, i.e., $[\![R]\!]_\xi^\mathfrak{A} \triangleright U$:*
- *If $f \notin mod(R)$ then $(f, a, b) \notin U$ for all a, b.*
- *If $g^\mathfrak{A} = g^{\mathfrak{A} \oplus U}$ for all $g \in dep(R, f)$*
  then $\mathfrak{A}, \xi \models \mathsf{asg}(R, f)$ iff $\mathfrak{A} \oplus U, \xi \models \mathsf{asg}(R, f)$.
- *If $(f, a, b) \in U$ for some a, b then $\mathfrak{A}, \xi\{f_{arg} \mapsto a\} \models \mathsf{asg}(R, f)$.*
- *If $\mathfrak{A}, \xi \models \mathsf{con}(R)$ then U is consistent.*

The lemma states that *mod*, *dep*, *asg*, resp. *con* are correct. The second bullet lifts (not) *asg* over the first rule in a sequential composition with updates U.

4   Relational Encoding

The relational encoding $\mathsf{rel}(R)(\underline{f}, \underline{f}')$ is a predicate over $\underline{f} \equiv mod(R)$ and primed versions $\underline{f}'$ of these functions. Its free variables are at most the free variables of R. Variable y as well as function variables f_1, f_2, F used below are fresh. We abbreviate $(\varphi \to t = u) \wedge (\neg\,\varphi \to t = v)$ with $t = (\varphi \supset u; v)$.

$$\mathsf{rel}(g(t) := u)(\underline{f}, \underline{f}') \quad \equiv \quad \forall\, y.\ g'(y) = (y = t \supset u; g(y)) \wedge \bigwedge_{f \in \underline{f}, f \neq g} f' = f$$

$$\mathsf{rel}(\rho(t; g(u)))(\underline{f}, \underline{f}') \quad \equiv \quad \exists\, y.\ y = t \wedge \mathsf{rel}(R_{z,x}^{g(u),y})(g, g') \wedge \bigwedge_{f \in \underline{f}, f \neq g} f' = f$$

$$\mathsf{rel}(R_1\ \mathbf{seq}\ R_2)(\underline{f}, \underline{f}') \quad \equiv \quad \exists\, \underline{f_1}.\ \mathsf{rel}(R_1)(\underline{f}, \underline{f_1}) \wedge \mathsf{rel}(R_2)(\underline{f_1}, \underline{f}')$$

$$\mathsf{rel}(R_1\ \mathbf{par}\ R_2)(\underline{f}, \underline{f}') \quad \equiv \quad \exists\, \underline{f_1}, \underline{f_2}.\ \mathsf{rel}(R_1)(\underline{f}, \underline{f_1}) \wedge \mathsf{rel}(R_2)(\underline{f}, \underline{f_2}) \wedge merge(\underline{f_1}, \underline{f_2}, \underline{f}')$$

$$\text{where } merge(\underline{f_1}, \underline{f_2}, \underline{f}') := \bigwedge_{f, f_1, f_2 \in \underline{f}, \underline{f_1}, \underline{f_2}} \forall\, y.\ f'(y) = (\mathsf{asg}(R_1, f)(y) \supset f_1(y); f_2(y))$$

$$\mathsf{rel}(\mathbf{if}\ \varphi\ \mathbf{then}\ R_1\ \mathbf{else}\ R_2)(\underline{f}, \underline{f}') \quad \equiv \quad (\varphi \supset \mathsf{rel}(R_1)(\underline{f}, \underline{f}'); \mathsf{rel}(R_2)(\underline{f}, \underline{f}'))$$

$$\mathsf{rel}(\mathbf{choose}\ x\ \mathbf{with}\ \varphi(x)\ \mathbf{in}\ R)(\underline{f}, \underline{f}') \quad \equiv \quad \exists\, x.\ \varphi(x) \wedge \mathsf{rel}(R)(\underline{f}, \underline{f}')$$

$$\mathsf{rel}(\mathbf{forall}\ x\ \mathbf{with}\ \varphi(x)\ \mathbf{in}\ R)(\underline{f}, \underline{f}') \quad \equiv \quad \exists\, \underline{F}.\ (\forall\, x.\ \varphi(x) \to \mathsf{rel}(R)(\underline{f}, \underline{F_x})) \wedge$$
$$\bigwedge_{f, F \in \underline{f}, \underline{F}} \forall\, y.\ \quad \big(\forall\, x.\ \varphi(x) \wedge \mathsf{asg}(R, f)(y) \to f'(y) = F_x(y)\big)$$
$$\wedge\ \big((\forall\, x.\ \varphi(x) \to \neg\,\mathsf{asg}(R, f)(y)) \to f'(y) = f(y)\big)$$

The rule for call renames variable x to y to avoid a conflict when x occurs in t. The definition solves the problem of parallel rules $R_1\ \mathbf{par}\ R_2$ from the introduction by first computing two individual results $\underline{f_1}$ and $\underline{f_2}$ for the two rules. Since we know from the fact that $\mathsf{con}(R)$ holds that each location (f, y) is assigned by at most one of the rules (so at most one of the predicates $\mathsf{asg}(R_i, f)(y)$ holds), the definition chooses the first rule if it assigned the location and otherwise the second one. The definition for **forall** generalizes from two results $\underline{f_1}, \underline{f_2}$ to a result F_x for every x satisfying φ. Note that since $x \in free(R)$ the result F_x may be different for every argument. In the presence of a clash, the formula is equivalent to false.

If we interpret $\mathfrak{A} \cup \mathfrak{A}'$ as the algebra that evaluates every unprimed function f over $\mathfrak{A}$ and every primed function f' over $\mathfrak{A}'$, we have:

Theorem 1. *Given a rule R with $mod(R) = \underline{f}$ and $\mathfrak{A}, \xi \models \mathsf{con}(R)$, then $\mathfrak{A} \cup \mathfrak{A}' \models \mathsf{rel}(R)(\underline{f}, \underline{f}')$ if and only if there is some consistent U such that $[\![R]\!]_\xi^\mathfrak{A} \triangleright U$ and $\mathfrak{A}' = \mathfrak{A} \oplus U$.*

5   Conclusion and Outlook

We have defined a clash-freedom check for ASM rules. All sequential rules check trivially. Typical parallel rules with disjoint tests (e.g. used in the WAM [1])

if $instruction = i_1$ **then** R_1 **par**   **if** $instruction = i_2$ **then** R_2

are also allowed. Lifting a rule $R(; lv)$ of one process p with process-local state lv to a parallel rule **forall** p **do** $R(; lv(p))$ for all processes p works, too (e.g. used for the threads of the Java ASM [8]).

We have verified the results in KIV by a predicate logic embedding of ASM rules (except for calls) and their semantics (see the URL [9]), similar to what we have done for the temporal logic RGITL [6] (including calls). This uncovered several mistakes in initial versions of the definitions.

The check presented in this paper could be improved in practice by using invariants of the ASM or preconditions of recursive rules as assumptions, e.g., the rule $f(x) := 1$ **par** $f(y) := 2$ is clash-free when the invariants imply $x \neq y$.

In parallel to our work, a relational encoding of ASMs to Event-B was developed in [5]. In contrast to ours, the clash-freedom check is exact and tolerates rules, which compute the same update several times. The approach has been evaluated with several examples, while we have only tried mimimal ones. The approach avoids the use of higher-order functions using set theory instead. On the other hand the approach is limited to ASM rules without recursion and sequential composition, so it is not sufficient to support the rules used in KIV.

This work is only the first step towards interactive proofs with a larger set of rules in KIV. Such deduction needs rules for symbolic execution, as the result of simply substituting the relation for the rule would be incomprehensible. For a clash-free rule R_1 **par** R_2 our result shows that it is valid to transform the rule to $\{\underline{f_1}, \underline{f_2} := \underline{f}, \underline{f}\}$ **seq** $R_1 \frac{f_1}{f}$ **seq** $R_2 \frac{f_2}{f}$ and compute the final result with *merge* from the definition of $\mathsf{rel}(R_1$ **par** $R_2)$. When **forall** iterates over a finite set, inductive arguments over its size should be possible.

References

1. Börger, E., Rosenzweig, D.: The WAM–definition, compiler correctness. In: Logic Programming: Formal Methods and Practical Applications, Studies in Computer Science and Artificial Intelligence, vol. 11, pp. 20–90. Elsevier (1995)
2. Börger, E., Stärk, R.F.: Abstract State Machines–a Method for High-level System Design and Analysis. Springer, Heidelberg (2003)
3. Ernst, G., Pfähler, J., Schellhorn, G., Haneberg, D., Reif, W.: KIV - overview and VerifyThis competition. Softw. Tools Techn. Transfer **17**(6), 677–694 (2015)
4. Ferrarotti, F., Schewe, K.-D., Tec, L., Wang, Q.: A logic for non-deterministic parallel Abstract State Machines. In: Gyssens, M. (ed.) FoIKS 2016. LNCS, vol. 9616, pp. 334–354. Springer, Heidelberg (2016). doi:10.1007/978-3-319-30024-5_18
5. Leuschel, M., Börger, E.: A compact encoding of sequential ASMs in Event-B. In: Butler, M., Schewe, K.D., Mashkoor, A., Biro, M. (eds.) ABZ 2016. LNCS, vol. 9675, pp. 119–134. Springer, Heidelberg (2016)

6. Schellhorn, G., Tofan, B., Ernst, G., Pfähler, J., Reif, W.: RGITL: a temporal logic framework for compositional reasoning about interleaved programs. Ann. Math. Artif. Int. (AMAI) **71**, 131–174 (2014)
7. Stärk, R.F., Nanchen, S.: A complete logic for Abstract State Machines. J. Univ. Comput. Sci. (J.UCS) **7**(11), 981–1006 (2001)
8. Stärk, R.F., Schmid, J., Börger, E.: Java and the Java Virtual Machine: Definition, Verification. Springer, Validation (2001)
9. A relational encoding for a clash-free subset of ASMs: Formalization and proofs.   https://swt.informatik.uni-augsburg.de/swt/projects/Refinement/ASM-clashfreedom.html

Towards an ASM Thesis for Reflective Sequential Algorithms

Flavio Ferrarotti[1](✉), Loredana Tec[1], and José María Turull Torres[2,3]

[1] Software Competence Center Hagenberg, 4232 Hagenberg, Austria
{flavio.ferrarotti,loredana.tec}@scch.at
[2] Universidad Nacional de La Matanza, Buenos Aires, Argentina
j.m.turull@massey.ac.nz
[3] Massey University, Palmerston North, New Zealand

Abstract. Starting from Gurevich's thesis for sequential algorithms (the so-called "sequential ASM thesis"), we propose a characterization of the behaviour of sequential algorithms enriched with reflection. That is, we present a set of postulates which we conjecture capture the fundamental properties of *reflective sequential algorithms* (RSAs). Then we look at the plausibility of an ASM thesis for the class of RSAs, defining a model of abstract state machine (which we call *reflective* ASM) that we conjecture captures the class of RSAs as defined by our postulates.

1 Reflective Sequential Algorithms

In this paper we are concerned with linguistic reflection [6], which can be defined as the ability of an algorithm to change itself.

In the field of computable functions this idea of reflection is as old as the field itself, think for instance of universal Turing machines. It has also been implemented in many programming languages. A prime example is LISP [5], where programs and data are represented uniformly as lists, and thus programs represented as data can be executed dynamically by means of an evaluation operator. Database theory is another field in which reflection has been deeply studied. It was shown that reflection can increase the expressive power of relational algebra [3] and relational machines [1]. Nowadays, most programming languages allow for some form of dynamic SQL, where the SQL queries are produced and evaluated dynamically during the program run-time, as opposed to static SQL where the queries are fixed at the time of compilation.

In the field of behavioural theory of algorithms however, linguistic reflection has not (up to our knowledge) been formally studied yet. This is surprising since dynamic self modifying code is a matter of increasing practical importance and key for the development of type-safe, dynamic agents, autonomous computing, and adaptive systems among others. The development of a good theoretical basis

The research reported in this paper results from the project *Behavioural Theory and Logics for Distributed Adaptive Systems* supported by the **Austrian Science Fund (FWF: [P26452-N15])**.

© Springer International Publishing Switzerland 2016
M. Butler et al. (Eds.): ABZ 2016, LNCS 9675, pp. 244–249, 2016.
DOI: 10.1007/978-3-319-33600-8_16

to describe, understand and prove properties of such systems, is then a pressing issue.

Our aim in this work is to contribute to the development of a behavioural theory of reflective algorithms. In particular, we are concerned with *reflective sequential algorithms* (RSAs), i.e., algorithms which are sequential in the precise sense of Gurevich's famous thesis [4], but which have the additional ability to change themselves.

In the remaining part of this section we propose to capture the class of RSAs by means of three postulates which naturally extend the sequential time, abstract state and bounded exploration postulates in Gurevich's thesis [4]. Then, in Sect. 2 we define a model of reflective ASM which we conjecture capture the class of RSAs as defined by our postulates. Section 3 concludes this short paper with two examples of RSAs which satisfy our postulates.

Similar to Gurevich's thesis for sequential algorithms [4], our first postulate states that every RSA works in sequential time. The key difference is that RSAs need to be able to change themselves. Thus, it seems natural to consider every state of a RSA as an *extended state* which includes (a representation of) a sequential algorithm (in the precise sense of Gurevich's thesis [4]) as part of it. In this way, transitions from one step to the next can also involve changes to the algorithm which now forms part of the state. Given a state $\mathbf{S}$ and a sequential algorithm A, we use $(\mathbf{S}, A)$ to denote an extended state which extends $\mathbf{S}$ with (a representation of) A.

Postulate 1 (Reflective Sequential Time Postulate). *A RSA $\mathcal{A}$ consists of the following:*

- *A non-empty set $\mathcal{S}_{\mathcal{A}}$ of extended states, where each state is extended with (a representation of) a sequential algorithm which forms part of the state.*
- *A non-empty subset $\mathcal{I}_{\mathcal{A}} \subseteq \mathcal{S}_{\mathcal{A}}$ of initial extended states such that for all $(\mathbf{S}_i, A_i), (\mathbf{S}_j, A_j) \in \mathcal{I}_{\mathcal{A}}$, $A_i = A_j$ (i.e., such that all initial extended states of $\mathcal{A}$ contain exactly the same sequential algorithm).*
- *A one-step transformation function $\tau_{\mathcal{A}} : \mathcal{S}_{\mathcal{A}} \to \mathcal{S}_{\mathcal{A}}$ such that $\tau_{\mathcal{A}}((\mathbf{S}_i, A_i)) = (\mathbf{S}_j, A_j)$ iff $\tau_{A_i}((\mathbf{S}_i, A_i)) = (\mathbf{S}_j, A_j)$, where τ_{A_i} denotes the one-step transformation function of the sequential algorithm A_i.*

The concept of run remains the same as in the thesis for sequential algorithms, except that we consider extended states instead of arbitrary states. That is, a *run* or *computation* of $\mathcal{A}$ is a sequence of extended states $(\mathbf{S}_0, A_0), (\mathbf{S}_1, A_1), (\mathbf{S}_2, A_2), \ldots$, where $(\mathbf{S}_0, A_0)$ is an initial extended state in $\mathcal{I}_A$ and $(\mathbf{S}_{i+1}, A_{i+1}) = \tau_{\mathcal{A}}((\mathbf{S}_i, A_i))$ holds for every $i \geq 0$.

While behavioural equivalent sequential algorithms have the same runs, this is not necessarily the case for RSA. In fact we can think of different runs $(\mathbf{S}_0, A_0), (\mathbf{S}_1, A_1), (\mathbf{S}_2, A_2), \ldots$, and $(\mathbf{S}_0', A_0'), (\mathbf{S}_1', A_1'), (\mathbf{S}_2', A_2'), \ldots$, where $\mathbf{S}_i = \mathbf{S}_i'$ and A_i is behavioural equivalent (in the classical sense) to A_i' for every $i \geq 0$. Since such runs clearly represent the same behaviour, we call them *essentially equivalent runs* and define *behavioural equivalent RSAs* as RSAs which have essentially equivalent classes of runs.

As in the sequential ASM thesis, our second postulate defines (extended) states as first-order structures. However, extended states are not just arbitrary first-order structures, since each extended state must also include (an encoding of) a sequential algorithm given by a finite text. It is important to note that the vocabulary of a RSA is *not* necessarily fixed. That is, we do not only allow RSAs to change themselves, but also to change their vocabularies.

Postulate 2 (Reflective Abstract State Postulate).

- *Extended states of RSAs are first-order structures.*
- *Every extended state $(\mathbf{S}, A)$ is formed by the union of an arbitrary first-order structure $\mathbf{S}$ and a finite first-order structure $\mathbf{S}_A$ which encodes the sequential algorithm A.*
- *The one-step transformation τ_A of a RSA $\mathcal{A}$ does not change the base set of any extended state of $\mathcal{A}$.*
- *The sets $\mathcal{S}_{\mathcal{A}}$ and $\mathcal{I}_{\mathcal{A}}$ of, respectively, extended states and initial extended states of a RSA $\mathcal{A}$, are both closed under isomorphisms.*
- *Any isomorphism between two extended states $(\mathbf{S}_1, A_1)$ and $(\mathbf{S}_2, A_2)$ of a RSA $\mathcal{A}$, is also an isomorphism from $\tau_{\mathcal{A}}((\mathbf{S}_1, A_1))$ to $\tau_{\mathcal{A}}((\mathbf{S}_2, A_2))$.*

Our next (key) definition of strong coincidence of two extended states over a set of ground terms, is based on the fact that by the sequential accessibility principle of the ASM thesis for sequential algorithms [4], the only way in which A can access an element a of the base set of the state $(\mathbf{S}, A)$ is by producing a ground term that evaluates to a in $(\mathbf{S}, A)$.

Definition 1. *Following the standard approach in reflective programming [6], for every extended state $(\mathbf{S}, A)$, we fix a total surjective function $raise_{(\mathbf{S},A)}$: $S_A \rightarrow Ground_A$ which maps (raises) elements from the domain S_A of the finite structure $\mathbf{S}_A$ that encodes A, to (the level of) well formed ground terms in the finite set $Ground_A$ formed by all the ground terms used by the sequential algorithm A to access elements of the extended state $(\mathbf{S}, A)$.*

Let $(\mathbf{S}, A)$ be an extended state, let Σ_S and Σ_A be the vocabularies of $\mathbf{S}$ and of the finite structure $\mathbf{S}_A$ which encodes the sequential algorithm A, respectively, let $val_{(\mathbf{S},A)}(t)$ denote the interpretation in $(\mathbf{S}, A)$ of a ground term t of vocabulary $\Sigma_S \cup \Sigma_A$, and let $val_{\mathbf{S}_A}(t)$ denote the interpretation in $\mathbf{S}_A$ of a ground term t of vocabulary Σ_A. We say that two extended states $(\mathbf{S}_1, A_1)$ and $(\mathbf{S}_2, A_2)$ strongly coincide on a set $W_S \cup W_A$ of ground terms of vocabulary $(\Sigma_{\mathbf{S}_1} \cup \Sigma_{A_1}) \cap (\Sigma_{\mathbf{S}_2} \cup \Sigma_{A_2})$ and $\Sigma_{A_1} \cap \Sigma_{A_2}$, respectively, iff the following holds:

- *For every $t \in W_S$, $val_{(\mathbf{S}_1,A_1)}(t) = val_{(\mathbf{S}_2,A_2)}(t)$.*
- *For every $t \in W_A$ and corresponding $a_1 = val_{\mathbf{S}_{A_1}}(t)$ and $a_2 = val_{\mathbf{S}_{A_2}}(t)$,*
  - *$raise_{(\mathbf{S}_1,A_1)}(a_1) = raise_{(\mathbf{S}_2,A_2)}(a_2)$, and*
  - *$val_{(\mathbf{S}_1,A_1)}(raise_{(\mathbf{S}_1,A_1)}(a_1)) = val_{(\mathbf{S}_2,A_2)}(raise_{(\mathbf{S}_2,A_2)}(a_2))$.*

We can now introduce our third and last postulate. It generalizes the bounded exploration postulate for sequential algorithms in [4] to RSAs. The key difference with the analogous postulate in the sequential ASM thesis, is that we use a stronger notion of coincidence. This is necessary because for each RSA $\mathcal{A}$, we want to have a *finite* bounded exploration witness set $W_{\mathcal{A}}$ which allows us to "extract" from every extended state $(\mathbf{S}_i, A_i)$ of $\mathcal{A}$, a corresponding bounded exploration witness W_{A_i} for A_i (in the sense of the sequential ASM thesis).

Let $(\mathbf{S}, A)$ be the extended state of a RSA $\mathcal{A}$, we use $\Delta(\mathbf{S}, A)$ to denote the unique set of updates produced by the sequential algorithm A in the extended state $(\mathbf{S}, A)$, which by virtue of Postulate 1 coincides with the set of updates produced by the RSA $\mathcal{A}$ in the extended state $(\mathbf{S}, A)$. The formal definition of update and update set produced by a sequential algorithm is exactly the same as in [2,4].

Postulate 3 (Reflective Bounded Exploration Postulate). *For every RSA $\mathcal{A}$, there is a finite set $W_S \cup W_A$ of ground terms (called* reflective bounded exploration witness*) such that $\Delta(\mathbf{S}_1, A_1) = \Delta(\mathbf{S}_2, A_2)$ whenever extended states $(\mathbf{S}_1, A_1)$ and $(\mathbf{S}_2, A_2)$ of $\mathcal{A}$ strongly coincide on $W_S \cup W_A$.*

A *reflective sequential algorithm* (RSA) is an algorithm satisfying the Reflective Sequential Time, Reflective Abstract State and Reflective Bounded Exploration Postulates. In our next section we introduce a model of ASM machine which we conjecture characterizes this class of RSAs.

2   Reflective ASMs

The set of ASM rules of the reflective ASMs, as well as the interpretation of these rules in terms of update sets, coincide with those of the sequential ASMs as defined in [4].

States of reflective ASMs are extended states. Each extended state $(\mathbf{S}, R)$ of a reflective ASM is formed by the union of an arbitrary first-order structure $\mathbf{S}$ and a finite first-order structure $\mathbf{S}_R$ which encodes the sequential ASM rule R as an abstract syntax tree T_R. $\mathbf{S}_R$ is formed by:

- A finite set V of nodes.
- A finite set L of labels which includes a different label for each ASM rule and each function symbol in the vocabulary of $(\mathbf{S}, R)$.
- A nullary function symbol *self* interpreted as the root node of T_R.
- Boolean binary function symbols *child* and *sibling* interpreted by the children and next sibling relationships of T_R, respectively.
- A function symbol *label* interpreted as a total labeling function of the nodes in V with labels from L.
- Nullary function symbols (constants) l_{par}, l_{if}, l_{update}, l_{import} interpreted by the labels in L corresponding to the ASM rules **par**, **if**, **update** and **import**, respectively.
- A different nullary function symbol (constant) l_f for each function symbol f in the vocabulary of $(\mathbf{S}, R)$, interpreted by the label in L corresponding to the function symbol f.

- A different nullary function symbol (constant) $node_{w_v}$ for each node $v \in V \setminus \{self\}$ which is interpreted by v. Here w_v is the word in the language defined by the grammar $P \to n \mid P.n$ where $n \in \mathbb{N}$, such that $w_v = n$ if v is the n-th child of $self$ and $w_v = w_{v'}.n$ if v is the n-th child of the node v'. For instance, if v is the node in T_R corresponding to the second child of the first child of $self$, then the constant $node_{1.2}$ is interpreted by v.

Let $\Delta(\mathbf{S}, R)$ denote the set of updates yielded by a sequential ASM rule R on an extended state $(\mathbf{S}, R)$. Let $(\mathbf{S}, R) + \Delta(\mathbf{S}, R)$ be the extended state obtained by applying the updates in $\Delta(\mathbf{S}, R)$ to $(\mathbf{S}, R)$. A *reflective ASM* $\mathcal{M}$ is formed by:

- A non-empty set $\mathcal{S}_{\mathcal{M}}$ of extended states which is closed under isomorphisms.
- A non-empty subset $\mathcal{I}_{\mathcal{M}} \subseteq \mathcal{S}_{\mathcal{M}}$ of *initial extended states* such that for all $(\mathbf{S}_1, R_1), (\mathbf{S}_2, R_2) \in \mathcal{I}_{\mathcal{M}}$, $R_1 = R_2$.
- A transition function $\tau_{\mathcal{M}}$ over $\mathcal{S}_{\mathcal{M}}$ such that $\tau_{\mathcal{M}}((\mathbf{S}, R)) = (\mathbf{S}, R) + \Delta(\mathbf{S}, R)$ for every $(\mathbf{S}, R) \in \mathcal{S}_{\mathcal{M}}$.

A *run* of a reflective sequential ASM is a finite or infinite sequence of extended states $(\mathbf{S}_0, R_0), (\mathbf{S}_1, R_1), (\mathbf{S}_2, R_2), \ldots$, where $(\mathbf{S}_0, R_0)$ is an initial extended state in $\mathcal{S}_{\mathcal{M}}$ and $(\mathbf{S}_{i+1}, R_{i+1}) = \tau_{\mathcal{M}}((\mathbf{S}_i, R_i))$ holds for every $i \geq 0$.

3   Examples

Let the sequential ASM rule in Fig. 1 be the rule encoded in the initial states of a reflective ASM $\mathcal{M}$. It follows that every run of $\mathcal{M}$ produces an infinite sequence of extended states $(\mathbf{S}_0, R_0), (\mathbf{S}_1, R_1), \ldots$ where each state $(\mathbf{S}_{i+1}, R_{i+1})$ is obtained by updating the sub-tree rooted at $node_{1.2}$ of the syntax tree of R_i so that it encodes the term $g + \underbrace{a + \ldots + a}_{(i+1)-\text{times}}$ instead of $g + \underbrace{a + \ldots + a}_{i-\text{times}}$, and by updating the location $(f, ())$ with the value of the term $g + \underbrace{a + \ldots + a}_{i-\text{times}}$ in $\mathbf{S}_i$.

If for every extended state $(\mathbf{S}, R)$ of $\mathcal{M}$, we fix the function $raise_{(\mathbf{S}, A)} : S_R \to Ground_R$ in Definition 1 to be such that, $raise_{(\mathbf{S}, A)}(a_i) = t_i$ iff the sub-tree of T_R rooted at a_i (T_R been the syntax tree encoded in $\mathbf{S}_R$) corresponds to the well formed ground term t_i in the set $Ground_R$ of ground terms in R, and $raise_{(\mathbf{S}, A)}(a_i) = \texttt{undef}$ otherwise. Then it is clear that the set $W = W_S \cup W_A$ where $W_S = \{\texttt{true}, \texttt{false}, l_+, l_a\}$ and $W_A = \{node_{1.2}\}$, is a reflective bounded exploration witness for $\mathcal{M}$.

As a second example, we consider the relational reflective machine (RRM) defined in [1] as a formal machine that computes only computable queries, as opposed to Turing machines. In RRMs the input (relational) database is stored in the so called relational store and it can be accessed only through first-order logic queries that are dynamic, i.e., are built during the execution of the machine. As part of its proof of completeness, such reflection power is used to find out the size of the domain of the database. To that end, the machine goes on building (and then executing) for each $n \geq 1$ the sentence $\exists x_1 \ldots x_n (\bigwedge_{1 \leq i \neq j \leq n} (x_i \neq x_j))$

par
  $f := g$
  $child(node_1, node_{1.2}) := \mathtt{false}$
  $sibling(node_{1.1}, node_{1.2}) := \mathtt{false}$
  import v_1, v_2 **do par**
    $l(v_1) := l_+$
    $l(v_2) := l_a$
    $child(node_1, v_1) := \mathtt{true}$
    $child(v_1, node_{1.2}) := \mathtt{true}$
    $child(v_1, v_2) := \mathtt{true}$
    $sibling(node_{1.1}, v_1) := \mathtt{true}$
    $sibling(node_{1.2}, v_2) := \mathtt{true}$
  endpar
endpar

Fig. 1. Sequential ASM rule in the initial states of $\mathcal{M}$.

until it becomes false, meaning that the previous value of n is the wanted size. Note that the kind of reflection that the RRM uses is a bit different to the one we propose in this work. We could call it "partial reflection", since the sequence of actions performed in each transition, except for the queries to the relational store, never changes. We could then think in a different definition of the reflective ASM to represent partial reflection, as follows. Essentially, we only add to the sequential ASM a rule **eval** t, which takes a term t as its argument, and interpret its value as the root of the syntax tree of a sequential ASM rule (other than **eval**) which is then executed.

References

1. Abiteboul, S., Papadimitriou, C.H., Vianu, V.: Reflective relational machines. Inform. Comput. **143**(2), 110–136 (1998)
2. Börger, E., Stärk, R.F.: Abstract State Machines: A Method for High-Level System Design and Analysis. Springer, Heidelberg (2003)
3. Van den Bussche, J., van Gucht, D., Vossen, G.: Reflective programming in the relational algebra. J. Comput. Syst. Sci. **52**(3), 537–549 (1996)
4. Gurevich, Y.: Sequential abstract-state-machines capture sequential algorithms. ACM Trans. Comput. Logic **1**(1), 77–111 (2000)
5. Smith, B.C.: Reflection and semantics in LISP. In: Proceedings of the 11th ACMSIGACT-SIGPLAN Symposium on Principles of Programming Languages POPL 1984, pp. 23–35. ACM (1984)
6. Stemple, D., Fegaras, L., Stanton, R., Sheard, T., Philbrow, P., Cooper, R., Atkinson, M., Morrison, R., Kirby, G., Connor, R., Alagic, S.: Type-safe linguistic reflection: a generator technology. In: Atkinson, M., Welland, R. (eds.) Fully Integrated Data Environments. Esprit Basic Research Series, pp. 158–188. Springer, Berlin, Heidelberg (2000)

A Model-Based Transformation Approach to Reuse and Retarget CASM Specifications

Philipp Paulweber[✉] and Uwe Zdun

Faculty of Computer Science, University of Vienna,
Währingerstraße 29, 1090 Vienna, Austria
{philipp.paulweber,uwe.zdun}@univie.ac.at

Abstract. The Abstract State Machine (ASM) theory is a way to specify algorithms, applications and systems in a formal model. Recent ASM languages and tools address either the translation of ASM specifications to a specific target programming language or aim at the execution in a specific environment. In this work-in-progress paper we outline a model-based transformation approach supporting (1) the specification of applications or systems using the Corinthian Abstract State Machine (CASM) modeling language and (2) retargeting those applications to different programming language and hardware target domains. An intermediate model is introduced, which not only captures software-based implementations, but also the generation of hardware-related code in the same model. This approach offers a new formal modeling perspective onto modular, reusable and retargetable software and hardware designs for the development of embedded systems. We provide a short overview of our CASM compiler design as well as the retargetable model-based approach to generate code for different target domains.

1 Introduction

Since 1995 where Gurevich has described the Abstract State Machine (ASM) theory [1], many approaches have been proposed to interpret, execute, translate, verify and validate ASM specifications (summarized by Börger [2]). Generally speaking all available (public) tools either aim to integrate an ASM language into a specific (software) platform system/framework or focus on a domain specific purpose. We want to enlarge the scope of ASM language tools and provide a general purpose modeling system for the Corinthian Abstract State Machine (CASM) modeling language (introduced by Lezuo et al. [3]). Such a system will enable us to specify arbitrary applications/systems in this language and translate them into one or multiple programming language and hardware target domains. To the best of our knowledge, such a generic translation does not yet exist.

Furthermore, not only is the focus of our investigation not limited to translations to several software environments, it also includes the idea to translate CASM specifications to different Hardware Description Language (HDL) contexts. This will enable us to even describe electronic circuit designs with CASM specifications and will result in a broad range of applications from specifying

© Springer International Publishing Switzerland 2016
M. Butler et al. (Eds.): ABZ 2016, LNCS 9675, pp. 250–255, 2016.
DOI: 10.1007/978-3-319-33600-8_17

small embedded applications up to Reduced Instruction Set Computing (RISC) microprocessors or even complete System-on-Chip (SoC) designs in a formal way.

1.1   Modeling Language and Compiler

The CASM modeling language was designed and used by Lezuo et al. [3] to describe the semantics of machine languages. Moreover, they performed compiler correctness proofs through the usage of the ASM machine models and compiled specifications written in this language into efficient C/C++ applications [4]. Unlike other ASM specification languages such as AsmL [5] or CoreASM [6], CASM currently consists of a small grammar and a static, strong type system, and it only supports a subset of rules from the CoreASM modeling language. The static, strong type system allows to optimize such specifications. Initially, the syntax of CASM followed CoreASM, but over time it diverged significantly (differences to other ASM modeling languages are described by Lezuo et al. [3]).

Due to the (currently) small grammar, the optimization potential and simplicity, the CASM modeling language is a good fit for our effort to retarget ASM specification. Before we go into details, let us review the design of the compiler infrastructure proposed by Lezuo et al. [4]. Figure 1 depicts the translation process.

The parsed CASM specifications are transformed into an Abstract Syntax Tree (AST), and after that type checks and type annotations are performed. Several static optimizations are performed to eliminate run-time overheads. All transformations which need run-time specific calculations and knowledge are redundantly implemented in the AST-based optimizations. The compiler directly emits C/C++ code in the next step, which then gets compiled and linked against the C/C++ run-time library. Important to mention here is that the generated code and the run-time are not synchronized in their implementation state.

1.2   Motivation and Goal

The design in Fig. 1 is not a retargetable infrastructure. That is, in this design, the existing code emitter and run-time implementation need to be checked for correctness, and it must be tested that the execution and calculation of the generated C program equals the specified CASM input specification. If we would retarget this design to different software or hardware environments, we would have to check for the code emitter and run-time implementation again for each

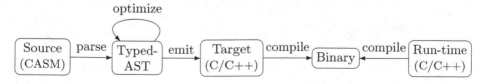

Fig. 1. CASM compiler with C/C++ back-end

new environment that the calculation behavior of the generated target equals the specified CASM input specification.

The emitting stage depicted in Fig. 1 is the main focus of our approach. Our solution to this *retargetable CASM specification problem* is to abstract the run-time and the emitted code in a specific calculation model. This will allow us to check the transformation from the CASM model to this specific computational model once. And for every new target environment (software or hardware) we add to the compiler, only the transformation has to be checked from the specific calculation model to the new target environment. Therefore, we can develop several different code emitter implementations hand-in-hand with *one* run-time implementation and *one* CASM transformation implementation.

This approach enables us to create and generate reusable and retargetable software or hardware artifacts. Those artifacts are self-contained because in our approach we even include the full CASM run-time in the generated artifacts. Hence, the generated artifacts of CASM input specifications can be deployed without further libraries or dependencies. The latter is very important when it comes to hardware-related generated code, because it will not only ease the integration in other hardware designs, but will also allow HDL compilers to fully optimize the generated HDL code on module level.

2   Retargetable Approach and Models

The design of our CASM implementation follows a strict model-based transformation approach to overcome the *retargetable CASM specification problem*. Figure 2 depicts our model-based transformation approach where we introduce two models – the Intermediate Representation (IR) and the Emitting Language (EL) model.

2.1   Intermediate Representation Model

The IR is a full CASM semantics aware model which will be used to analyze and optimize the input specification. An instance of this model is created during the AST to IR transformation (first transformation step in Fig. 2). The IR

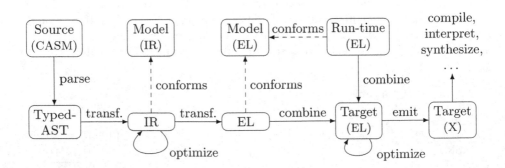

Fig. 2. CASM compiler with model-based transformation

model consists of two important characteristics – parallel/sequential Control Flow Graph (CFG) (introduced by Lezuo et al. [4]) and explicitly modeled operations which are not covered in the AST representation from Lezuo et al. [4] e.g. the location of a ASM state function. The proposed ASM specific *lookup and update elimination* optimizations by Lezuo et al. [4] are planned to be implemented at this level. Software back-ends will profit from those optimizations to be able to execute the specifications much faster (as shown in [4]). Furthermore, we strongly believe the hardware back-ends will benefit from the proposed optimizations too. Because the generated HDL code will result in a less complex digital design by reducing the number of performed calculations just like it applies to the generated software code.

2.2   Emitting Language Model

The EL model is the main contribution in this paper. An instance of this model is created during the IR to EL transformation of the IR instance (depicted in Fig. 2). It allows us to express the CASM run-time and the CASM input specification in a CASM semantics unaware fashion. Thereby we are forced to find generic abstract language constructs for the EL model which allow us to express calculations, procedures and sequential and parallel execution behavior. Figure 3 depicts the class diagram of the EL model.

The EL model is designed to make the mapping to different software/hardware targets easier, but this generic abstraction does not come without limitations. For example the only data type allowed in the EL model is a bit-precise integer value (*Bit-type*) to enable a clean translation to HDL data types. To represent complex or compound data a structure concept is available in the EL model as well to create records of several bit-precise integer values.

The overall model construct is a *Module* which can contain besides *Constants*, *Variables*, *CallableUnits* also explicitly defined *Memory* blocks. The *Memory* blocks are used to properly allocate the appropriate amount of wiring and memory storage in the generated HDL designs. The difference between a *Memory* and *Variable* storage is that *Variables* are translated to HDL designs as plain registers and only permit a single write access. *Memory* blocks permit multiple

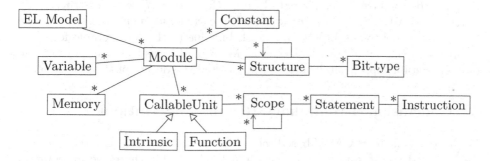

Fig. 3. Emitting language model class diagram

write access. We assume in the EL model that each write access is mutually exclusive. The latter is important, because the model allows the construction of mixed parallel and sequential *Statement* blocks.

CallableUnits are divided into two procedural constructs – *Functions* and *Intrinsics*. Software back-end languages like C, Python etc. use this differentiation to emit efficient target language code, which can be used by the target compiler/interpreter to optimize the execution of the program. Hardware back-end languages can derive a differentiation between behavioral descriptions and computational logic blocks. At this point, another important EL model characteristic is that a *CallableUnit* does not have a "return" value. All incoming and outgoing data of a *CallableUnit* has to be explicitly defined through "in" and "out" parameters. Hence, software back-ends will use this to generate "call-by-reference" constructs and hardware back-ends generate direct component wiring.

All *CallableUnits* can contain mixed parallel and sequential *Scopes* to define a concurrent and sequential calculation hierarchy. Every *Scope* in the EL model can contain several *Statements*. A *Statement* can either be a "trivial", "branch" or "loop" behavioral container. Every *Statement* consists of a list of *Instructions*, which form the leaf nodes in the EL model and perform the actual operations.

Furthermore, due to the flexibility of the EL model and the possibility of unbounded in time of rule evaluations in the sense of CASM, we decided to translate EL instances in the HDL back-ends to asynchronous digital designs. Hence, every *Function, Intrinsic, Statement* etc. from the EL model follows a request-acknowledge handshake protocol. Currently we only focus, besides the software C back-end, for the hardware back-ends on the generation of Very High Speed Integrated Circuit Hardware Description Language (VHDL) code with an assumed annotated timing information. The generated designs are validated in a HDL simulator environment. But in the future the generated code shall be synthesizeable to Field Programmable Gate Array (FPGA) boards as well.

2.3   Compiler Design

From the software design point of view of the compiler, both presented models (IR and EL) follow a Single Static Assignment (SSA) based internal representation. They use a similar class design and analyze/transformation pass design proposed by the Low Level Virtual Machine (LLVM) compiler infrastructure by Lattner and Adve [7]. The latter was used in early experiments to translate the CASM IR model to the LLVM IR model, but due to the retargetable focus for assembly code it turned out that the LLVM IR model was to low-level to realize our retargetable approach. Therefore, we started the design of the EL model.

3   Conclusion, Preliminary Results and Outlook

We have outlined our CASM based retargetable compiler infrastructure and the model-based transformation approach which will enable the reuse, integration and execution of a single CASM specification in different software and hardware environments through the usage of the EL model.

The current development status of the compiler and the models are in an early state. Major compiler infrastructure and transformation passes are implemented to parse, dump and transform CASM input specifications. We were able to retarget a small CASM filter application to a valid C program and VHDL digital design (not synthesizeable yet). The example application consists of three functions, one rule and two parallel update terms.

The overall goal we want to achieve in our future work is to create at least for four language domains a translation back-end implementation. CASM specifications shall be translated to C11 (native), Python (script), JavaScript (web) and VHDL (hardware). A possible field of application would then be the construction of a new RISC microprocessor design in CASM. The proposed retargetable approach of our modeling system generates then directly an Instruction Set Simulator (ISS) for software debugging, an ISS for integration in a website (e.g. for presentation and testing purposes), and a valid synthesizeable hardware implementation.

References

1. Gurevich, Y.: Evolving Algebras 1993: Lipari Guide Specification and Validation Methods, pp. 9–36. Oxford University Press, Inc., New York (1995)
2. Börger, E.: The abstract state machines method for high-level system design and analysis. In: Boca, P., Bowen, J.P., Siddiqi, J. (eds.) Formal Methods: State of the Art and New Directions, pp. 79–116. Springer, London (2010)
3. Lezuo, R., Barany, G., Krall, A.: CASM: implementing an abstract state machine based programming language. In: Software Engineering (Workshops), pp. 75–90 (2013)
4. Lezuo, R., Paulweber, P., Krall, A.: CASM - optimized compilation of abstract state machines. In: SIGPLAN/SIGBED Conference on Languages, Compilers and Tools for Embedded Systems (LCTES), pp. 13–22. ACM (2014)
5. Gurevich, Y., Rossman, B., Schulte, W.: Semantic essence of ASML: extended abstract. In: Boer, F.S., Bonsangue, M.M., Graf, S., Roever, W.-P. (eds.) FMCO 2003. LNCS, vol. 3188, pp. 240–259. Springer, Heidelberg (2004)
6. Farahbod, R., Gervasi, V., Glässer, U.: CoreASM: an extensible ASM execution engine. Fundamenta Informaticae **77**(1–2), 71–104 (2007)
7. Lattner, C., Adve, V.: LLVM: a compilation framework for lifelong program analysis and transformation. In: Code Generation and Optimization, pp. 75–86. IEEE (2004)

Modeling a Discrete Wet-Dry Algorithm for Hurricane Storm Surge in Alloy

John Baugh$^{(\boxtimes)}$ and Alper Altuntas

Civil, Construction, and Environmental Engineering,
North Carolina State University, Raleigh, NC, USA
{jwb,aaltunt}@ncsu.edu

Abstract. We describe an Alloy model that helps check the correctness of a discrete wet-dry algorithm used in a system for hurricane storm surge prediction. Derived from simplified physics and encoded with empirical rules, the algorithm operates on a finite element mesh to allow the propagation of overland flows. Our study is motivated by complex interactions between the algorithm and a recent performance enhancement to the system that involves mesh partitioning. We briefly outline our approach and describe safety properties of the extension, as well as directions for future work.

1 Introduction

The tools and techniques most often associated with scientific computing are those of numerical analysis and, for large-scale problems, structured parallelism to improve performance while limiting complexity. Beyond those conventional tools, we also happen to see a role for state-based methods and present one such application here using the Alloy language [4]. Our immediate concern is the correctness of an extension made by our group to ADCIRC [5], a finite element code widely used by the U.S. Army Corps of Engineers and others to simulate storm surge. ADCIRC itself has been extensively validated against actual flooding conditions, with simulation times of about 1 000 CPU hours.

To get a sense of the problem, the mesh in Fig. 1 depicts a shoreline extracted from a larger domain with 620 089 nodes and 1 224 714 triangular elements that encompasses the western North Atlantic Ocean, the Caribbean Sea, and the Gulf of Mexico. Forced with winds and tides, three primary routines drive the physics of the model and are executed in succession at each time step. The first finds the free surface elevation for each node in the domain. Next the wet-dry status of each node is determined via a series of checks involving water surface elevation, velocity, and prior wet-dry states. Finally, velocities at each node are determined by solving the shallow water equations.

Our extension, now included in ADCIRC, is an exact reanalysis technique that enables the assessment of local *subdomain* changes with less computational effort than would be required by a complete resimulation [1]. Figure 2 shows a domain Ω partitioned at interface Γ into a subdomain Ω_I, representing the

M. Butler et al. (Eds.): ABZ 2016, LNCS 9675, pp. 256–261, 2016.
DOI: 10.1007/978-3-319-33600-8_18

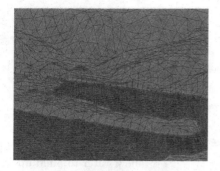

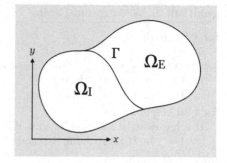

Fig. 1. Finite element mesh   **Fig. 2.** Region of interest Ω_I

interior of a geographic region of interest, and Ω_E. The technique starts with a simulation on Ω that produces elevations, velocities, and wet-dry states that are used as *boundary conditions* along interface Γ in subsequent low-cost simulations on Ω_I. We refer to the first simulation as a *full run* and the latter as a *subdomain run*. A correctness condition requires that boundary conditions be enforced in such a way that results obtained in both cases match within subdomain Ω_I.

A particularly tricky interface condition arises from ADCIRC's discrete wet-dry algorithm, which operates on a finite element mesh to accommodate advancing and receding flood waters [3]. We begin by describing a spatial representation in Alloy that forms the basis for state used by the algorithm.

2   Statics: Representing a Mesh

Finite element methods work by discretizing a continuous domain and approximating a solution with piecewise polynomials. The resulting mesh of elements and nodes can be thought of as a triangulation of a surface. We begin with a representation of mesh topology and later add physical attributes:

```
sig Mesh {                        abstract sig Triangle {
    triangles: some Triangle          edges: Vertex → Vertex,
}                                     adj: set Triangle
abstract sig Vertex {}            }
```

Facts are defined to ensure that every triangle has three directed edges and is oriented, i.e., its edge set forms a ring. Distinct triangles with common anti-parallel edges define the *adj* relation and, correspondingly, the dual of a mesh:

fact { **all** t, t': Triangle | t **in** t'.adj **iff one** ~(t.edges) & t'.edges }

Using *adj*, we ensure that a mesh is connected and oriented; edges are required to be unique. Other facts prevent the possibility of local "cut points" in the mesh as well as overlapping triangles, with the latter following from Euler's formula, $T - E + V = 1$, where T triangles, E (undirected) edges, and V vertices exist. These are specified in terms of helper functions and predicates that define and distinguish between interior and border vertices and edges.

Algorithm 1. Wetting and Drying

0: **for** e **in** *elements* **do**   ▷ start with all elements being wet
    make e wet
1: **for** n **in** *nodes* **do**   ▷ make nodes with low water column height dry
    if W_n **and** $H_n < H_0$ **then**
        $W_n \leftarrow$ false, $W_n^t \leftarrow$ false
2: **for** e **in** *elements* **do**   ▷ propagate wetting
    if e has exactly 2 wet nodes **and** $V_{ss}(e) > V_{min}$ **then**   unless slow flow
        Let j be the remaining dry node $\mid W_j^t \leftarrow$ true
3: **for** e **in** *elements* **do**   ▷ allow water to build up
    find nodes i and j of e with highest water surface   on downhill slopes
    if H_i, $H_j < 1.2H_0$ **then**
        make element e dry
4: **for** n **in** *nodes* **do**   ▷ make landlocked nodes dry
    if W_n^t **and** n on only inactive elements **then**
        $W_n^t \leftarrow false$
5: **for** n **in** *nodes* **do**   ▷ set the final wet-dry state for nodes
    $W_n \leftarrow W_n^t$

---

3   Dynamics: Wetting and Drying

The purpose of the wet-dry routine, characterized in Algorithm 1, is to determine which nodes participate in the calculation of physical properties in the next time step. Its output is W_n for each node n, which is set true when the node is wet. Within the algorithm, both nodes and elements have intermediate wet-dry states, determined by current physical properties, that are set and unset, e.g., W_n^t, which is true when a node is "temporarily" wet. Additionally, as part of the algorithm, an element is said to be *active* if it is wet and has three temporarily wet nodes, and a node is *landlocked* if it is incident only to inactive elements.

The algorithm has both spatial and temporal dimensions, with the former being maintained by vertices and triangles that are now extended to include features of the problem domain:

```
sig Node extends Vertex {          sig Element extends Triangle {
    W, Wt: Bool → State,               wet: Bool → State,
    H: one Height                      slowFlow: one Bool,
}                                      lowNode: one Node
sig State {}                       }
```

In addition to discrete wet-dry states $n.W$ and $n.Wt$ for node n, and $e.wet$ for element e, we incorporate physical attributes and tests on them by making use of predicate abstraction: a water column height $n.H$ may be low ($H_n < H_0$), medium ($H_0 \le H_n < 1.2H_0$), or high ($H_n \ge 1.2H_0$), as used in parts 1 and 3; flow across an element $e.slowFlow$ is true when $V_{ss}(e) \le V_{min}$, as used in part 2; and $e.lowNode$ is an element's node with the *lowest* water surface, supporting the test in part 3 of the algorithm. To accommodate local state changes within a mesh, a *State* atom is added in the last column of the W, Wt, and *wet* relations.

We allow the algorithm to begin with arbitrary $n.W$ states, as though they had been produced in a prior time step, and check correctness at the end. Each part of the algorithm is modeled by a predicate defining the state change:

> **pred** part2 [s, s': State] {
>   noElementChange[s, s']
>   **all** n: Node | n.W.s' = n.W.s
>   **and** (make_wet[n, s] **implies** n.Wt.s' = True **else** n.Wt.s' = n.Wt.s) }

where *noElementChange*[*s*, *s'*] specifies the frame condition, and *make_wet*[*n*, *s*] defines the conditions in part 2 that cause a node to become wet, namely:

> **pred** make_wet [n: Node, s: State] {
>   **some** e: Element | e.slowFlow = False **and** loneDryNode[n, e, s] }
>
> **pred** loneDryNode [n: Node, e: Element, s: State] {
>   n **in** dom[e.edges] **and** n.W.s = False **and** wetNodes[e, s] = 2 }
>
> **fun** wetNodes [e: Element, s: State]: Int {
>   #(dom[e.edges] <: W).s.True }

Other parts of the algorithm are similarly defined.

4   Full and Subdomain Runs

With the above, we are able to represent a mesh and the dynamic behavior of ADCIRC's wet-dry algorithm by chaining together its parts and thereby constraining intermediate states to form a trace. What is left is to distinguish between full and subdomain runs (denoted F and S), which we achieve by extending *State* so that a unique trace can be generated for each type of run:

> **sig** F, S **extends** State {}

Then, by making use of predicate abstraction, we can have a single mesh instance do double duty and serve the needs of both. Here is how. We recognize that, within Ω_I, the two types of runs perform identical computations, though on their own state variables. Where the two cases differ is along interface Γ.

For a full run, we represent only Ω_I and use nondeterminism along Γ to model arbitrary behavior external to it, as depicted in Fig. 3 (in white). Boundary nodes are defined to realize that capability:

> **sig** Boundary **extends** Node {
>   allowsWetting, allowsDrying: **one** Bool }

where $n.allowsWetting$ is true when a node n on Γ in a full run is incident to an *imaginary* element e in Ω_E that has exactly two wet nodes and $V_{ss}(e) > V_{min}$, and $n.allowsDrying$ is likewise true when such an element is active.

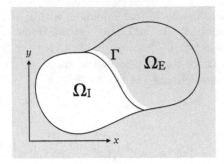

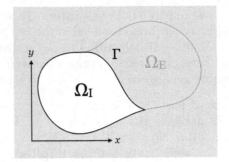

Fig. 3. Mesh for Full Run

Fig. 4. Mesh for Subdomain Run

The conditions defined by the two fields can then be used in parts 2 and 4 of the algorithm, respectively, to account for interactions with Ω_E in a full run. Within part 2, for instance, *make_wet* now becomes:

```
pred make_wet [n: Node, s: State] {
    (some e: Element | e.slowFlow = False and loneDryNode[n, e, s])
    or (s in F and n in Boundary and n.allowsWetting = True) }
```

Intuitively, the updated predicate makes clear that subdomains require state from a prior full run on Γ if they are to produce final wet states that match their full domain counterparts.

Naturally, for a subdomain run, we represent only the portion of the domain over which an ADCIRC simulation is performed, i.e., the geographic region of interest, Ω_I, as shown in Fig. 4 (in white). Absent any forcing along the interface, results internal to it clearly diverge from those produced by a full run. The following assertion confirms this by producing a counterexample:

```
assert sameFinalStates {
    all n: Node | n.W.FD/last = n.W.SD/last }
```

where *FD* is an ordering on F, and *SD* is an ordering on S, so $n.W.FD/last$ denotes the final wet state of a node n in a full run, for instance.

4.1   Enforcing Boundary Conditions

For actual simulations in ADCIRC, we can store the intermediate states of boundary nodes produced during a full run, and then use them as boundary conditions in a subdomain run. Doing so makes subsequent low-cost simulations possible, since all computations external to a geographic region of interest, Ω_I, are avoided. In practice, that cost is only a fraction of a percent of the time required for full runs [1].

The value of state-based modeling in Alloy is in gaining confidence that the boundary conditions are right, since it facilitates experimentation with (a) the amount of state along interface Γ needed from a full run, and (b) the manner in which that state is enforced in subdomain runs. To satisfy the *sameFinalStates* assertion, for instance, we pull wet-dry states out of a full run and apply them to a subdomain run. This modification is made to the last conjunct in the *part2* predicate—call it x—so that it becomes:

(s **in** S **and** n **in** Boundary **implies** n.Wt.s' = n.W.FD/last **else** x)

With a similar change in part 4, we are able to show that enforcing *intermediate* wet-dry states on subdomain boundary nodes with the corresponding *final* wet-dry states obtained from a full run is sufficient to satisfy safety properties. Thus, for actual simulations in ADCIRC, we can record a minimal amount of data from a full run—the final wet-dry states on Γ—and during a subdomain run force those states in parts 2 and 4 of the wet-dry algorithm.

5   Conclusion and Future Work

As far as we are aware, ours is the first study to model in Alloy some of the discrete computational aspects of a finite element solver for systems of partial differential equations. Among several related studies, however, are one that uses Larch and CCS on an illustrative numerical algorithm [2] and another that combines symbolic execution and model checking for numerical subroutines [6].

Our look at model checking began with a question ADCIRC developers raised but were unable to answer without resorting to experiments: how is it that an element with three wet nodes can apparently be dry? We quickly put together Promela/SPIN and FSP/LTSA models that produced traces from a limited set of constructively-defined topologies. A strength of Alloy, of course, is model *generation,* allowing mesh topologies to be defined by declarative properties instead of trying to devise an algorithm to produce them. Given the prevalence of network-like structures of various types in science and engineering, we imagine that Alloy might find further uses there. Our own efforts are focused on a reimplementation of ADCIRC that incorporates adaptivity both for reanalysis and mesh refinement. We expect to make use of Alloy's support for experimenting with abstractions, building object models, and finding representation invariants.

References

1. Baugh, J., et al.: An exact reanalysis technique for storm surge and tides in a geographic region of interest. Coast. Eng. **97**, 60–77 (2015)
2. Chadha, H., Baugh, J., Wing, J.: Formal specification of concurrent systems. Adv. Eng. Softw. **30**, 211–224 (1999)
3. Dietrich, J.C., Kolar, R.L., Luettich, R.A.: Assessment of ADCIRC's wetting and drying algorithm. Dev. Water Sci. **55**, 1767–1778 (2004)
4. Jackson, D.: Software Abstractions: Logic, Language, and Analysis (revised edition). MIT Press, Cambridge (2012)
5. Luettich, R.A., Westerink, J.J.: Formulation and Numerical Implementation of the 2D/3D ADCIRC Finite Element Model Version 44.xx. http://www.adcirc.org
6. Siegel, S., et al.: Combining symbolic execution with model checking to verify parallel numerical programs. ACM TOSEM **17**(2), 1–34 (2008). Article no. 10

'The Tinker' for Rodin

Yibo Liang, Yuhui Lin$^{(\boxtimes)}$, and Gudmund Grov

Heriot-Watt University, Edinburgh, UK
{yl9,Y.Lin,G.Grov}@hw.ac.uk

Abstract. *PSGraph* [3] is a graphical proof strategy language, which uses the formalisation of labelled hierarchical graphs to provide support for the development and maintenance of large and complex proof tactics. PSGraph has been implemented as the *Tinker* system, which previously supported the *Isabelle* and *ProofPower* theorem provers [4]. In this paper we present a *Rodin* version of Tinker, which allows Rodin users to encode, analyse and debug their proof strategies in Tinker.

1 PSGraph and Tinker

PSGraph [3] is a graphical proof strategy language, where proof tactics are represented as directed hierarchical graphs, which provides an intuitive representation to understand and work with proof strategies. The nodes in PSGraph contain tactics, provided by the underlying theorem prover, or nested graphs, and are composed by labelled wires. The labels are called *goal types* which are predicates describing expected properties of sub-goals. Each sub-goal is a special *goal node* on the graph, which "lives" on a wire. Evaluation is handled by applying a tactic to a goal node that is on one of its input wires. The resulting sub-goals are sent to the out wires of the tactic node. To add a goal node to a wire, the goal-type must be satisfied. This mechanism is used to ensure that goals are sent to the right place.

In Fig. 1 we show an example adapted from [3]. This is a proof strategy called *rippling*, which has also been used to automate invariant proofs in Event-B [5]. Here, there are two tactics nodes, *induct* and *ripple*, connected by edges with suitable goal types to guide the sub-goals to the correct target. In this example, a goal node, labelled by a is the input of the first tactic node (*induct*). Applying the *induct* tactic to a generates three new sub-goals, where a and b are *base* cases, while d is a *step* case. The step cases will be sent to the *ripple* tactics. Over two iterations, the *ripple* tactic will generate two new sub-goals, f and d, which will be sent to the output of the graph. Major features of PSGraph is: this ability to control the goal flow with goal types; step through evaluation during debugging as Fig. 1 illustrates and abstraction of the order and number of sub-goals (e.g. both b and c were sent to the same edge).

PSGraph is formalised using *string diagrams* [1], which supports dangling edges (i.e. an edge without a source/target): an edge without a source becomes

This work has been supported by EPSRC grants EP/J001058, EP/K503915, EP/M018407 and EP/N014758.

M. Butler et al. (Eds.): ABZ 2016, LNCS 9675, pp. 262–268, 2016.
DOI: 10.1007/978-3-319-33600-8_19

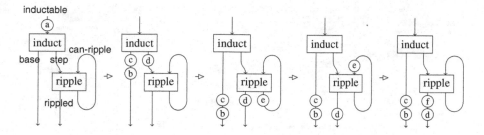

Fig. 1. An example of using PSGraph [3]

the input; and an edge without a target becomes an output. By exploiting the goal types, string diagrams support a novel type-correct composition of tactics[1]. For example, we can only connect the (output) *base* edge of Fig. 1 with a graph that has an input *base* goal type.

PSGraph has been implemented in the *Tinker* tool [4,7], which also provide support for developing, debugging and maintaining proof strategies. We will illustrate this in the tool in Sect. 2.

Previously Tinker only supported *Isabelle* and *ProofPower*. In this paper, we present a *Rodin* version of Tinker, which consists of the Tinker tool and a 'Tinker for Rodin' plugin. This will enable Rodin users to encode their proof strategies in Tinker with the following features: (1) users can draw proof strategies as flow graphs; (2) users can reuse and modify existing proof strategies by drawing; (3) users can step through how sub-goals flow through the graph during a proof, including debugging features such as breakpoints and the ability to modify the proof strategy. In Sect. 2 we will show how Tinker can be used to support developing and debugging proof strategies with a case study. This is followed by a discussion of the architecture and implementation details in Sect. 3. We conclude and briefly discuss further work in Sect. 4.

2   Developing and Debugging Proof Strategies in Rodin

To illustrate usage of the tool, consider proof obligations (POs) `X_act/inv19` (1) and `X_act/inv21` (2) taken from an encoding of a landing gear case study [8].

$$\texttt{partition}(\texttt{EV}, \{prs_ev\}, \{dpr_ev\}, \{opn_ev\}, \{cls_ev\}, \{ext_ev\}, \{rtr_ev\}, \{no_ev\}), \dots$$
$$\vdash N(ext \mapsto which\_ev) \in \{prs\_ev, opn\_ecv\} \land ext = \texttt{FALSE}$$
$$\Rightarrow X(rtr\_ev = \texttt{TRUE}) \tag{1}$$

$$\texttt{partition}(\texttt{EV}, \{prs_ev\}, \{dpr_ev\}, \{opn_ev\}, \{cls_ev\}, \{ext_ev\}, \{rtr_ev\}, \{no_ev\}), \dots$$
$$\vdash N(ext \mapsto which\_ev) \in \{rtr\_ev, ext\_ecv\} \Rightarrow X(opn\_ev = \texttt{TRUE}) \tag{2}$$

---

[1] Composition of two graphs is formalised as a categorical push-out [1].

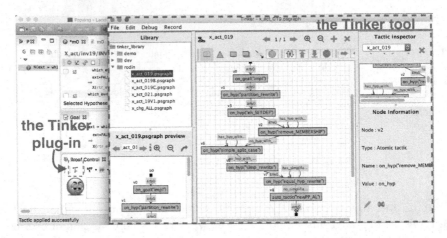

Fig. 2. The Tinker plug-in and Tinker GUI

Both of these require interactive proofs. The proof strategy for `X_act/inv19` is as follows:

- Eliminate the implication ($\Rightarrow$) in the goal;
- Unfold the definitions of partition (`partition`);
- Rewrite hypotheses with the set equality hypothesis unfolded from `partition`;
- Unfold the definitions of membership ($\in$);
- Apply a case split on any disjunctive hypothesis;
- Simplify hypotheses with the default rewrite rules of Rodin;
- Simplify hypotheses by rewriting with equational hypotheses;
- Apply the auto prover on the remaining sub-goals.

To develop this proof strategy, users can open the Tinker tool (as seen in Fig. 2) and draw it in a click and drag style. The PSGraph we have developed is shown in Fig. 3 (left), where each green box is a proof tactic provided by Rodin:

on_goal means that the given rewrite rule(s) should be applied to the goal, e.g. `on_goal(impI)` eliminates the implication ($\Rightarrow$) in the goal.

on_hyp means that the given rewrite rule(s) should be applied to the hypotheses, e.g. `on_hyp(remove_membership)` rewrites any hypotheses of the shape

$$x \in \{S_1, \ldots, S_n\}$$

to

$$x = S_1 \vee \cdots \vee x = S_n$$

auto_tactic to apply the automatic tactic specified by the argument, e.g.

`auto_tactic(newPP_AL)` calls the interface of Rodin's `newPP_ALL`.

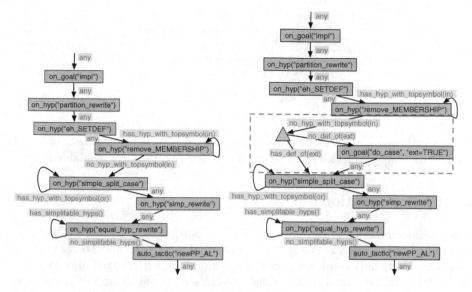

Fig. 3. Proof strategy of X_act/inv19 (left) and X_act/inv21 (right)

A key feature of Tinker is that the edges of the graph are labeled by *goal types*, which are predicates that are used to guide sub-goals to the correct target. This is also used to pinpoint where a proof fails during debugging. Here, we illustrate some of the goal types that we have developed for Rodin proofs:

any is the default goal-type which allows any goal to proceed.
(has/no)_hyp_with_topsymbol succeeds if there is (or is not) a hypothesis in which the top level symbol is the operation specified by the argument.
(has/no)_simplifable_hyps succeeds if there exists (or not exists) any equational hypothesis to rewrite the hypotheses.

The available tactics and goal-types are hard-coded in the current implementation, meaning users cannot define new ones. We discuss this limitation further in Sect. 4.

Tinker allows users to save the existing proof strategies into a library and import those from the library to develop new ones. To illustrate, the proof of X_act/inv21 only differs from X_act/inv19, in that a case split on ext=TRUE is required before applying case split on the disjunctive hypothesis. Figure 3 (right) shows the proof strategy which is developed based on the X_act/inv19 one, with the changes highlighted. Tinker could then have been used to help generalising these strategies into a single more high-level strategy.

To apply a proof strategy to a PO, users can open the PO in Rodin and then click the button T from the *prove control view* in Rodin. After selecting the proof strategy to be applied, Tinker will be launched to allow users to step through proofs with the features, such as interactive controlled inspection, debugging

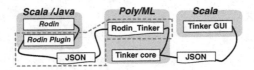

Fig. 4. Architecture

using breakpoints and a logging mechanism [7]. A screenshot is shown in Fig. 2. The demo of applying the proof strategies to the two POs are available in [9].

3   Implementation

The Rodin version of Tinker consists of three parts: the **GUI**, **CORE** and the **Rodin plug-in**. Each is shaded in a separated grey box, and the new parts, which are contributions of this paper, are highlighted with dotted lines in Fig. 4. The GUI is implemented in Scala and provides a graphical way of developing, and debugging proof strategies. The core is implemented in Poly/ML, and holds the key functionalities such as goal-type checking and evaluation. The core is theorem prover independent, as most functionality is implemented using ML `functors`. Each theorem prover, i.e. Isabelle, ProofPower and Rodin, has a prover configuration `structure` that implements a provided `signature`, as indicated by `Rodin_Tinker` for Rodin in Fig. 4.

Isabelle and ProofPower are both encoded in Poly/ML which made it easy to integrate with the core. For Rodin, we had to develop a JSON-based communication protocol between the actual Rodin plug-in and the Tinker core. The Rodin plug-in therefore acts as the communication agent between the *Rodin Proof Obligation Manager (POM)* and the core. It is responsible for calling the correct tactic in the POM and send the updated proof status back to the core. To illustrate, the tactic application

```
on_goal(impI)
```

is translated to a JSON message containing the command

```
APPLY_TACTIC
```

and the parameters `ON_GOAL` and `impI`. This is sent to the Rodin plug-in, which will interpret the message to call the following interface in Rodin to eliminate the implication in the goal:

```
Tactics.impI().apply(pnode, pm)
```

where `pnode` is the proof tree node at which this tactic should be applied and `pm` is the proof monitor in Rodin.

4   Conclusion and Future Work

We have presented a Rodin version of the Tinker tool. This is the first Tinker supported theorem prover that does not strictly follow the LCF approach [2] and that required an additional communication protocol. The plug-in allows Rodin user to develop and debug tactics as graphs, and the approach has been shown to scale to an industrial setting [6].

We believe that the integration of Rodin and Tinker has great potential. To illustrate, a tactic node in a PSGraph can be used to select relevant hypothesis, which are then used as parameters for subsequent tactic node.s For example, we can apply a tactic

bind_hyp_with_topsymbol (in, ?x)

which selects all the hypotheses satisfying that the top symbol is ∈, and then bind the list of the hypotheses to an environment variable ?x. We can then apply a tactic with the variable, e.g.

on_hyp (remove_MEMBERSHIP, ?x)

which will apply the tactic to eliminate ∈ for the list hypotheses bound in ?x. This feature, which seems very useful for Rodin which can have a large number of hypothesis, is supported in the Isabelle and ProofPower version, and we need to update the Rodin prover configuration **structure** to support it. We also would like to support configurable tactic and goal type features that will allow users to define new tactics and goal types in a easy manner without resorting to the source code. We would also like to investigate richer proof strategies in Rodin. For example, the second author's PhD thesis used rippling (in Isabelle) to automate Rodin POs [5]. We have previously developed a simplified version of rippling in Tinker [4] which we would like to incorporate with the Rodin version.

References

1. Dixon, L., Kissinger, A.: Open graphs and monoidal theories. CoRR, abs/1011.4114 (2010)
2. Gordon, M.J.: Edinburgh LCF: A Mechanised Logic of Computation. LNCS. Springer, Heidelberg (1979)
3. Grov, G., Kissinger, A., Lin, Y.: A graphical language for proof strategies. In: McMillan, K., Middeldorp, A., Voronkov, A. (eds.) LPAR-19 2013. LNCS, vol. 8312, pp. 324–339. Springer, Heidelberg (2013)
4. Grov, G., Kissinger, A., Lin, Y.: Tinker, tailor, solver, proof. In: UITP, ENTCS, vol. 167, pp. 23–34. Open Publishing Association (2014)
5. Lin, Y.: The Use of Rippling to Automate Event-B Invariant Preservation Proofs. Ph.D. thesis (2015)
6. Lin, Y., O'Halloran C. Grov, G.P.G.: A super industrial application of PSGraph. In: ABZ (2016). (to appear)
7. Lin, Y., Le Bras, P., Grov, G.: Developing & debugging proof strategies by tinkering. In: TACAS (2016). to appear

8. Su, W., Abrial, J.-R.: Aircraft landing gear system: approaches with event-b to the modeling of an industrial system. In: Boniol, F., Wiels, V., Ait Ameur, Y., Schewe, K.-D. (eds.) ABZ 2014. CCIS, vol. 433, pp. 19–35. Springer, Heidelberg (2014)
9. Liang, Y., Lin, Y., Grov, G.: Tinker - ABZ 16 paper ressources. http://ggrov.github. io/tinker/abz2016/. Accessed 3 Feb 2016

A Graphical Tool for Event Refinement Structures in Event-B

Dana Dghaym[1]($^{\boxtimes}$), Matheus Garay Trindade[2], Michael Butler[1],
and Asieh Salehi Fathabadi[1]

[1] University of Southampton, Southampton, UK
{dd4g12,mjb,asf08r}@ecs.soton.ac.uk
[2] Federal University of Santa Maria, Santa Maria, Brazil
mtrindade@inf.ufsm.br

Abstract. The Event Refinement Structures (ERS) approach provides a graphical extension of the Event-B formal method to represent event decomposition and control-flow explicitly. In this paper we present an improved version of the ERS plug-in, which provides a graphical environment for the ERS approach within the Event-B tool, Rodin. The improved ERS plug-in is based on the available frameworks that are developed to support Event-B with an EMF framework, language extensions and generic diagram extensions.

1 Introduction

The ERS plug-in provides an automatic generation of part of the Event-B model related to the ordering of events and their relationships at different refinement levels. The ERS plug-in can generate additional variables, events, guards, actions and invariants to an Event-B machine.

The ERS language is defined using the Eclipse Modelling Framework (EMF) [1] meta-model, and then transformed into an Event-B EMF meta-model [2]. In the earlier version of the tool [4], an ERS diagram was defined in an EMF tree structure. The transformation from the ERS language to Event-B was performed using the Epsilon Transformation Language (ETL) [3].

In the updated version of the ERS plug-in, we provide a graphical environment for ERS, and we apply a different approach to transform from the ERS language to Event-B. The new approach followed is based on the generic Diagram Extensions framework for Event-B [5]. The framework used, is built specifically to support Event-B providing lots of helpful functionalities. In addition to supporting model transformation to Event-B, the generic Diagram Extensions framework provides graphical and validation support. Unlike, ETL which is a generic model to model transformation language, and supporting a graphical interface and validation requires other tools adding more learning efforts. Moreover, at the time we used ETL, we had technical problems related to debugging and code auto-completion features, while the current approach simply uses Java.

© Springer International Publishing Switzerland 2016
M. Butler et al. (Eds.): ABZ 2016, LNCS 9675, pp. 269–274, 2016.
DOI: 10.1007/978-3-319-33600-8_20

2   Event Refinement Structures (ERS) Approach

The ERS approach provides a tree-like graphical representation of the events, with an explicit representation of the events ordering and the refinement relationships. We illustrate the ERS approach using a small example of the order workflow, Fig. 1. The root of the tree, *Order_Workflow(o)*, represents the name of the flow-diagram and the parameter *"o"* indicates multiple instances. Multiple instances means different instances of a workflow may be executed in an interleaved manner. If no parameter was provided, then the ERS diagram will represent a single instance of the flow.

The leaves, *Make_Order, Receive_Order* etc., will be transformed into events in the Event-B machine. The ordering of the leaf events is from left-to-right, so *Make_Order* can execute first, followed by *Receive_Order*, etc. To describe the control-flow of the events using Event-B, variables with the same name of the leaf events are generated. We refer to these variables as control variables. The type of the control variables is boolean in the case of single instance modelling, and a set in the case of multiple instances modelling. These control variables are used in the Event-B model to specify the control-flow of the events using invariants and guards.

In ERS, the dashed lines indicate that the events are newly added events, and they are not refining events. Therefore, the abstract level of Event-B, first row, can only have dashed lines. ERS also allows the addition of different combinators between events, represented within an oval shape. In Fig. 1, we used two different combinators, the *and-combinator* and the *xor-combinator*. The *xor-combinator* indicates the exclusive choice between events, in this case either *Accept_Order* or *Reject_Order*, but not both, can execute before enabling the event *Close_Order*.

In the first refinement, second row, *Accept_Order* is the only refined event. The solid line means a direct refinement of an event, indicated using the keyword *"refines"* in Event-B. ERS requires that an event can be only refined by one event, this is referred to as "single solid line rule". The only case in ERS, where an event can be refined by more than one event, is by introducing refinement using the *xor-combinator*, which only allows the execution of one event

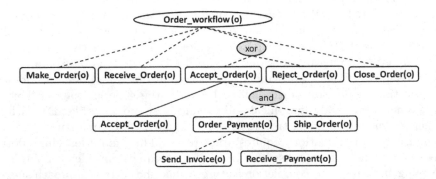

Fig. 1. Overall ERS diagram of the order workflow

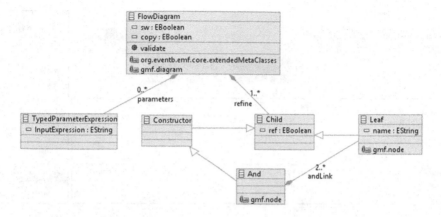

Fig. 2. Part of the ERS meta-model

without requiring mutually exclusive guards. The event *Accept_Order* is decomposed into the sequence of events *Accept_Order* followed by the interleaved execution of the events *Order_Payment* and *Ship_Order*, as a result of applying the *and-combinator*. Similarly at the second level of refinement, *Order_Payment* is decomposed into the sequence of events *Send_Invoice* followed by *Receive_Payment*, where *Receive_Payment* is the directly refining event and *Send_Invoice* is a newly added event.

In Fig. 2, we present part of the ERS meta-model, showing the *and-combinator*. The meta-model describes the different classes of an ERS diagram, such as the *FlowDiagram, Leaf, Child*. Each class can have its own attributes, for example a *Child* has the boolean attribute *ref* to determine whether it is refining or not, this is reflected by the solid and dashed lines in the diagram.

Relationships and associations between classes can be represented by the links between them. A link with solid diamond at the end represents a containment association between the two classes, for example a *FlowDiagram* can contain one or more *Children* as indicated by the upper and lower bounds of the association (1..*), and it can refer to a *Child* using the *refine* relation. A link with a triangular end represents a specialisation relation, for example a *Leaf* and a *Constructor* are special types of *Child* and they inherit all its properties.

3   The ERS Plug-In

Using the ERS plug-in, we first need an Event-B machine, then we can add an ERS flow-diagram to the machine. We can start from an empty machine, but we need to define the *"sees"* relationship to a context, if parameters are needed. Figure 3 presents an image of the tool interface for the abstract level of the ERS diagram in Fig. 1. *Overview* and *Parameters*, panels on the left, are properties of the ERS flow-diagram that can be updated by the user. The palette on the right shows the different combinators available for an ERS diagram such

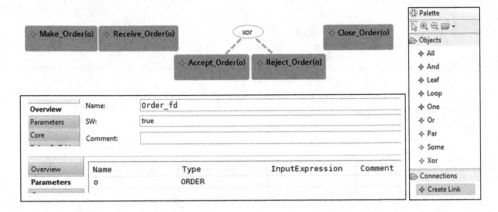

Fig. 3. The ERS diagram of the order workflow at the abstract level

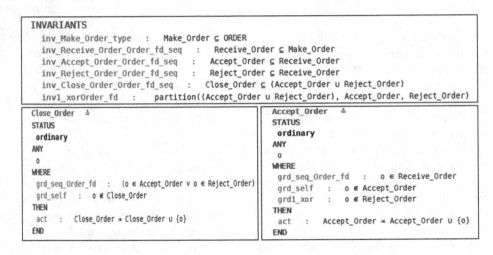

Fig. 4. Part of the generated event-B at the abstract level of the order workflow

as the *xor*, while the leaf, e.g. *Make_Order(o)*, will be translated to an event in Event-B. Figure 4 shows part of the generated Event-B from the ERS diagram of Fig. 3. In the ERS plug-in we also support some validation of the diagram, such as the single solid line rule mentioned above. We only show the invariants and some of the events generated, but the complete generated Event-B also includes control variables to support the control-flow of events, initialisations of the control variables and events for every single leaf. In ERS if the leaf does not already exist in the machine, a new event will be generated with the same name of the leaf. Otherwise, we add the generated guards and actions to the existing events.

When refining a machine, the ERS diagram at the previous level will be copied to the refined machine. Then the user can right click and refine the required leaves. After refining an ERS diagram, all the old generated parts of the Event-B will be deleted and updated according to the new refinement. For example after the first refinement, the sequencing guard of the *Close_Order* event will be updated to $((o \in Order\_Payement \land o \in Ship\_Order) \lor o \in Reject\_Order)$ as a result of applying *and-refinement* to the preceding event.

Generating the Event-B elements from the ERS diagram is based on the generator framework which is part of the generic Diagram Extension framework [5,6]. Each rule transforming an ERS element to an Event-B element implements the *Irule* and defines the methods *enabled, dependenciesOK*, and *fire*. The *enabled* method checks when the translation rule should be applied, e.g. the rule transforming a leaf to a variable is enabled if the ERS source element is a leaf that is not a child of a loop. The *dependenciesOK* method checks if there are any dependencies required like other elements that need to be generated first, e.g. generating a sequencing guard to an event requires the event to be generated first or already existing in the machine, so if dependencies are not satisfied firing the rule will be postponed until all dependencies are satisfied. The *fire* method is the main method where the mapping of ERS elements to Event-B elements takes place. The *fire* method returns a list of *GenerationDescriptors* describing what should be generated, e.g. an ERS *leaf* is mapped to an *event* in Event-B.

Every element generated from the ERS diagram is read-only except for the generated events. They are editable so that users can add application-specific guards and actions using an Event-B editor. The *GenerationDescriptor* has the option whether to mark a generated element as editable or not. In ERS, when refining a leaf, all the elements that are not generated, e.g. the manually added guards and actions, will be passed over to the solid leaf.

4   Conclusions

ERS is a graphical approach that explicitly describe the control-flow of the Event-B events and helps to structure refinement. UML-B [7] is one of the important approaches that provides Event-B with a graphical front-end. UML-B supports class and state-machine diagrams. Similar to ERS, the UML-B state-machine can explicitly describe control-flow in Event-B and supports hierarchical decomposition. The main difference between the two approaches is that UML-B state-machines focus on the state transitions, whereas ERS diagrams focus on the events. depending on the problem one can decide which approach is more appropriate. In many cases both approaches can complement each other, especially after the introduction of the new integrated form of UML-B, iUML-B [8].

The new ERS plug-in provides a graphical environment for the ERS approach, and also supports the validation of the ERS diagrams. Applying the tool in modelling a complex case study resulted in having more consistent and systemic models, faster modelling and the graphical front-end made understanding and validating the model easier.

Using frameworks that directly support Event-B made the development of the plug-in easier and more systematic, by providing different functionalities for defining transformation rules, Event-B generators and graphical editors. In addition to providing various convenience classes and methods for Event-B, like making and finding Event-B elements. In the future, we would like to add more validations to the diagrams, enhance some of the translation rules and allow the users to add application-specific guards, actions and invariants from within the ERS graphical environment without the need to switch to the Event-B editor.

References

1. Steinberg, D., Budinsky, F., Paternostro, M., Merks, E.: EMF: Eclipse Modeling Framework, 2nd edn. Addison-Wesley Professional, Reading (2008)
2. Snook, C., Fritz, F., Illisaov, A.: An EMF framework for Event-B. In: Workshop on Tool Building in Formal Methods, ABZ Conference, Orford (2010)
3. Kolovos, D., Rose, L., Garcia-Dominguez, A., Paige, R.: The epsilon Book (2014). http://www.Eclipse.org, http://www.eclipse.org/epsilon/doc/book/
4. Salehi, F.A., Butler, M., Rezazadeh, A.: Language and tool support for event refinement structures in Event-B. Formal Aspects Comput. 27(3), 499–523 (2015)
5. Savicks, V., Snook, C.: A framework for diagrammatic modelling extensions in rodin. In: Rodin Workshop, Fontainbleau (2012)
6. Wiki.event-b.org: Generic Event-B EMF extensions - Event-B (2016). http://wiki.event-b.org/index.php/Generic_Event-B_EMF_extensions
7. Snook, C., Butler, M.: UML-B and Event-B: an integration of languages and tools. In: The IASTED International Conference on Software Engineering - SE2008 (2008)
8. Snook, C.: iUML-B statemachines: new features and usage examples. In: Proceedings of the 5th Rodin User and Developer Workshop (2014)

Rodin Platform Why3 Plug-In

Alexei Iliasov, Paulius Stankaitis, David Adjepon-Yamoah,
and Alexander Romanovsky[✉]

Newcastle University, Newcastle upon Tyne, UK
alexander.romanovsky@ncl.ac.uk

Abstract. We briefly present the motivation, architecture and usage experience as well as proof statistics for a new Rodin Platform proof back-end based on the Why3 umbrella prover. Why3 offers a simple and versatile notation as a common interface to a large number of automated provers including all the leading SMT-LIB and TPTP compliant tools. The plug-in can function either in a local mode when all the provers are installed locally, or remotely as a cloud service. We discuss the experience of building the tool, the current status and the potential advantages of a cloud-hosted proof infrastructure.

1 Overview

The Rodin Platform offers a fairly capable development and proof support for the Event-B specification language. Some of the automated provers are a part of the Platform and there is a number of add-on provers that significantly improve proof success. Two more important one are the Atelier-B ML prover and the SMT plug-in [3] that offers a bridge to a number of SMT-LIB compliant provers.

In addition to SMT-LIB interface, the majority of the prominent automated provers support the TPTP [6] interface that originated as a common notation for prover competitions.

Recently some important work has been done to bring a large number of TPTP and SMT-LIB provers under the roof of a common, versatile notation - the Why3 verification platform [1]. At the basic level Why3 offers a common interface to over a dozen of automated provers; it also has its own high-level specification notation to reason about software correctness though we do not make any use of it in this work and rather rely on Why3 to offer a bridge to tools like Z3 [4], SPASS [7], Vampire [5] and Alt-Ergo [2].

A theorem prover is a computationally and memory intensive program typically run for rather short periods of time (the vast majority of proofs is done within two seconds) with long idling periods in between. Proof success and perceived usability depend on the capability of an execution platform. Such requirement is best met by the cloud technology.

Doing proofs on a cloud opens possibilities that we believe were previously not explored, outside, perhaps, prover contests. The cloud service keeps a detailed record of each proof attempt along with (possibly obfuscated) proof obligations, supporting lemmas and translation rules. There is a fairly extensive library of

© Springer International Publishing Switzerland 2016
M. Butler et al. (Eds.): ABZ 2016, LNCS 9675, pp. 275–281, 2016.
DOI: 10.1007/978-3-319-33600-8_21

Event-B models constructed over the past 15 years and these are a ready of source of proof obligations. Some of these come from academia and some from industry. We are now starting to put models through our prover plug-in in order to collect some tens of thousands of proof obligations. One immediate point of interest is whether one can train a classification algorithm to make useful prediction of relative prover performance. If such a prediction can yield statistically significant results, prover call order may be optimized to minimize resource utilization while retaining or improving average proof time.

In order to convert Event-B into Why3 verification we had provide a mapping for various Event-B operators, especially its set-theoretic treatment of functions and relations. Unlike say, Isabelle/HOL, Why3 does not rely on a small proof kernel and allows one to make axiomatic definitions. It is a much quicker way to define an embedding of a logic but there is always a danger of making it unsound. In a simpler case, an unsound axiomatisation may be detected by proving of a tautological falsity but there are more intricate situations where unsound definitions show up only in specific circumstances (that is, unsound part is guarded by an implication or requires instantiation of some bound variables). A database of proof attempts makes detecting suspect changes much easier as we can go through historic proofs at any time to see if the outcome changes. We also perform negated proofs on thousands of saved proof efforts.

Table 1. Performance benchmark for the cloud-based proving service.

Model	Total POs	Open, built-in	Open, built-in + SMT	Open, Why3
Multi-core runtime	625	281	62	18
Paxos	348	121	9	4
Fisher's alg	82	14	2	0
Train control system	133	36	5	32

Table 1 shows the comparative performance of the plug-in for some models. Why3 plug-in at the moment is slower than the SMT plug-in but generally more capable though we have one example where its performance is much inferior to the SMT plug-in. The plug-in is open source and is available from the authors on request; we plan to release it with a public cloud service in the coming few months.

2   Translation

The translation of an Event-B proof sequent into a Why3 theory is split into following four activities: a lightweight syntactic translation; construction of a theory from an Event-B sequent and translated formulas; filtering of sequent hypotheses and support lemmas, Why3 axiomatisation of the Event-B mathematical language. Most of the effort goes in the last part so that the programmatic bit of translation is relatively lightweight and generic.

The syntactic translator is written in Tom/Java and simply pretty-prints an AST of an Event-B formula as an S-expression (which is, in essence, the input syntax of Why3) with a static mapping of Event-B operator names. Thus a formula $f \backslash \{t\}$ becomes (diff f (singleton t)). There two non-trivial mappings: the folding of left- or right- associative multi-operators into equivalent binary forms, and the detection of enumerated set definitions (a native, algebraic definition of enumerated sets significantly improves prover performance).

An Event-B proof sequent is mapped into a Why3 theory. The Why3 language has a different treatment of types - type variables are explicit and are separate from the notion of a set - hence every carrier sets is defined twice: as a type variable and as a maximal set. For instance, carrier set CORES and enumerated set STATUS are translated into:

```
type tp_CORES
type tp_STATUS = T_ON | T_OFF

constant id_STATUS : (set tp_STATUS)
constant id_CORES : (set tp_CORES)

axiom hyp1 :(maxset id_STATUS)
axiom hyp2 :(maxset id_CORES)
```

where

```
axiom maxset_axm:
    forall s:set 'a, x:'a .  (maxset s) -> mem x s
```

Free identifiers occurring in a sequent become constants of a Why3 theory; hypotheses are theory axioms and the sequent goal is mapped into a theory goal (i.e., a lemma).

The filtering of hypotheses and support conditions is essential to enable proof within reasonable time. It is discussed in Sect. 3.

Most of the translation effort goes into the construction and fine-tuning of Why3 support theories. We define each Event-B operator in a separate theory and give the bare minimum axiomatic definition that must be checked by hand. For instance, the following is the cardinality operator defined inductively:

```
theory Cardinality
  use import Set
  ..
  use import ElementAddition

  function card (set 'a) : int

  axiom card_def1:
    forall s: set 'a.
```

```
        finite s /\ is_empty s -> (card s) = 0

  axiom card_def2:
    forall x : 'a, s : set 'a.
      finite s /\ not mem x s -> card (add x s) = 1 + card s
end
```

This is all one needs to know but not really enough to carry out proofs. Thus we construct and prove a fairly long list of support conditions. These are deposited in a separate theory (to facilitate filtering). The following gives an example of such support lemmas:

```
theory Cardinality_support
  use import Set
  ...
  use import Union
  ...
  lemma lemma_card_def5:
    forall s, t: set 'a.
      finite t /\ (forall x : 'a. mem x s -> mem x t) ->
        card s <= card t
  ...
  lemma lemma_card_def10:
    forall s : set 'a, t : set 'b, f : rel 'a 'b.
      finite t /\ (mem f s >->> t) ->
        card s = card t
  ...
  lemma lemma_card_def11:
    forall s : set 'a, t : set 'b, f : rel 'a 'b.
      card s = card t /\ (mem f s >-> t) ->
        mem f s -->> t
  ...
end
```

3   Hypotheses and Lemma Filtering

The initial experiments have shown that a minimal axiomatisation support is not sufficient to discharge a sizeable proportion of proof obligations. Provable lemmas were added to assist with specific cases but then it become clear that a large number of support conditions slow down or even preclude a proof. On top of that, the auto tactic language of Rodin offers a very crude hypotheses selection mechanism that for larger models tends to include tens if not hundreds of irrelevant statements. It was thus deemed essential to attempt to filter out unnecessary axiomatisation definitions, Why3 support lemmas and proof obligation hypotheses.

The Rodin mechanism for hypotheses is based on matching conditions with common free identifiers. To complement this mechanism we do filtering on the structure of a formula. It is also a natural choice since support lemmas do not have any free identifiers.

Directly comparing some two formulae is expensive: a straightforward algorithm (tree matching) is quadratic unless memory is not an issue. We use a computationally cheap proxy measure known as the Jaccard similarity which, as the first approximation, is defined as $JS(P,Q) = \mathrm{card}(P \cap Q)/\mathrm{card}(P \cup Q)$.

The key is in computing the number of overall and common elements and, in fact, defining what an "element" means for a formula. One immediate issue is that P and Q are sets and a formula, at a syntactic level, is a tree. One common way to match some two sequences (e.g., bits of text) using the Jaccard similarity is to use *shingles* of elements to attempt to capture some part of the ordering information. A shingle is a tuple preserving order of original elements but seen as an atomic element. Thus sequence $[a, b, c, d]$ could be characterised by two 3-shingles $P = \{[a, b, c], [b, c, d]\}$ (here $[b, c, d]$ is but a structured name) and matching based on these shingles would correctly show that $[a, b, c, d]$ is much closer to $[a, b, c, d, e]$ than to $[d, c, b, a]$. Trees are slightly more challenging. On one hand, a tree may be seen (but not defined uniquely) as a set of paths from a root to leaves and we could just do matching on a set of sequences and aggregate the result. This is not completely satisfactory as tree structure is not accounted for. So we add another characterisation of tree as a set of sequences of the form $[p, c_1, \ldots, c_2]$ where p is a parent element and $c_1, \ldots, c_2$ are children. This immediately gives a set of n-shingles that might need to be converted into shorter m-shingles to make things practical.

As an example, consider the following expression $a * (b + c/d) + e * (f - d * 2)$. We are not interested in identifiers and literals so we remove them to obtain tree $+(*(+(/))(*(-*)))$ which has the following 3-shingles based on paths, $[*, +, /], [+, *, +], [+, *, -], [*, -, *]$, and only 1 3-shingle, $[+, *, *]$, based on the structure. The shingles are quite cheap to compute (linear to formula size) and match (fixed cost if we disregard low weight shingles, see below). Let $\mathrm{sd}(P)$ and $\mathrm{sw}(P)$ be set of depth and structure shingles of formula P. Then the similarity between some P and Q is computed as

$$s(P,Q) = \sum_{i \in I_1} w_{\mathrm{d}}(i) + c \sum_{i \in I_2} w_{\mathrm{w}}(i) \qquad I_1 = \mathrm{sd}(P) \cap \mathrm{sd}(Q), I_2 = \mathrm{sw}(P) \cap \mathrm{sw}(Q)$$

where $w_*(i) = \mathrm{cnt}(i)^{-1}$ and $\mathrm{cnt}(i)$ is number of times i occurs in all hypotheses and support lemmas. Very common shingles contribute little to the similarity assessment and may be disregarded so that there is some k such that $\mathrm{card}(I_1) < k, \mathrm{card}(I_2) < k$.

4   Prover Scenarious

The cloud service accepts as inputs sequents $\mathcal{S}$ of the form: (p, t) where p defines a proof scenario stipulating which provers need to be run and t is a Why3 theory containing a single goal.

The server executes a proof scenario p to obtain a proof result. A proof scenario is a function from an input sequent to a proof result:

$q \in \mathcal{S} \longrightarrow \{\mathsf{unknown}, \mathsf{valid}, \mathsf{invalid}\}$ and is defined via the following proof scenario primitives:

$$
\begin{aligned}
p(t) := \ & \mathsf{pr}(t) && (\text{prover call, positive}) \\
| \ & \mathsf{pr}(\neg t) && (\text{prover call, negative}) \\
| \ & p(t) \rhd w && (\text{deadline}), w \in \mathbb{R}_+ \\
| \ & p_1(t) \wedge p_2(t) && (\textit{and composition}) \\
| \ & p_1(t) \vee p_2(t) && (\textit{or composition}) \\
| \ & \neg p(t) && (\text{result negation})
\end{aligned}
$$

The negation operator $\neg$ on proof results turns valid into invalid, invalid into valid and does not affect unknown and failed.

The *or* composition $(p_1 \vee p_2)(t)$ is opportunistic: it may return any result r such that $r \in \{p_1(t), p_2(t)\} \backslash \{\mathsf{unknown}\}$, and when no such result exists, returns unknown. The *and* composition $(p_1 \wedge p_2)(t)$ evaluates $\max(\{p_1(t), p_2(t)\})$ where unknown $<$ valid $<$ invalid.

The compositions are distributive and commutative so that provers invocations may be scheduled rather flexibly or invoked at the same time. In practical terms, the *or* composition runs until any prover returns a definite results and the *and* composition runs all the provers until it sees invalid result.

The multiplicity of (independently developed) back-end verification tools may be relied on to increase the confidence in a proof result by applying adjudicating on the results of prover calls: $SA(t) \equiv \mathsf{pr}_1(t) \wedge \mathsf{pr}_2(t) \wedge \mathsf{pr}_3(t) \wedge \ldots$.

An important case is proving both positive and negative forms of an input sequent: $\mathsf{pr}_1(t) \wedge \neg \mathsf{pr}_1(\neg t)$. Negation may also be employed opportunistically with the parallel composition: $\mathsf{pr}_1(t) \vee \neg \mathsf{pr}_1(\neg t)$.

Provers may be run with a timeout. A practical example is to run a less capable but often faster prover in parallel with a slower prover: $\mathsf{pr}_1(t) \rhd w_1 \vee \mathsf{pr}_2(t) \rhd (w_1 + w_2), w_1, w_2 \in \mathbb{R}_+$.

An efficient implementations of both sequential and parallel compositions must rely on concurrent invocation of some or all of the composed prover calls.

References

1. Bobot, F., Filliâtre, J.-C., Marché, C., Paskevich, A.: Why3: shepherd your herd of provers. In: Boogie 2011: First International Workshop on Intermediate Verification Languages, pp. 53–64, August 2011
2. Conchon, S., Contejean, É., Kanig, J., Lescuyer, S.: CC(X): semantical combination of congruence closure with solvable theories. In: Post-proceedings of the 5th International Workshop on Satisfiability Modulo Theories (SMT 2007), vol. 198, issue 2 of Electronic Notes in Computer Science, pp. 51–69. Elsevier Science Publishers (2008)
3. Guyot, Y., Voisin, L., Deharbe, D., Fontaine, P.: Integrating SMT solvers in Rodin. Sci. Comput. Program. **94**(Part 2), 130–143 (2014)
4. Bjørner, N.S., de Moura, L.: Z3: an efficient SMT solver. In: Ramakrishnan, C.R., Rehof, J. (eds.) TACAS 2008. LNCS, vol. 4963, pp. 337–340. Springer, Heidelberg (2008)

5. Kovács, L., Voronkov, A.: First-order theorem proving and VAMPIRE. In: Sharygina, N., Veith, H. (eds.) CAV 2013. LNCS, vol. 8044, pp. 1–35. Springer, Heidelberg (2013)
6. TPTP: Thousands of Problems for Theorem Provers. www.tptp.org/
7. Weidenbach, C., Dimova, D., Fietzke, A., Kumar, R., Suda, M., Wischnewski, P.: SPASS version 3.5. In: Schmidt, R.A. (ed.) CADE-22. LNCS, vol. 5663, pp. 140–145. Springer, Heidelberg (2009)

Semi-Automated Design Space Exploration for Formal Modelling

Gudmund Grov[1][(✉)], Andrew Ireland[1], Maria Teresa Llano[2],
Peter Kovacs[1], Simon Colton[2], and Jeremy Gow[2]

[1] School of Mathematical and Computer Sciences,
Heriot-Watt University, Edinburgh, Scotland, UK
{G.Grov,A.Ireland,PK157}@hw.ac.uk

[2] Goldsmiths College, University of London, London, UK
{m.llano,s.colton,j.gow}@gold.ac.uk

Abstract. Refinement based formal methods allow the modelling of systems through incremental steps via abstraction. Discovering the right levels of abstraction, formulating correct and meaningful invariants, and analysing faulty models are some of the challenges faced when using this technique. We propose *Design Space Exploration* that aims to assist a designer by automatically providing high-level modelling guidance.

Keywords: Design · Abstraction · Event-B · Theory formation

1 Introduction

During the development of software intensive systems, the mathematical rigour of *formal methods* brings unique benefits. Specifically, the precision of a formal notation enables design decisions to be clearly communicated and formally verified. However, the use of a formal notation alone is not sufficient to achieve these benefits. Developing design models at the "right" level of abstraction is a creative process, requiring significant skill and experience on the part of the designers. Typically within industrial-scale projects, a design will be modelled at too concrete a level, with the details obscuring the clarity of key design decisions, making it harder to determine if the customer's requirements have been satisfied. In addition, starting with too concrete a design may prematurely "lock" the design team into a particular solution and increase the complexity of the associated formal verification task, i.e. proving properties of the design. Addressing these problems would significantly leverage the creativity of a designer.

We aim at developing a tool that analyses the work of a designer behind the scenes, and automatically suggests design alternatives for Event-B models [1] – alternatives which improve the clarity and correctness of a design. Moreover, a

This work has been supported by EPSRC platform grants EP/J001058/1 and EP/N014758/1, and FP7 WHIM project 611560. We are grateful for feedback on our approach by Jean-Raymond Abrial.

© Springer International Publishing Switzerland 2016
M. Butler et al. (Eds.): ABZ 2016, LNCS 9675, pp. 282–289, 2016.
DOI: 10.1007/978-3-319-33600-8_22

tool that explains for each alternative *what* issue it is addressing and *how* it will effect the design as a whole. The tool will be *semi*-automatic in that while the analysis and synthesis outlined above will be automatic, the designer will remain in full control of the design process. We believe that we can achieve this goal by combining common patterns of modelling with techniques from automated reasoning, in particular automated theory formation. This paper takes the first steps towards such tool. As a working example, consider the requirements given below of a simplified protocol for transferring money between bank accounts:

R1: the sum of money across all accounts should remain constant;
R2: transactions can only be completed if the source account has enough funds;
R3: if an amount m is debited from a source account, the target account should be credited by m;
R4: progress should always be possible (no deadlocks).

A designer might choose to represent the protocol as follows in Event-B:

$$start(a1, a2, m) \overset{def}{=} \textbf{when } a1 \notin active$$
$$\textbf{then } pend := pend \cup \{((a1, a2), m)\} \; \| \; active := active \cup \{a1\}$$
$$debit(a1, a2, m) \overset{def}{=} \textbf{when } ((a1, a2), m) \in pend \wedge bal(a1) \geq m$$
$$\textbf{then } bal(a1) := bal(a1) - m \; \| \; pend := pend \setminus \{((a1, a2), m)\} \; \|$$
$$trans := trans \cup \{((a1, a2), m)\}$$
$$credit(a1, a2, m) \overset{def}{=} \textbf{when } ((a1, a2), m) \in trans$$
$$\textbf{then } bal(a2) := bal(a2) + m \; \| \; trans := trans \setminus \{((a1, a2), m)\} \; \|$$
$$active := active \setminus \{a1\}$$

The chosen approach involves three steps, each of which is represented through an event that is parametrised by the names of the source ($a1$) and target ($a2$) accounts, along with the value of money (m) associated with the transfer. Step one (event *start*) initiates a transfer by adding the transaction to a *pending* set (*pend*), and uses a set (*active*) to ensure that an account can only be the source of one transfer at a time. Note that $\|$ denotes parallel execution. The second step (event *debit*) removes the funds from the source account if sufficient funds exist – *bal* denotes a function that maps an account to its balance. If successful, the transaction is removed from the *pending* set and is added to the *transfer* set. The final step (event *credit*) completes the transaction by adding the funds to the target account, as well as updating the *trans* and *active* sets accordingly. Finally, requirement **R1** is formalised as an invariant, I1: $\Sigma_{a \in dom(bal)} bal(a) = C$ where C is a constant that represents the sum of money across all accounts.

This design abstraction only represents a starting point for the modelling process. A designer will next refine their design ideas through a series of progressively more concrete design abstractions. This gives leverage over the inherent complexity of the design process, enabling the designer to incrementally achieve a customer's requirements. Crucially each refinement step must be formally proved correct. This process is called *correctness-by-construction*. A longer version of this paper is available on ArXiV [5].

2   Towards Design Space Exploration

Key to the style of modelling outlined above is *abstraction* – the ability to create a design at the right level of detail; and to "glue" it to any abstract model through a set of gluing invariants. Trial-and-error is very much part of the expert methodology, where low-level proof failures are examined, and design alternatives in terms of abstractions are experimented with manually (see [2]). Within Design Space Exploration, our goal is to automate much of the low-level grind associated with the trial-and-error nature of formal modelling, and provide a designer with *high-level* modelling advice in real-time.

In particular, we aim to generate alternative models at a higher level of **abstraction** than the original model to deal with a flaw. The intuition is that the flaw is a result of being too concrete. Moreover, within a correct abstraction, the designer has the additional burden of correctly defining the system behaviour and supplying numerous auxiliary invariants that are required for the formal verification process. To support this, we will suggest **adaptations** of the initial model at the same level of abstraction. This could be for instance in terms of additional invariants, or even changes to the behaviour of the system. As can be seen in the next section, unconstrained generation of new models will result in an enormous search space which will be infeasible in practice. Instead, the approach we are proposing has two phases, **analysis** and **generation**, which will iterate until a satisfactory solution is found, possibly including user input.

Analysis Phase. Automated Theory Formation (ATF) is a technique that invents concepts to describe and categorise examples from the input domain, makes conjectures which relate the concepts, and seeks proofs and counterexamples to determine the truth of the conjectures. The HR ATF system [3] will be used in the **analysis** phase to explore given Event-B models and highlight problematic areas. A major challenge will be to find heuristic techniques that effectively prune the design space so that a designer is presented with a useful set of modelling alternatives. This analysis will aim to pin-point both *where* and *what* the problem may be in order to guide the generation phase, and to identify the most interesting solutions. Our approach will be a significant evolution of our previous work on using HR for Event-B [4,9], where we will explore unrestricted theories and include event information in order to explore hypotheses related to the events. We explore simulation traces derived from simulating models, to identify conjectures that are associated with failed steps from the simulation trace. This strategy has proven successful as evidenced in [4], and is extended here by including event information. This will indicate that a variable or an event are associated with failures in the model and therefore should be the focus for the generation phase, as will be illustrated in Sect. 3. This section also illustrates how HR can be used to exploit erroneous user given invariants in order to suggest adaptations of them. We will also search for invariants that are required in order to prove the consistency between the abstract and concrete models; i.e. gluing invariants, which we have already explored in [9]. Finally, we will exploit HR's support for the generation of near conjectures, i.e. conjectures that are true for a percentage threshold of the examples they have. Building upon this functionality, we will explore how this can be tailored to the needs of formal modelling. That is, although formal methods are typically based on definite answers, e.g. a property is either true or false, we believe that a weaker notion of truth is called for when exploring design alternatives, what we call *near-properties*; i.e. properties that are true for most, but not all, behaviours, e.g. *"event X always violates invariant I, but it is always re-established by event Y"*. Paying attention to such properties can lead to insights and in particular suggest solutions which lie just beyond the fringe of what is currently *true* about a design.

Generation Phase. The results of the analysis phase are then used in the model **generation** phase, where alternative abstractions and adaptations of the model are generated. The system must be able to 'explore' design alternatives also for new and previously unseen scenarios. The component that performs the actual generation of new abstractions and adaptations can therefore not be too prescriptive, as was the case with our reasoned modelling critics [6]. For his (unpublished) honours dissertation, one of the authors (Kovacs) has made the first step towards such a component by implementing a generic framework for model generation as a plug-in to the Rodin tool-set [7]. The key feature of this plug-in is that it has a layered design: at the bottom is a set of low-level but generic 'atomic operators' that make small changes to a model, e.g. 'delete variable' and 'merge events'. These atomic operators can then be combined in order to generate new models, and constrained to reduce the number of possible models generated. It is up to the system to find the right combination of operators and to constrain them in the best possible manner. Thus, a "complete" set of atomic operators would allow the generation of all possible alternative models. This gives flexibility to our proposed approach to Design Space Exploration, enabling us to handle new and unforeseen circumstances. Due to space constraints, the details of this tool has been omitted and we refer the interested reader to [5,7]. In Sect. 3 we give examples of how this framework is used.

Common Patterns of Modelling. As will be illustrated in Sect. 3, common modelling patterns will play a central role in finding the right combination of operators. These will be at a very high-level to enable flexibility in terms of their application and therefore enable us to provide assistance in situations where there are no applicable design patterns. The analysis will be used to suggest suitable patterns and guidance as to how they can be implemented. To support this, we have already identified several *refinement patterns* [4] in previous work; however as we cannot refine away flaws, this will be applied in inverse, essentially turning them into *abstraction patterns*. Some abstraction patterns have also been identified and represented using the operator framework in [7]. The experiments in the next section are utilising two patterns: (1) "undoing" bad behaviour by introducing a special **error** (or exception) **case**; and (2) **abstracting away** the problem when it can be pinpointed between certain events. This amounts to "atomising" sequential events into a single event.

3   Illustrative Examples and Initial Experiments

In terms of realising our vision we have undertaken experiments at the level of analysing design models as well as mechanising generation. We present these experiments next. The selection of operators and the integration of the two phases is currently manual; our ultimate goal is to automate the full development chain.

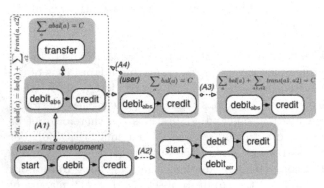

Starting from the initial development, abstraction (A1) and adaptation (A2) are suggested to deal with violation of requirement R4. Given that I1 is a near-invariant, a new invariant is suggested in (A3); or an abstraction (A4) with the required gluing invariant.

Fig. 1. A diagrammatic summary of a small design space exploration.

Consider again the user provided model of a money transfer protocol given in Sect. 1. As it stands, the model is flawed since **R4** is violated when all accounts have started a transaction but none of the source accounts have sufficient funds. Moreover, event *debit* violates invariant I1 since the amount removed from the source account is not accounted for in the invariant, which breaks requirement **R2**. Our aim in such situations will be to offer the designer modelling alternatives that address the flaws. Figure 1 summarises the alternatives generated through our approach, and below we outline how this was achieved. More details can be found in the long version of the paper [5].

Abstraction A1. The first step of the analysis is to generate simulation traces by running the ProB simulator [8], which will also check if the invariants hold. This is input for HR which will use the concept *good* for states in which ProB did not find any invariant violations. HR is then used to search for properties that involve the concept ¬*good*, and this analysis suggest that the generation of bad states are associated to event *debit* and variable *active*.

We can apply the "abstract away" pattern to this violation. One implementation of this pattern is to remove the variable that two (sequential) events use to communicate an intermediate result, and then combine this sequence into an atomic event. A naive application of this pattern in our operator framework will generate 12 alternatives, however by constraining the generation to always include the event *debit* and variable *active*, this is reduced to 2 alternatives (thus pruning the search space by 83%), one of them being the desired abstraction:

$$debit_{abs}(a1, a2, m) \overset{def}{=} \textbf{when } a1 \notin active \land bal(a1) \geq m$$
$$\textbf{then } active := active \cup \{a1\} \ || \ bal(a1) := bal(a1) - m$$
$$|| \ trans := trans \cup \{((a1, a2), m)\}$$

Adaptation A2. An alternative analysis is to apply the error-case pattern. Intuitively, this means introducing a new "error-handling" event that will "undo" some previous state changes when the desired path is not applicable. This can be implemented so that it reverses a previous action in cases when an event of the desired path stays disabled. This require transformations to negate an event's guard, reverse an action of an event and combine the guards of one event with the actions of another. Here a naive implementation will generate 10 alternatives, while if we apply the same constraints as in (A1) then this is reduced to 7, including the generation of the error-handling event:

$$debit_{err}(a1, a2, m) \overset{def}{=} \textbf{when } ((a1, a2), m) \in pend \land bal(a1) < m$$
$$\textbf{then } pend := pend \setminus \{((a1, a2), m)\} \ ||$$
$$active := active \setminus \{a1\}$$

$debit_{err}$ handles the case when the source account does not have enough funds.

Adaptation A3. Let's assume the user selects **A1**. Through analysis of this alternative, invariant (I1) is still violated and HR is re-applied. Through manual inspection of the result of HR, we can see that we are in a "bad state" when *trans* and *active* are not empty, i.e. when there are transactions currently in progress. As a results HR is re-applied to search for conjectures that involve the concepts *trans* and *active* as well as the invariant itself; i.e. $C = \Sigma_{a \in dom(bal)} bal(a)$. HR is then able to generate an *adaptation* of the invariant I1 that addresses the violation by $debit_{abs}$. Note that this adaptation is achieved by including the "internal state" *trans* within the invariant. The Event-B representation of the invariant, which replaces I1, is:

$$I2: \Sigma_{a \in dom(bal)} bal(a) + \Sigma_{(a1, a2) \in dom(trans)} trans(a1, a2) = C$$

Abstraction A4. Although correct, invariant I2 is not a natural representation of **R1**, as compared with near-invariant I1. The designer may wish to explore an alternative abstraction in which I1 is an invariant. Our final alternative **A4** represents such an abstraction. Based on the output given by HR for alternative **A1**, we can re-apply our "abstract away" pattern, albeit with a slightly modified implementation that deletes two variables. Unconstrained, this operator will generate 6 possible alternatives, while a constrained application, which takes into account the analysis, only generates 2 alternatives, one of them being the desired *transfer* event[1]:

$$transfer(a1, a2, m) \overset{def}{=} \textbf{when } abal(a1) \geq m \wedge a1 \neq a2$$
$$\textbf{then } abal(a1) := abal(a1) - m \;\|$$
$$abal(a2) := abal(a2) + m$$

Finally, in order to prove the consistency between the abstract and concrete models, a gluing invariant is required. Therefore, we enter again in an analysis phase where HR is used to form a theory of the refinement step and search for the invariant. HR is able to figure out the relation between the abstract variable *abal* and the concrete representation; i.e. variables *bal* and *trans*. Part of our future work will be focused on tailoring HR for the formal methods context so that invariants such as the gluing invariant required in this refinement step can be formed.

4   Conclusion and Future Work

Focusing on Event-B, we have introduced our approach to *Design Space Exploration* for formal modelling, supported by an initial implementation with partly automated experiments. Currently, the sub-components of our approach are partly automated, while their integration is manual. HR has to be manually guided and we have to manually inspect its output as well as select and combine the relevant operators to perform the generations. Our goal is to fully automate all parts, and provide users with a list of new (and ideally ordered by perceived relevance) modelling alternatives. The approach is *semi*-automatic in that the user will decide on how to use the alternatives. In this paper we have provided the first step towards realizing our goal and have shown the feasibility of the overall approach. However, there is still a long way to go: we have already discussed the desirable features for the analysis phase; in addition, we need to identify a sufficiently small, yet complete, set of atomic operators, constraints and combinators, in order to be able to generate all necessary alternatives in the generation phase. It is crucial that these are controlled to avoid generating duplicates. The phases must then be integrated to be able to automate the selection and combination of operators based upon the analysis. The level of support we aim to provide is very ambitious. If successful, our approach will increase the productivity and accessibility of Event-B, but more importantly, it will provide valuable insights into how formal methods can be deployed more widely.

---

[1] Technically, the Event-B syntax of the action should be: $abal := abal \Leftarrow \{a1 \mapsto abal$ $(a1) - m, a2 \mapsto abal(a2) + m\}$.

References

1. Abrial, J.-R.: Modeling in Event-B - System and Software Engineering. Cambridge University Press, Cambridge (2010)
2. Butler, M., Yadav, D.: An incremental development of the mondex system in Event-B. Formal Aspects Comput. **20**(1), 61–77 (2008)
3. Colton, S.: Automated Theory Formation in Pure Mathematics. Springer, Heidelberg (2002)
4. Grov, G., Ireland, A., Llano, M.T.: Refinement plans for informed formal design. In: Derrick, J., Fitzgerald, J., Gnesi, S., Khurshid, S., Leuschel, M., Reeves, S., Riccobene, E. (eds.) ABZ 2012. LNCS, vol. 7316, pp. 208–222. Springer, Heidelberg (2012)
5. Grov, G., Ireland, A., Llano, M.T., Kovacs, P., Colton, S., Gow, J.: Semi-Automated Design Space Exploration for Formal Modelling. arXiv:1603.00636
6. Ireland, A., Grov, G., Llano, M., Butler, M.: Reasoned modelling critics: turning failed proofs into modelling guidance. SCP **78**(3), 293–309 (2013)
7. Kovacs, P.: Automating abstractions in formal modelling, Heriot-Watt University, Undergraduate Honors Thesis (2015). http://bit.ly/1JnL0Ts
8. Leuschel, M., Butler, M.: ProB: A model checker for B. In: Araki, K., Gnesi, S., Mandrioli, D. (eds.) FME 2003. LNCS, vol. 2805, pp. 855–874. Springer, Heidelberg (2003)
9. Llano, M.T., Ireland, A., Pease, A.: Discovery of invariants through automated theory formation. Formal Aspects Comput. **26**(2), 203–249 (2012)

Handling Continuous Functions in Hybrid Systems Reconfigurations: A Formal Event-B Development

Guillaume Babin[✉], Yamine Aït-Ameur, Neeraj Kumar Singh, and Marc Pantel

IRIT / INPT-ENSEEIHT, Université de Toulouse,
2 rue Charles Camichel, Toulouse, France
guillaume.babin@irit.fr, {yamine,nsingh,marc.pantel}@enseeiht.fr

Abstract. This paper presents a substitution mechanism for systems having a continuous behavior. It shall preserve the safety property stating that the output of both systems remain in a safety envelope. The whole approach is formalized using Event-B, and relies on the Rodin tools and a theory of Reals provided by the Rodin Theory Plug-in to check the internal consistency with respect to safety properties, invariants and events.

1 Introduction

This paper relies on stepwise refinement in Event-B [2] to contribute to the formalization and verification of controllers in Cyber Physical Systems, relying on a generic substitution mechanism. In this work, we show how to apply the defined substitution as a reconfiguration mechanism to handle hybrid systems characterized by continuous functions which can be solutions of differential equations In this case as we model elements from the physical world, system substitutions are not instantaneous. We extensively use our previous work [4] on discrete controllers synthesis from continuous behavior descriptions. The primary use of the models is to assist in the construction, clarification, and validation of the continuous behavior controller requirements to build a digital controller in the presence of system reconfiguration or system substitution. In this development, we use the Rodin Platform [3,9] to manage model development, refinement, proofs checking, verification and validation.

2 The Event-B Method

An Event-B model [2] is defined by *contexts* and *machines*. It encodes a state-transitions system which consists of: *variables* to represent the state; and *events* to represent the transitions (defined by before-after predicates). A model also contains *invariants* to represent its relevant properties. A *variant* may ensure convergence properties when needed. An Event-B machine can *see* contexts that

© Springer International Publishing Switzerland 2016
M. Butler et al. (Eds.): ABZ 2016, LNCS 9675, pp. 290–296, 2016.
DOI: 10.1007/978-3-319-33600-8_23

contain the relevant *sets*, *constants* and *axioms*. Refinement allows to add more behavior properties and system requirements by refining the abstract model to a more concrete one. New variables and new events may be introduced at the refinement level. Once a machine is defined, generated proof obligations need to be proven in order to demonstrate the preservation of the invariants.

Use of Reals in Event-B. In our work, we are interested in formalizing and analyzing system specifications while using reals. Therefore, we rely on the theory for *reals*, written by Abrial and Butler[1]. It provides a dense mathematical $REAL$ datatype with arithmetic operators, axioms and proof rules.

3   Our Previous Work

This section recalls the seminal results we obtained for both system substitution and discretization of continuous functions. The work presented in this section is detailed in [4] and all the Event-B models related to the discretization of continuous functions are available in [1].

Fig. 1. Behavior of studied systems   **Fig. 2.** System substitution

Studied Systems. They are formalized as state-transition systems. The behavior of such systems can be characterized by two states: the initial one (usually called *boot*) and the nominal one (*progress*). According to Fig. 1, after initialization, a system enters into the *booting* state (state 1). Then, after a *start* transition, the system progresses (state 2, known as *running* state). If the system stops, it switches to state 3.

System Substitution [5,6]. One of the key properties studied in system engineering is the capability of a system to react to changes. It may occur in different situations (e.g. failures, change of quality of service, context evolution, maintenance, etc.). System substitution can be defined as replacing a system by another one while preserving the required behavior.

In [5,6], we have developed a formal model (pattern) for system substitution described in the Event-B modeling language. This pattern is depicted in Fig. 2. When a failure occurs, the running system is halted (*fail* transition), it is repaired in state 3 where the state of the substitute system is restored from the one of the

---

[1] http://wiki.event-b.org/index.php/Theory_Plug-in#Standard_Library.

halted system. Finally, the control is restored to the substitute system (transition *repaired*). The correctness of this substitution has been studied in different cases (equivalent, degraded or upgraded ones). The defined substitution mechanism (Fig. 2) preserves the behavior of the original specification and restores correctly the state of the halted system.

Discretization of Continuous Functions [4]. To control a system, in particular for system reconfiguration, it is required to observe (monitor the feedback behavior of the function) and to control (maintain or change system mode) the system. Such observation and control is performed by a software requiring the discretization of continuous functions. When using computers to implement such controllers, time is observed according to specific clocks and frequencies. Therefore, it is mandatory to define a correct discretization of time that preserves the observed continuous behavior introduced previously. This preservation entails the introduction of other requirements on the defined continuous function. Note that, in practice, these requirements are usually satisfied by the physical plant.

In [4], we reported the stepwise formal development in Event-B of a correct discretization of a Lipschitz continuous function f characterizing a hybrid system Sys_f. The development consists in the following steps.

- **Specification: the mode controller.** It models systems whose behaviors are described in the state-transitions system of Fig. 1. The safety requirement ensures that the observed values remain in the safety envelope defined by the interval $[m, M]$.
- **Introduction of continuous behaviors.** This refinement consists in introducing a continuous function f defined on the domain of positive real numbers (for dense time). The observed values of f will belong to the defined safety envelope $[m, M]$.
- **The second refinement: correct discretization of continuous behaviors.** Discretization requires the introduction of a margin allowing the controller to anticipate (predictive control) the next observable behavior before incorrect behavior occurs. Let $z > 0$ be this margin. We use the Lipschitz continuous function property of f to define the amount of time between two consecutive states observed by the discrete controller. We introduced macro time steps of duration δt (discrete control sampling time interval). As a consequence, we define z such that $z \geq \max_{t \in \mathbb{R}^+} |f(t) - f(t + \delta t)|$. In order to make it well-defined, δt must be small enough so that the property $m + z < M - z$ holds. As a consequence, the set $\mathbb{D}$ of observation instants can be defined as, $\mathbb{D} = \{t_i \mid t_i \in \mathbb{R} \wedge i \in \mathbb{N} \wedge t_0 = 0 \wedge t_{i+1} = t_i + \delta t\}$
It can be rewritten as $\mathbb{D} = \{t_i \mid t_i \in \mathbb{R} \wedge i \in \mathbb{N} \wedge t_0 - 0 \wedge t_i = i \times \delta t\}$. As a result of the definitions of z and δt, if $f(n \times \delta t)$ is in $[m + z, M - z]$ then we can safely predict using the Lipschitz condition that the next value of f observed by the controller, $f((n + 1) \times \delta t)$, is in $[m, M]$.

4 Hybrid System Substitution in the Presence of Continuous Behaviors

We consider two continuous functions f and g characterizing the behavior of two hybrid systems Sys_f and Sys_g. We also assume that these two systems maintain their feedback information value in the safety envelope $[m, M]$. As a consequence, these two systems can substitute each other since they fulfill the same safety requirement. The studied scenario consists in substituting Sys_f after a failure by Sys_g (see requirements in Table 1).

Figure 3a and b show the substitution scenario in both continuous and discrete cases. The X axis describes time passing and the vertical dashed lines model state transitions according to the behavior depicted in Fig. 2. Observe that during repair (state 3 of Fig. 2), the function f (associated with Sys_f) decreases while the function g (associated with Sys_g) is booting. The invariant states that the function $f + g$ belongs to the safety envelope $[m, M]$ during the repair. Finally, the system returns to *progress* (state 2) using Sys_g as the running system.

Table 1. Requirements in the abstract specification.

At any time, the feedback information value of the controlled system shall be less or equal to M in any mode.	Req.1
At any time, the feedback information value of the controlled system shall belong to an interval $[m, M]$ in *progress* mode.	Req.2
The system feedback information value can be produced either by f, g or $f + g$ (f and g being associated to Sys_f and Sys_g).	Req.3
The system Sys_f may have feedback information values outside $[m, M]$.	Req.4
At any time, in the *progress* mode, when using Sys_f, if the feedback information value of the controlled system equals to m or to M, Sys_f must is stopped.	Req.5

4.1 Refinement Strategy

The substitution process is defined for replacing Sys_f by Sys_g, both systems being described by the state-transitions system of Fig. 2. Following the approach defined in [10], the adopted refinement strategy consists of an abstract specification and three refinements: *Definition of a Mode controller (M0), Introduction of the safety envelope (M1), Handling continuous behavior and dense time (M2)* and *Discretization of the continuous behavior (M3)*.

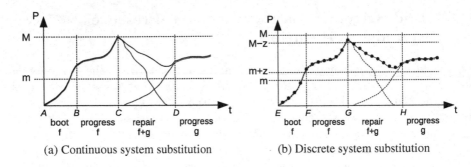

(a) Continuous system substitution   (b) Discrete system substitution

Fig. 3. Examples of the evolution of the function f

4.2   The Event-B Models

We describe the stepwise formal development of the models corresponding to the refinement strategy sketched above. All the Event-B models are available in [1].

– **The required contexts.** Contexts define the relevant concepts needed for our developments. The context C_reals defines a set of theorems for positive reals, continuous and monotonic functions. This context uses the $REAL$ type for real numbers defined in $RealTheory$ (using the Theory plug-in for Rodin). The context C_modes defines the constants $MODE\_X$ representing the different system modes (F, G and R for Sys_f, Sys_g and $Repair$ modes) belonging to the $MODES$ set. The next two contexts ($C\_envelope$ and C_margin) deal with the definition of the safety envelope. They define the intervals of safe values: $[m, M]$ in the continuous case, and $[m + z, M - z]$ with margin z in the discrete case.

– **The root machine: Definition of a Mode controller.** Machine $M0$ describes the reconfiguration state-transitions system depicted in Fig. 2. The modes are used in the events guards that allow the system to switch from one state to another. At initialization, Sys_f is started ($MODE\_F$), it becomes active when the *active* variable is true (Sys_f ended the booting phase). When a failure occurs, progress of Sys_f is stopped. The controller enters in the reparation mode $MODE\_R$. Once the reparation is completed, the mode is switched to $MODE\_G$ and Sys_g enters into a progress state.

– **First refinement: Introduction of the safety envelope.** Machine $M1$ refines $M0$. It preserves the behavior defined in $M0$ and introduces two kinds of events: environment events (event name prefixed with ENV) and controller events (event name prefixed with $CTRL$) [10]. The ENV events produce the system feedback observed by the controller. Three continuous variables f, g and p are introduced. f and g record the feedback information values of Sys_f and Sys_g individually, while p records the feedback value of both systems before, during and after substitution. p corresponds to f of Sys_f in $MODE\_F$, g of Sys_g in $MODE\_G$ and $f + g$ of combined Sys_f and Sys_g in $MODE\_R$ corresponding to the reparation after failure. Once the system has booted, p must belong to the safety envelope in all the cases. The $CTRL$ events correspond to refinements of the abstract events of *M0*. They modify the control variable *active* and *md* (the mode). The *ENV* events observe real values corresponding to the different situations where Sys_f and Sys_g are running or when Sys_f failed and Sys_g is booting. This last situation corresponds to the reparation case.

– **Second refinement: Continuous behavior and dense time.** Machine *M2* (refining *M1*) describes the continuous behavior (Fig. 3a). Once the modes and the observed values are correctly set, the next refining events are straightforward. They correspond to a direct reuse of the development of a correct discretization of a continuous function as done in [4]. Indeed, continuous variables f_c, g_c, p_c and md_c corresponding to the functions f, g, p and md to the modes in *M1* are introduced. A real positive variable *now* is defined to represent the current time. The gluing invariants (for example $p = p_c(now)$) connect the variables of *M1* with the continuous variables at time *now*. In the same way, each event of *M1* is refined. A non-deterministic time step *dt* is introduced and the continuous functions are updated by the *ENV* events on the interval $[now, now + dt]$ while *now* is updated to $now := now + dt$. The control *CTRL* events observe the value $p_c(now)$ to decide whether specific actions on mode md_c variable are performed or not. A detailed description of this refinement is given in [4].

– **Third refinement: Discretization of the continuous behavior.** Following [4], the discrete behavior is described in a Machine *M3* (Fig. 3b). As mentioned in the context *C_margin*, the margin *z* is defined, such as $0 < z \wedge m + z < M - z$ and $M - m > 2 \times z$. This margin defines, at the discrete level, the new safety envelope as $[m + z, M - z] \subset [m, M]$. The new discrete variables f_d, g_d, p_d and md_d of *M3* are glued to f_c, g_c, p_c and md_c of *M2*. They correspond to discrete observations feedback of f_c, g_c, p_c and md_c. The discretization step is defined as δt. Each environment event corresponding to a continuous event is refined by three events. One discrete event starting a time interval δt, one for time intervals included in δt and the next discrete event (third one) at end of δt. The second event may occur several times, a variant enforces it to be Zeno free. The discrete controller only observes the events at time $n \times \delta t$.

5   Conclusion

Modeling hybrid systems using Event-B was studied by several other authors [7,8,10]. In this paper, we have extended our work on system substitution to handle systems characterized by continuous models. First, we modeled system substitution at a continuous level, then we provided a discrete model for substitution which preserves the original continuous behavior. This work reused previous results we obtained on the correct system reconfiguration modeling and the correct discretization of continuous behaviors. By correctness, we mean the preservation of system information feedback in a safety envelope. The whole approach is supported by proof and refinement based on the Event-B method. Refinement proved useful to build a stepwise development which allowed us to gradually handle the requirements. Moreover, the availability of a theory of reals allowed us to introduce continuous behaviors which usually raise from the description of the physics of the controlled plants. All these models have been developed within the Rodin platform [3] and the developed formal models can be downloaded from [1].

References

1. Models. http://babin.perso.enseeiht.fr/r/ABZ_2016_Models/
2. Abrial, J.R.: Modeling in Event-B: System and Software Engineering, 1st edn. Cambridge University Press, New York (2010)
3. Abrial, J.R., Butler, M., Hallerstede, S., Hoang, T.S., Mehta, F., Voisin, L.: Rodin: an open toolset for modelling and reasoning in event-B. International Journal on Software Tools for Technology Transfer **12**(6), 447–466 (2010). http://dx.doi.org/10.1007/s10009-010-0145-y
4. Babin, G., Aït-Ameur, Y., Nakajima, S., Pantel, M.: Refinement and proof based development of systems characterized by continuous functions. In: Li, X., et al. (eds.) SETTA 2015. LNCS, vol. 9409, pp. 55–70. Springer, Heidelberg (2015). doi:10.1007/978-3-319-25942-0_4
5. Babin, G., Ait-Ameur, Y., Pantel, M.: A generic model for system substitution. In: Romanovsky, A., Ishikawa, F. (eds.) Trustworthy Cyber Physical Systems Engineering. CRC Press Taylor & Francis Group (2016)
6. Babin, G., Ait-Ameur, Y., Pantel, M.: Correct instantiation of a system reconfiguration pattern: a proof and refinement-based approach. In: 2016 IEEE High Assurance Systems Engineering Symposium, HASE 2016, Orlando, FL, USA, January 7–9, 2016. IEEE Computer Society Press (2016)
7. Banach, R.: Pliant modalities in hybrid Event-B. In: Liu, Z., Woodcock, J., Zhu, H. (eds.) Theories of Programming and Formal Methods. LNCS, vol. 8051, pp. 37–53. Springer, Heidelberg (2013)
8. Butler, M., Abrial, J.R., Banach, R.: From Action Systems to Distributed Systems: The Refinement Approach, chap. Modelling and Refining Hybrid Systems in Event-B and Rodin, p. 300. Taylor & Francis, February 2016. http://www.taylorandfrancis.com/books/details/9781498701587/
9. Jastram, M.: Rodin User's Handbook (Oct 2013). http://handbook.event-b.org
10. Su, W., Abrial, J.R., Zhu, H.: Formalizing hybrid systems with Event-B and the Rodin platform. Sci. Comput. Program. **92**(2), 164–202 (2014). http://www.sciencedirect.com/science/article/pii/S0167642314002482

UC-B: Use Case Modelling with Event-B

Rajiv Murali[✉], Andrew Ireland, and Gudmund Grov

School of Mathematical and Computer Sciences,
Heriot-Watt University, Edinburgh, UK
{rm339,a.ireland,g.grov}@hw.ac.uk

Abstract. Use cases are a popular but informal technique used to define and analyse system behaviour. We introduce UC-B a plug-in for the Rodin platform (Event-B tool) that supports the authoring and management of use case specifications with both informal and formal components. The formal component is based on Event-B's mathematical language. Once the behaviour of the use case is specified, UC-B automatically generates a corresponding Event-B model. The resulting model is then amenable to the Rodin verification tools that enable system level properties to be verified. By underpinning informal use case modelling with Event-B we are able to provide greater precision and formal assurance during the early stages of design.

Keywords: Event-B · Rodin · UML · Use cases

1 Introduction

UML use cases [1] are a popular but informal method for capturing behavioural requirements for software systems. They often appear in two complementary forms: (1) a *use case diagram* that provides an easy-to-understand illustration of the *subject, actors* and individual *use cases*, and (2) an informal textual *use case specification* that outlines a *contract* and *scenarios* for each use case. Errors introduced during use case modelling may later manifest themselves in the design or implementation phases where the cost of fixing them is significantly more expensive.

As shown in Fig. 1, UC-B[1] is a plug-in for the Rodin platform [2] that supports a more formal approach to use case modelling. The tool allows users to provide formal counterparts to the informal use case specifications, i.e. pre- and post-conditions along with triggers and actions associated with the various flows (scenarios). The generic structure of the use case is then used to automatically generate an Event-B [3] model reflecting the natural levels of abstractions in the use case specification [4]. The generated Event-B model is then amendable to the verification tools provided by the Rodin platform. Furthermore, UC-B also supports the notion of an *accident case* which provides a way of representing potential accident scenarios and enables safety concerns to be explicitly taken

---

[1] Tool information on UC-B: https://sites.google.com/site/rajivmkp/uc-b.

© Springer International Publishing Switzerland 2016
M. Butler et al. (Eds.): ABZ 2016, LNCS 9675, pp. 297–302, 2016.
DOI: 10.1007/978-3-319-33600-8_24

into consideration during use case modelling [4]. An accident case is a sequence of actions that a system and/or other actors can perform that results in an accident or loss to some stakeholder if the sequence is allowed to complete. An important role of an accident case is communication – it allows designers to explain how the system behaviour will deal with hazards identified by safety engineers.

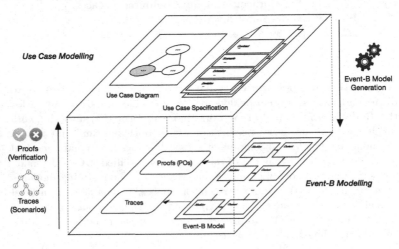

Fig. 1. The Rodin platform with UC-B.

This paper is structured as follows: Sect. 2 describes the process of using UC-B. Section 3 describes the tool architecture and future work, while our conclusion is presented in Sect. 4.

2   Using UC-B

Consider Fig. 2 that presents a use case diagram for a *water tank system*. For simplicity we have included a single regular use case, i.e. MaintainH. UC-B provides its own data model that supports the creation of a use case model (UC-B model) on the Rodin platform. This UC-B model is allowed to contain use cases, actors and a subject that corresponds to those in the use case diagram (see Fig. 3). The actors and subject are

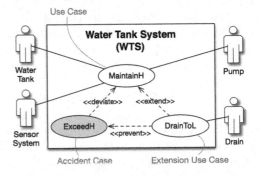

Fig. 2. Use case diagram.

represented as *agents* in UC-B as described in Sect. 2.1. Each use case and agent are allowed to have name, label and informal description. The labels are required to be unique as they are later used to help provide traceability between the use cases and its generated Event-B model.

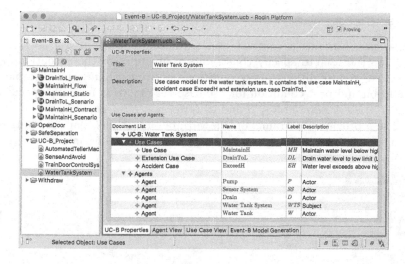

Fig. 3. UC-B model on Rodin: Use cases, actors and subject.

2.1 Agents

Each agent is allowed to specify a distinct list of *variables*, *constants* and *carrier sets* that are associated to the domain of the actor or subject it represents. These provide the basis for the formal aspect of the use case specification.

In Fig. 4, the agent Water Tank introduces the variable *waterlevel* that represents the dynamic water level in the tank. Each variable specifies a predicate and assignment that denotes its type and initialisation respectively. For example, the water level has the type $waterlevel \in L..H$ and initialisation $waterlevel := H$. The constants L, LT, HT and H denote the low limit, low threshold, high threshold and high limit of the water tank. Each constant specifies a predicate that denotes its type, e.g. the type for L is $L = 0$ and LT is $LT > L$. UC-B relies on the Rodin platform to check for well-formedness of predicates and assignments.

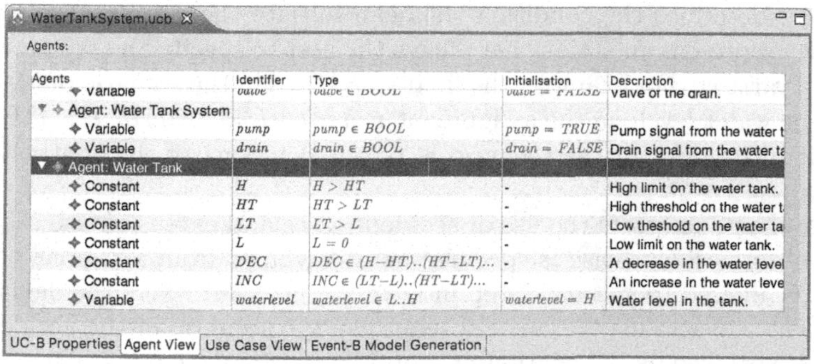

Fig. 4. Agents with constants and variables.

2.2   Use Cases

Each use case contains a contract and scenarios as seen in Fig. 5. The contract allows the user to specify constraints that apply to the execution of the use case, i.e. pre-conditions, post-conditions and invariants. For example, the pre-condition and post-condition for use case MaintainH is formally specified as $waterlevel > HT \land waterlevel \leq H$ and $waterlevel \geq L \land waterlevel \leq HT$ respectively. These provide constraints on the variable $waterlevel$ that are required to be satisfied before and after the execution of the use case MaintainH.

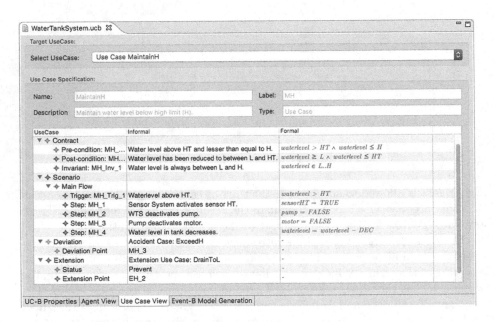

Fig. 5. Use case specification of MaintainH with contract and scenario.

The scenario is represented by a trigger and a sequence of steps (flow). The trigger describes the condition that can initiate the execution of the flow. A step may represent an action that allows the user to specify an assignment that modifies a variable. For example, the final step MH_4 in the scenario of MaintainH reduces the water level via the assignment $waterlevel := waterlevel - DEC$. The behaviour introduced in the scenario is required to satisfy the contract of the use case.

The tool also supports the use of accident cases and extension use cases. The scenario of an accident case is specified as a *deviation* from a regular use case. The user is allowed to specify a step in flow of a use case where the deviation to the accident scenario may occur. For example, the scenario of the accident case ExceedH is specified as a deviation at step MH_2. If the scenario of the accident case is allowed to complete the system results in an accident, i.e. the contract of the use case is expected to be violated.

In use case modelling, an extension use case is often used to describe how a system should respond in exceptional circumstances. The behaviour of an extension use case can be introduced between the steps of a scenario via an *extension-point*. UC-B allows the user to specify a step before which the behaviour of the extension use case can be introduced. In the example, the extension use case DrainToL is introduced before the final step of the accident scenario EH_2. A full description of this example is given in [4].

2.3 Event-B

As noted above, once the use cases have been specified, UC-B supports the automatic generation of a corresponding Event-B model. Figure 6 illustrates the structure of the Event-B model generated for MaintainH. All static aspects in the use case model are introduced in the context MaintainH_Static. This context is *seen* by all machines generated for the use case.

The contract associated with the MaintainH use case is represented in the initial Event-B machine, i.e. MaintainH_Contract. The initial machine is refined by MaintainH_Scenario which represents the main scenario associated with MaintainH. Since the refinement is formally verified, the MaintainH scenario is guaranteed to satisfy the given contract. Any deviations and extension-points in the use case scenario result in accident and extension use cases to be taken into account in the Event-B model respectively. The accident scenario for the accident case ExceedH is introduced in the machine MaintainH_Scenario as a deviation from the ideal scenario. Extension use cases result in a further refinement, e.g. Drain-ToL_Scenario. The refinement style of modelling promoted by Event-B reduces the complexity of proofs and thus increases proof automation.

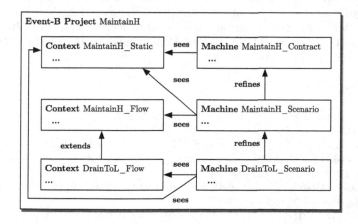

Fig. 6. Structure of Event-B model for MaintainH.

3   Architecture and Further Developments

Figure 7 shows the relationship between UC-B and the Rodin platform. The UC-B meta-model is structured using the Eclipse Modelling Framework [5]. APIs provided by Rodin enable UC-B to detail the content of the UC-B model using Event-B's mathematical language and support the subsequent generation of Event-B models. Further development is being undertaken to integrate UC-B with the Papyrus [6]

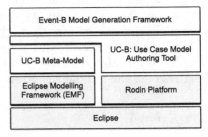

Fig. 7. UC-B architecture.

Eclipse-based tool that supports standard UML2 modelling environment. The integration is aimed at relating use case specification detailed in UC-B with other downstream UML diagrams.

4   Conclusion

Building upon the Rodin platform for Event-B, we have described the UC-B tool which supports the authoring and management of use case specifications. UC-B focuses in particular on the textual aspects of use cases. Specifically users are required to specify a formal counterpart to the informal text that is normally provided. The pay back comes through the automatic generation and verification of a corresponding Event-B model. UC-B also introduces the notion of an accident case which provides a mechanism for explicitly representing and reasoning about potential accidents during use case modelling.

Acknowledgements. The first author was supported by an Industrial CASE studentship which was funded by EPSRC and BAE Systems (EP/J501992), while the second and third authors were partially supported by EPSRC grants EP/J001058 and EP/N014758/1. We also would like to thank Benjamin Gorry, Rod Buchanan and Paul Marsland from BAE Systems.

References

1. Booch, G., Rumbaugh, J., Jacobson, I.: Unified Modeling Language. Addison-Wesley, Boston (1997)
2. Abrial, J.R., Butler, M., Hallerstede, S., Hoang, T.S., Mehta, F., Voisin, L.: Rodin: an open toolset for modelling and reasoning in Event-B. Int. J. Softw. Tools Technol. Transfer **12**(6), 447–466 (2010)
3. Abrial, J.R.: Modeling in Event-B: System and Software Engineering. Cambridge University Press, Cambridge (2010)
4. Murali, R., Ireland, A., Grov, G.: A rigorous approach to combining use case modelling and accident scenarios. In: Havelund, K., Holzmann, G., Joshi, R. (eds.) NFM 2015. LNCS, vol. 9058, pp. 263–278. Springer, Heidelberg (2015)
5. Steinberg, D., Budinsky, F., Merks, E., Paternostro, M.: EMF: Eclipse Modeling Framework. Pearson Education, New Jersey, US (2008)
6. Gérard, S., et al.: Papyrus uml, 8 (2012). http://www.papyrusuml.org

Interactive Model Repair by Synthesis

Joshua Schmidt$^{(\boxtimes)}$, Sebastian Krings, and Michael Leuschel

Institut für Informatik, Universität Düsseldorf,
Universitätsstr. 1, 40225 Düsseldorf, Germany
joshua.schmidt@uni-duesseldorf.de,
{krings,leuschel}@cs.uni-duesseldorf.de

Abstract. When using B or Event-B for formal specifications, model checking is often used to detect errors such as invariant violations, deadlocks or refinement errors. Errors are presented as counter-example states and traces and should help fixing the underlying bugs. We suggest automating parts of this process: Using a synthesis technique, we try to generate more permissive or restrictive guards or invariants. Furthermore, synthesized actions allow to modify the behaviour of the model. All this could be done with constant user feedback, yielding an interactive debugging aid.

1 Introduction and Motivation

Writing a formal model is a complicated and time consuming task. Often, one iterates between changing a specification and proof or model checking. Once an error has been detected, the model has to be adapted to make it disappear.

To some extent, the correction phase can be automated: Using examples of positive or negative behaviour we can synthesize corrected guards, invariants or actions. Such examples can be collected during model checking or directly provided by the user.

For simplicity, we will focus on Event-B below. Our approach has been implemented for Event-B and classical B and could be extended to various other languages supported by PROB. As we do not have a user interface in place, our prototype is not yet available to the general public. We intent to ship it as a standalone tool or bundled with one of the next releases of PROB.

2 Synthesis Technique

Our synthesis approach is based on the one by Jha et. al. [7] and is implemented inside PROB [10,11] using its capabilities as a constraint solver as outlined in [9].

The main idea is the composition of program components, represented as formulas describing input and output. Each component defines a single line of the program written in three-address code: For instance, arithmetic operations can be encoded by components that map two input values i_1, i_2 to an output value o_1. In case of addition, a constraint would ensure that $o_1 = i_1 + i_2$ holds.

M. Butler et al. (Eds.): ABZ 2016, LNCS 9675, pp. 303–307, 2016.
DOI: 10.1007/978-3-319-33600-8_25

By setting up constraints for each I/O example one defines valid connections between program parameters, input and output values and components as well as in between components. In the example above, this could result in connecting o_1 to the input of another component in order to synthesize more involved operations.

Other constraints encode the position or line of components in the code block. Additionally, well-definedness constraints are added to enforce a syntactically correct program. This includes ensuring type compatibility, i. e., we define connections between locations referring to the same type and explicitly add constraints preventing connections between differently typed ones.

Once a candidate program has been found, we search for another semantically different solution. That is, we search for a set of input values where the output of the solutions differs. We ask the user to choose amongst the solutions based on this distinguishing input. We iterate through further solutions in the same fashion. The ongoing search for distinguishing inputs provides us with additional I/O examples. Eventually this leads to a unique solution. Once it is found, we return the program synthesized so far.

The synthesis technique in [7] relies on two oracles. One is used to compute the output of synthesized events while the other is used to assert the correctness of a solution. We implement them as follows:

The *I/O oracle* used to compute output based on given input is replaced by the model checker and the user. For a given input, we use ProB to compute the matching output. Essentially, this amounts to computing a single animation step from the given input state to the output state using the synthesized event. Afterwards, we check the invariant on the output state: If it holds, we assume the state pair is correct and the target state is used as the output. Otherwise, the user has to decide: The event can be disabled on the input state, the invariant can be relaxed or a new output state can be provided.

The *validation oracle* is used to check if a synthesized event is correct. To provide it, we ask ProB to find two states s_1, s_2 such that the synthesized event connects $s_1 \mapsto s_2$ and s_2 violates the invariant. If a solution is found validation fails. If ProB finds a contradiction validation succeeds. Of course, a timeout might occur. In this case we have to rely on other validation options like the provers ml/pp of Atelier B [4] or the SMT solvers [5]. As a last resort, we can again query the user.

3   Interactive Workflow

The process as outlined in Fig. 1 is guided and enforced by ProB. The workflow itself is quite mature and has been fully implemented within the system. Given that our implementation is prototypical we have not considered a user interface by now. Currently, we only provide access via the developer command-line interface of ProB.

Repair is performed successively, i. e., we loop until no error can be found anymore and the user is satisfied with the model. Each step starts with regular

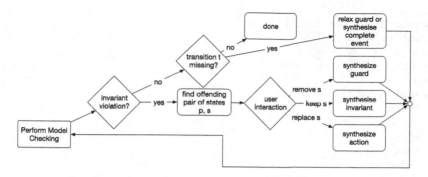

Fig. 1. Interactive workflow

explicit-state model checking as supported by PROB. There are two possible outcomes:

- An invariant violation has been found. The user can decide to make the last transition, leading from a state satisfying the invariant to one violating it, impossible by synthesizing a stronger guard. Alternatively, the system can generate a weaker invariant or a new action in case the user edited the output state of a transition.
- The model has been checked and no error was found.[1] We query the user if a state transition is missing. In case any action is able to reach the missing state, we can synthesize a relaxed guard for the event if necessary. Otherwise, we can synthesize a new action, possibly with already appropriate guards.

4   Running Example

In Fig. 2 you find the Event-B model of a simple vending machine. The machine accepts coins and gives out a can of soda per coin. There are two errors we intend to fix as outlined in Sect. 3:

- The invariant is violated if $soda = 0$, i. e., selling the last can is not allowed.
- get_soda can be executed if $coins = 0$. The guard is too permissive.

First, PROB discovers the violating state $S \triangleq coins = -1 \land soda = 2$. The user now has to decide, whether $coins = 0 \land soda = 3 \xrightarrow{get\_soda} S$ is a valid transition. We select to discard the transition resulting in a negative example. Other positive examples are collected from the state space. After a few seconds, the missing guard $coins > 0$ is synthesized.

We update the machine and model check it again. Now PROB finds $coins = 0 \land soda = 0$ to violate the invariant. This time we decide this state is valid, i. e., we need to change the invariant. The system now tries to find a replacement for $soda > 0$ using positive and negative examples found during model checking. The solution, $soda > -1$ is then synthesized and gets conjoined with $coins >= 0$.

```
machine SimpleVendingMachine
variables soda coins
invariants
  soda > 0 & coins >= 0
events
  event INITIALISATION
    then
       soda, coins := 3, 0
  end
  event get_soda
    where
       soda > 0 & coins >= 0
    then
       soda, coins := soda − 1, coins − 1
  end
  event insert_coin
    then coins := coins + 1
  end
end
```

Fig. 2. Vending machine

```
event SYNTHESIZED_EVENT
  where
     soda > 2 & coins > 1
  then
     soda, coins := soda − 3, −2 + coins
end
```

Fig. 3. Synthesized event

After updating the machine, no further invariant violation is found. To proceed further, we can synthesize a new event. For instance, we could synthesize a "three for the price of two" event by providing the two example transitions

$$coins = 2 \wedge soda = 3 \xrightarrow{?} coins = 0 \wedge soda = 0, \text{ and}$$

$$coins = 4 \wedge soda = 4 \xrightarrow{?} coins = 2 \wedge soda = 1.$$

For these inputs PROB synthesizes the event in Fig. 3. Of course there is no guarantee that proper guards will be synthesized simultaneously, as it highly depends on the given I/O examples. Another iteration might be necessary.

---

[1] Either the check has been exhaustive or a timeout occurred.

5   Discussion and Conclusion

Compared to other approaches like [1,2] we synthesize entirely new predicates or actions based on input and output values instead of transforming an existing specification. Synthesis has been used for repair as well, see for instance [6]. An interactive approach has been suggested in [8].

In CEGAR [3] spurious counter examples are used to refine a model checking abstraction. Our synthesis tool is guided by real counter examples and provides an interactive debugging aid for model repair. Moreover, we not only rely on the model checker to find counter-examples but also make use of PROB as a constraint-solver. This leads to more flexibility in model repair, i. e., we are able to avoid or allow specific states and even extend a machine in case model checking has been exhaustive.

We believe that an interactive modeling assistant like the one we outlined above will have its merits both for teaching as well as for professional use. First tests using our prototypical implementation seem promising.

References

1. Bartocci, E., Grosu, R., Katsaros, P., Ramakrishnan, C.R., Smolka, S.A.: Model repair for probabilistic systems. In: Abdulla, P.A., Leino, K.R.M. (eds.) TACAS 2011. LNCS, vol. 6605, pp. 326–340. Springer, Heidelberg (2011)
2. Chatzieleftheriou, G., Bonakdarpour, B., Smolka, S.A., Katsaros, P.: Abstract model repair. In: Goodloe, A.E., Person, S. (eds.) NFM 2012. LNCS, vol. 7226, pp. 341–355. Springer, Heidelberg (2012)
3. Clarke, E., Grumberg, O., Jha, S., Lu, Y., Veith, H.: Counterexample-guided abstraction refinement. In: Emerson, E.A., Sistla, A.P. (eds.) CAV 2000. LNCS, vol. 1855, pp. 154–169. Springer, Heidelberg (2000)
4. ClearSy. Atelier B, User and Reference Manuals. Aix-en-Provence, France (2014). http://www.atelierb.eu/
5. Déharbe, D., Fontaine, P., Guyot, Y., Voisin, L.: SMT solvers for rodin. In: Derrick, J., Fitzgerald, J., Gnesi, S., Khurshid, S., Leuschel, M., Reeves, S., Riccobene, E. (eds.) ABZ 2012. LNCS, vol. 7316, pp. 194–207. Springer, Heidelberg (2012)
6. Gvero, T., Kuncak, V.: Interactive synthesis using free-form queries. In: Proceedings ICSE, pp. 689–692 (2015)
7. Jha, S., Gulwani, S., Seshia, S.A., Tiwari, A.: Oracle-guided component-based program synthesis. In: Proceedings ICSE, pp. 215–224 (2010)
8. Kneuss, E., Koukoutos, M., Kuncak, V.: Deductive program repair. In: Kroening, D., Păsăreanu, C.S. (eds.) CAV 2015. LNCS, vol. 9207, pp. 217–233. Springer, Heidelberg (2015)
9. Krings, S., Bendisposto, J., Leuschel, M.: From failure to proof: the ProB disprover for B and Event-B. In: Calinescu, R., Rumpe, B. (eds.) SEFM 2015. LNCS, vol. 9276, pp. 199–214. Springer, Heidelberg (2015)
10. Leuschel, M., Butler, M.: ProB: a model checker for B. In: Araki, K., Gnesi, S., Mandrioli, D. (eds.) FME 2003. LNCS, vol. 2805, pp. 855–874. Springer, Heidelberg (2003)
11. Leuschel, M., Butler, M.: ProB: an automated analysis toolset for the B method. Int. J. Softw. Tools Technol. Transf. **10**(2), 185–203 (2008)

SysML2B: Automatic Tool for B Project Graphical Architecture Design Using SysML

David Mentré[(⊠)]

Mitsubishi Electric R&D Centre Europe, Rennes 35708, France
d.mentre@fr.merce.mee.com

Abstract. We present an approach to transform SysML structural diagrams, BDD and IBD with constraints, into a B Method project skeleton. This project can then be directly used for implementation development through usual B refinement mechanism. We prototyped this approach.

1 Introduction

The B Method [1] has been industrially used for more than 18 years now since its first application on Paris subway line 14 [2]. It is mainly used in the railway domain, especially for radio-based train control and signalling systems. Common drawbacks of this formal approach are the difficulty to correctly define the architecture of a B project and the effort needed to understand the very mathematical notation of the B Method [3].

As a first step to palliate this issue, we propose to use a graphical semi-formal language, in our case SysML, to *graphically design the architecture of the B software as a SysML model* and then, using a translation tool, to *automatically transform this model into a B project* loadable into the Atelier B development tool. Our central idea is to use this model as the pivot point between system architects and B formal software experts using a formalism that everybody understands and, *through formal annotations on the model, system assumptions and guarantees regarding the software can be expressed* on the B software side. Once the B project skeleton is automatically produced, the actual behaviour of the B software is developed using the usual B refinement mechanism.

Moreover this approach also targets an *easier integration with the global system architecture* and the *encoding of B Method architecture best practices into a tool*, thus reducing the global effort and cost.

2 Transformation of SysML Language into B Project

Our graphical modelling language is SysML 1.3. We do not handle the whole SysML language but a reduced subset supporting our needs and fully defined in a separate report [4]. We focus on the *architecture* of the system and its contained software. Therefore in SysML we only consider Block Description Diagram (BDD) and Internal Block Diagram (IBD), as well as Package diagram for better structuring (but without translating it). Other SysML diagrams are

© Springer International Publishing Switzerland 2016
M. Butler et al. (Eds.): ABZ 2016, LNCS 9675, pp. 308–311, 2016.
DOI: 10.1007/978-3-319-33600-8_26

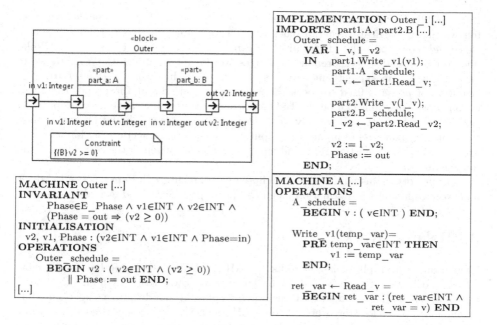

Fig. 1. Transformation of outer and a SysML blocks into B Method

not supported. Each SysML element in a diagram cannot have the name of a B Method keyword. From the BDD and IBD, each block is translated into a B machine, e.g. block Outer in Fig. 1. Input and output ports as well as local variables are translated into variables local to the considered B machine, e.g. v1 and v2 in Outer. Operations are also generated to write (resp. read) to variables corresponding to input (resp. output) ports, e.g. Write_v1 and Read_v operations in machine A.

Each input and output ports can be typed with a Flow Specification, a SysML data structure which can contain other data types. We are supporting SysML Enumeration, DataType (SysML record) and primitive types Boolean and Integer. Each field of a Flow Specification or a DataType is translated into a separate variable, as records should be avoided in B models. Nesting of DataType into another DataType is supported. An Enumeration is translated into a B set. Primitive types Boolean and Integer are translated into B $\mathbb{B}$ and INT. SysML block's constraints can be used on input ports, output ports or local variables to express requirements or guarantees applied on a block. Those constraints are translated into B **INVARIANT**, e.g. the constraint on v2 in Outer. To allow block reuse, essential for scalability (e.g. use of libraries), we are using SysML Instance Specification to instantiate a given block several times, each instance having different block attribute values using SysML slots.

The SysML block hierarchy is translated into a tree of B **IMPORTS**, each machine importing the machine corresponding to the contained SysML blocks, e.g. part1.A and part2.B in Outer_i. As seen in operation Outer_schedule, the connections between blocks are translated into a B operation that uses read and write operations of imported machine (aka contained blocks) to transfer information from one machine to the other (aka one block to the other). The order of called operations in the schedule operation follows the topological sorting of block connections shown in the IBD. For simplicity we consider only pure data-flow links between blocks but adding loops with PRE-like operator as found in synchronous languages would not be difficult. We made some experiments by marking them with a stereotype but this is not very user friendly. A notion of *phase* is used to check the generated invariants at the proper time, e.g. output port constrains can be checked only when output values have been propagated (v2 invariant in Outer). Obviously, connection between blocks should be well-formed (no loop, matching use of input and output ports, correct connections within the hierarchy). Detailed translation rules and more examples are given in the complete on-line documentation[1].

Our prototype was built upon the Eclipse environment, with Papyrus SysML editor and Acceleo model transformation language. It is open-source and available on-line[2].

3 Related Work

Transforming a graphical language like SysML or UML into B Method is not new. Snook and Butler [13] worked on a translation from UML to B Method. While their motivation is similar to ours, they transform a whole UML model into a single B module which is not scalable. Moreover they tag UML elements with many B concepts, while we keep the default SysML profile to avoid the need of any specific B knowledge. Lano et al. [10] worked on the translation of UML-RSDS models into B. Laleau and Polack [9] suggested a transformation process for Information Systems (IS) modeled in UML. Meyer and Souquières [12] originally worked on the translation of Object Modeling Technique (OMT) models. However, all these approaches only worked with B abstract modules and **INCLUDES** links, while we focused and developed modules and **IMPORTS** links as used in industrial projects. Regarding SysML, Laleau et al. [11] focus on the SysML requirement diagram, transforming it into Event-B and thus do not consider the B architecture. Kinoshita et al. [8] have our target but prefer to annotate the SysML diagrams with B concepts. Their tool uses **INCLUDES** which is not scalable and necessitates manual annotations compared to our automatic approach. We previously worked on aligning SysML and B Method concepts [6,7], translating behaviour encoded into SysML state machines. Our current work focuses on the architecture, avoiding the use of any B concepts.

4 Conclusion

This paper presents a transformation of the SysML language into the skeleton of a B project. We voluntarily kept a strict subset of SysML for a wider understanding of the architecture. This first prototype could be improved. To express formal properties, we are using the B notation which is not user friendly. We could use instead a C-like syntax, like OMG's ALF used in our previous experiment [7] or a simplification of ACSL formal language [5]. But it remains to see how to augment this language with abstractions needed for high-level formal specification. We also noticed that Papyrus is

---

[1] www.github.com/openETCS/toolchain/wiki/User-Documentation.

[2] www.github.com/openETCS/toolchain/tree/master/tool/bundles/org.openetcs. sysml2b.transformation.

a very complex tool to use: a simplified GUI is certainly needed for engineer acceptance. As said previously, handling of loops could be improved. Regarding traceabilty, the generated B model could have comments identifying the generated architecture part from further refinements.

Acknowledgement. The author thanks T. Bardot for his work on the prototype, his very helpful remarks and discussions when designing the approach; and the reviewers for their helpful comments.

References

1. Abrial, J.R.: The B-Book: Assigning Programs to Meanings. Cambridge University Press, Cambridge (1996)
2. Abrial, J.R.: Formal methods in industry: achievements, problems, future. In: SE 2006 (2006)
3. Badeau, F., Amelot, A.: Using B as a high level programming language in an industrial project: Roissy VAL. In: Treharne, H., King, S., Henson, M., Schneider, S. (eds.) ZB 2005. LNCS, vol. 3455, pp. 334–354. Springer, Heidelberg (2005)
4. Bardot, T.: SysML modeling rules for SysML to B. Technical report, MitsubishiElectric R&D Centre Europe (2014). www.github.com/openETCS/ toolchain/blob/master/ToolDescription/MERCE-SysML-to-B/2014_12_04-SysMLModelingRules-for-SysMLtoB_MFR14-ARC-839--MFR14-ECD-530.pdf
5. Baudin, P., Filliâtre, J.C., Marché, C., Monate, B., Moy, Y., Prevosto, V.: ACSL: ANSI/ISO C Specification Language
6. Bousse, E.: Requirements management led by formal verification. Master's thesis, INSA Rennes (2012). www.people.irisa.fr/Benoit.Combemale/research/2012/ bousse_erwan_report.pdf
7. Bousse, E., Mentré, D., Combemale, B., Baudry, B., Katsuragi, T.: Aligning SysML with the B Method to Provide V&V for Systems Engineering. In: MoDeVVa 2012 (2012)
8. Kinoshita, S., Nishimura, H., Takamura, H., Mizuguchi, D.: Describing software specification by combining SysML with the B method. In: ISSREW 2014 (2014)
9. Laleau, R., Polack, F.A.C.: Coming and going from UML to B: a proposal to support traceability in rigorous IS development. In: Bert, D., Bowen, J.P., C. Henson, M., Robinson, K. (eds.) B 2002 and ZB 2002. LNCS, vol. 2272, pp. 517–518. Springer, Heidelberg (2002)
10. Lano, K., Clark, D., Androutsopoulos, K.: UML to B: formal verification of object-oriented models. In: Boiten, E.A., Derrick, J., Smith, G.P. (eds.) IFM 2004. LNCS, vol. 2999, pp. 187–206. Springer, Heidelberg (2004)
11. Matoussi, A., Gervais, F., Laleau, R.: A goal-based approach to guide the design of an abstract Event-B specification. In: ICECCS (2011)
12. Meyer, E., Souquières, J.: A systematic approach to transform OMT diagrams to a B specification. In: Wing, J.M., Woodcock, J. (eds.) FM 1999. LNCS, vol. 1708, pp. 875–895. Springer, Heidelberg (1999)
13. Snook, C., Butler, M.: UML-B: formal modelling and design aided by UML. ACM Trans. Softw. Eng. Methodol. **15**, 92–122 (2006)

Mechanized Refinement of Communication Models with TLA$^+$

Florent Chevrou$^{(\boxtimes)}$, Aurélie Hurault, Philippe Mauran,
and Philippe Quéinnec

IRIT – Université de Toulouse, 2 rue Camichel, 31000 Toulouse, France
florent.chevrou@enseeiht.fr
http://www.irit.fr

Abstract. In distributed systems, asynchronous communication is often
viewed as a whole whereas there are actually many different interaction
protocols whose properties are involved in the compatibility of peer compositions. A hierarchy of asynchronous communication models, based on
refinements, is established and proven with the TLA$^+$ Proof System.
The work serves as a first step in the study of the substituability of the
communication models when it comes to compatibility checking.

1 Introduction

Properties of distributed systems are directly impacted by the interaction protocol in use. Unlike in synchronous communication, the decoupling of send
and receive events in asynchronous communication allows for many ordering
strategies and thus, communication models. Yet, these models are seldom distinguished. For instance, the multiple variations of FIFO communication are seen
to be used interchangeably despite of their fundamental differences. In [CHQ15],
the consequences on the compatibility of the composition of peers under these
circumstances have been highlighted thanks to the modeling of such systems and
classic communication models in TLA$^+$ [Lam02]. Knowing which substitutions of
communication models preserve compatibility is of great interest. Some models
have simpler specifications which ease formal studies and proofs of compatibility
in practical cases. As a first step of this work, we propose here to exhibit the
refinements between each of the models. The hierarchy of refinements is a key
result in the further study of the communication models when compatibility of
peers is involved. The models and the structure of their TLA$^+$ module are introduced in Sect. 2. In Sect. 3 the approach behind the proofs of refinement with
the TLA$^+$ Proof System is exposed along with the obtained results.

2 Specification

We consider point-to-point message-passing communication through channels.
Messages consist of a unique id and metadata. Histories of past sent messages
are part of the metadata to allow for the specification of ordering properties.

© Springer International Publishing Switzerland 2016
M. Butler et al. (Eds.): ABZ 2016, LNCS 9675, pp. 312–318, 2016.
DOI: 10.1007/978-3-319-33600-8_27

Table 1. Specification of the communication models. The "Send" and "Receive" columns only contain the model-specific guards on the *send* and *receive* actions (where *m* denotes the message to be received), as in the TLA$^+$ modules. When applicable, the last column informally symbolizes an implementation based on queues.

Model	Specification	Send	Receive	Queues
M_{RSC}	Messages are immediately delivered after their send [CBMT96].	$net = \emptyset$	$\top$	size 1 ×1
M_{n-n}	Global ordering. Messages are delivered in their send order.	$\top$	$\neg(\exists m_2 \in net :$ $mid(m_2) \in mhg(m))$	n, n ×1
M_{1-n}	Messages from the same peer are delivered in their send order.	$\top$	$\neg(\exists m_2 \in net :$ $mp(m_2) = mp(m)$ $\wedge\ mid(m_2) \in mhl(m))$	1, n ×n
M_{n-1}	On a given peer, messages are received in their send order.	$\top$	$\neg(\exists m_2 \in net :$ $mc(m_2) \in listened$ $\wedge\ mid(m_2) \in mhg(m))$	n, 1 ×n
M_{causal}	Messages are delivered according to the causality of their emission [Lam78].	$\top$	$\neg(\exists m_2 \in net :$ $mc(m_2) \in listened$ $\wedge\ mid(m_2) \in mhc(m))$	
M_{1-1}	Messages between two designated peers are delivered in their send order.	$\top$	$\neg(\exists m_2 \in net :$ $mp(m_2) = mp(m)$ $\wedge\ mc(m_2) \in listened$ $\wedge\ mid(m_2) \in mhl(m))$	1, 1 ×n^2
M_{async}	Fully asynchronous. No order on message delivery is imposed.	$\top$	$\top$	×1 bag

This allows for homogeneous descriptions of the models even though a particular model might not make use of the whole information. The content of a message is irrelevant to the specification of ordering properties although it can be taken into account in practical implementations. As messages are exchanged on channel and there is no explicit peer destination, multiple senders and receivers can interact with the same channel. The state variables in the TLA$^+$ module of a communication model are:

- *net* the network: a set that contains messages in transit.
- *hg* the global history contains the ids of all the messages the peers have sent.
- *hl* the local histories: $hl[p]$ is a set that holds the ids of messages sent by p.
- *hc* the causal histories: $hc[p]$ is a set that contains the ids of the messages in the causal past of p built according to Lamport's causal relation [Lam78].

A message m on the network is a tuple $\langle id_m, c_m, p_m, hl_m, hc_m, hg_m \rangle$ where id_m is the message's unique id, c_m the channel on which it has been sent, p_m the sender, and hl_m, hc_m, hg_m snapshots of $hl(p)$, $hc(p)$, and hg at send event. We define mid, mc, mp, mhl, mhc, mhg the associated accessors (e.g. $mc(m) = c_m$).

Communication models are specified by two actions: *send* and *receive*. The $send(peer, chan)$ action consists in sending a new message from peer *peer* on

───── MODULE *fifo*11 ─────

EXTENDS *Naturals, Defs*
$Init \triangleq id = 1 \land net = \{\} \land hl = [i \in Peer \mapsto \{\}] \land hc = [i \in Peer \mapsto \{\}] \land hg = \{\}$

$send(peer, chan) \triangleq$ **The peer "peer" sends a new message on channel "*chan*"**
$\quad \land id' = id + 1$
$\quad \land \text{LET } m \triangleq \{id\} \times \{chan\} \times \{peer\} \times \{hl[peer]\} \times \{hc[peer]\} \times \{hg\}\text{ IN}$
$\qquad net' = net \cup m$
$\quad \land hl' = [hl \text{ EXCEPT } ![peer] = @ \cup \{id\}]$
$\quad \land hc' = [hc \text{ EXCEPT } ![peer] = @ \cup \{id\}]$
$\quad \land hg' = hg \cup \{id\}$

$deliveryOk(m, listened) \triangleq \neg(\exists m2 \in net :$ **Ordering property.** There is no tran-
$\qquad \land mp(m) = mp(m2)$ siting message $m2$ from the same peer,
$\qquad \land mc(m2) \in listened$ whose channel is listened, and in the lo-
$\qquad \land mid(m2) \in mhl(m))$ cal history of m (thus previously sent by
the same peer).

$receive(peer, chan, listened) \triangleq$
$\quad \exists m \in net :$ The peer "peer" receives a message on "*chan*",
$\qquad \land chan = mc(m)$ while listening to a set of channels "listened"
$\qquad \land deliveryOk(m, listened)$
$\qquad \land net' = net \setminus \{m\}$
$\qquad \land \text{UNCHANGED } \langle id, hl, hg \rangle$
$\qquad \land hc' = [hc \text{ EXCEPT } ![peer] = @ \cup mhc(m) \cup \{mid(m)\}]$

$NextSend \triangleq \exists p \in Peer : \exists c \in Channel : send(p, c)$
$NextRecv \triangleq \exists p \in Peer : \exists c \in Channel : \exists l \in \text{SUBSET } Channel : receive(p, c, l)$
$Next \triangleq NextSend \lor NextRecv$
$Spec \triangleq Init \land \Box[Next]_{vars}$

Fig. 1. TLA$^+$ specification of M_{1-1}. The TLA$^+$ module *Defs* defines sets and type invariants for the state variables, invariants on histories (inclusion between the different histories) and the uniqueness of message identifiers.

channel *chan*. It is always enabled except in the *RSC* (Realizable with Synchronous Communication [CBMT96]) model where an empty network is expected. The *receive(peer, chan, listened)* action consists in receiving a message on peer *peer*, retrieved from channel *channel*, while being interested in channels in the set *listened*. It is enabled when a message m with a matching channel is in transit and no other message on a listened channel should be received first according to the ordering property of the communication model. For each communication model, the ordering policy and this last condition are introduced in Table 1. Figure 1 shows a comprehensive TLA$^+$ module of the FIFO 1-1 model.

Receiving a message on a peer consists in removing it from the network and updating that peer's causal past accordingly. Sending a message consists in building the tuple $\langle id, chan, peer, hl[peer], hc[peer], hg \rangle$, adding it in the network, and adding the message id to hg, $hl[peer]$, and $hc[peer]$.

Given a set of peers $Peer$ and a set of channels $Channel$, the specification of a communication model is $Spec_M \triangleq Init_M \wedge \Box[Next_M]_{vars_M}$ where $vars_M$ groups the state variables of M, $Init_M$ specifies their initial values and $[Next_M]_{vars_M}$ accounts for all the possible *send* and *receive* actions along with stuttering on the state variables. $Next_M \triangleq NextSend_M \vee NextReceive_M$ where $NextSend_M \triangleq \exists p \in Peer : \exists c \in Channel : send_M(p, c)$ and $NextReceive_M \triangleq \exists p \in Peer : \exists c \in Channel : \exists l \subseteq Channel : receive_M(p, c, l)$.

3   Refinement

Proofs of refinement between the communication models have been carried with the TLA$^+$ Proof System [CDLM10]. The resulting hierarchy is summed up in Fig. 2. This adds to existing results about the comparison of models as in [KS11] and [CBMT96].

In TLA$^+$, M_2 refines M_1 iff $Spec_{M_2} \Rightarrow Spec_{M_1}$ where the state variables of M_1 are mapped to the variables of M_2 when instantiating the module M_1. All our models have the same state variables and actions that evolve accordingly. For some models, some history variables are constructed but unused (e.g. the causal history in M_{n-n}, or all the histories in M_{RSC}) and play the role of shadow variables. This simplify the refinement proofs, as the mapping relation is the identity. Proving that $Spec_{M_2} \Rightarrow Spec_{M_1}$ here consists in refining each action:

$\forall p \in Peer : \forall c \in Channel : send_{M_2}(p, c) \Rightarrow send_{M_1}(p, c)$
$\forall p \in Peer : \forall c \in Channel : \forall l \subseteq Channel : receive_{M_2}(p, c, l) \Rightarrow receive_{M_1}(p, c, l)$

The proofs require highlighting inductive invariants for each model, especially to refine the *receive* actions since they differ the most (see Table 1). Among the inductive invariants that are introduced to guide the refinement proofs, most are common to all the models. The uniqueness of the messages (different message ids) and relations between the different histories are such invariants. For instance $\forall p \in Peer : hl[p] \subseteq hc[p] \subseteq hg$: the sent messages of peer p is a subset of the known messages of this peer (the causal history of peer p), which is a subset of all sent messages. The same applies to histories carried by messages in transit ($\forall m \in net : mhl(m) \subseteq mhc(m) \subseteq mhg(m)$). Some invariants are specific to a communication model. For instance, in M_{1-n}, messages in transit that are causally related are from the same peer ($\forall m_1, m_2 \in net : mid(m_2) \in mhc(m_1) \Rightarrow mp(m_1) = mp(m_2)$). This hypothesis is crucial to prove that M_{1-n}

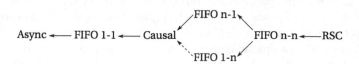

Fig. 2. Refinement of the communication models. An arrow means "refines". A dashed line means the proof is still in progress.

─────────── MODULE *refinement_11_causal* ───────────

EXTENDS *Defs*
causal $\triangleq$ INSTANCE *causal*
*fifo*11 $\triangleq$ INSTANCE *fifo*11

───

THEOREM *RaffSend* $\triangleq$ $\forall p \in Peer : \forall c \in Channel$:
  causal! *send*(p, c) $\Rightarrow$ *fifo*11! *send*(p, c)BY DEF *fifo*11! *send*, *causal*! *send*
THEOREM *RaffRecv* $\triangleq$ $\forall p \in Peer : \forall c \in Channel : \forall l \in$ SUBSET *Channel* :
  causal! *invHistories* $\wedge$ *causal*! *receive*(p, c, l) $\Rightarrow$ *fifo*11! *receive*(p, c, l)
  BY   DEF *fifo*11! *receive*, *causal*! *receive*, *causal*! *deliveryOk*, *fifo*11! *deliveryOk*, ...

───

THEOREM *Refinement* $\triangleq$ *causal*! *Spec* $\Rightarrow$ *fifo*11! *Spec*
  ⟨1⟩a. *causal*! *Init* $\wedge$ $\Box$([*causal*! *invHistories* $\wedge$ *causal*! *Next*]$_{causal}$! *vars*) $\Rightarrow$ *fifo*11! *Spec*
    ⟨2⟩1. *causal*! *Init* $\Rightarrow$ *fifo*11! *Init*BY DEF *causal*! *Init*, *fifo*11! *Init*
    ⟨2⟩2. *causal*! *invHistories* $\wedge$ *causal*! *Next* $\Rightarrow$ *fifo*11! *Next*
      BY *RaffSend*, *RaffRecv* DEF *causal*! *Next*, *fifo*11! *Next*, *causal*! *NextSend*,
      *fifo*11! *NextSend*, *causal*! *NextRecv*, *fifo*11! *NextRecv*
    ⟨2⟩3. [*causal*! *invHistories* $\wedge$ *causal*! *Next*]$_{causal}$! *vars* $\Rightarrow$ [*fifo*11! *Next*]$_{fifo11}$! *vars*
      BY ⟨2⟩2 DEF *causal*! *vars*, *fifo*11! *vars*
    ⟨2⟩ QED BY *PTL*, ⟨2⟩1, ⟨2⟩3 DEF *fifo*11! *Spec*
  ⟨1⟩.QED BY *PTL*, ⟨1⟩a, *causal*! *Invariant* DEF *causal*! *Spec*

Fig. 3. TLA$^+$ proof that M_{causal} refines M_{1-1}. The propositional temporal logic tactic PTL is used to step from one transition (*Next*) to the specification ($\Box[Next]$).

refines M_{causal} (*receive* action). Similarly, the proof of the refinement of M_{n-n} by M_{RSC} (*send* action) requires an invariant that is specific to M_{RSC} and states that *net* contains at most one message.

We had to carefully separate the proof steps regarding individual actions from the ones regarding the complete specification. The former are formulae of first-order logic with quantifiers and are handled by SMT backends (CVC3 and Z3 in our case); the latter deal with temporal logic ($\Box$ operator) and are handled by the LS4 backend, a propositional temporal logic prover (Fig. 3). The inductive invariants which are required to prove the refinements are large formulae (10 state variables and up to 20 quantifiers) and need several proof steps. However, they were gradually built and were easily decomposed in successive strengthening (type invariants, invariants on peers, invariant on messages) to allow for incremental proofs. Once this natural decomposition was done, the TLAPS backends have shown to be efficient enough to directly prove the formulae, without having to go down to reasoning by cases.

Our main difficulty was with the representation of messages: a message is a tuple of six elements (message id, channel, sender, histories). In the current state of TLAPS, the handling of tuples $\langle e_1, \ldots, e_n \rangle$ is sometimes awkward. They are internally considered as functions of domain $1..n$, in accordance with their TLA$^+$ semantics. But a product of sets is also a set of tuples, and we were unable to switch between both points of view. For instance, we had to assume a lemma similar to $\{1\} \times \{2\} = \{\langle 1, 2 \rangle\}$ (more precisely, that the product of N singleton sets ($N > 0$) is a set with a unique tuple).

At this point, all the refinements are proved except for two secondary invariants, only required for the refinement of M_{causal} by M_{1-n}. These two invariants have been manually provedusing induction but their TLAPS proof is still elusive. All the TLA$^+$ modules that specify the communication models and the proofs of refinement are available at http://queinnec.perso.enseeiht.fr/ABZ2016.

4   Related Work

Asynchronous communication models in distributed systems have been studied and compared in [KS11] (notion of ordering paradigm) and [CBMT96] (notion of distributed computation classes). In our work, we consider additional distributed communication models, namely M_{n-n}, M_{1-n} and M_{n-1}, which are of interest since they are not totally ordered. M_{n-1} for instance, the FIFO order with instantaneous delivery, is often used in the literature without distinction from the classic FIFO order. Our approach to isolate the communication model as a transition system is reminiscent of Tel's textbook [Tel00], but his focus is on describing distributed algorithms, whereas ours is on comparing the models.

5   Conclusion

This paper explains how proofs of refinement between communication models have been conducted with the TLA$^+$ Proof System. A unified TLA$^+$ specification of classic communication models along with common and model-specific invariants is the key to achieve these formal proofs. Ongoing work consists in studying other descriptions of the models. For example, they can be specified as properties on distributed executions (sequences of communication events). Practical implementations based on queues and counters are also of interest. The verification of refinement relations between these models is in progress.

References

[CBMT96] Charron-Bost, B., Mattern, F., Tel, G.: Synchronous, asynchronous, and causally ordered communication. Distrib. Comput. **9**(4), 173–191 (1996)

[CDLM10] Chaudhuri, K., Doligez, D., Lamport, L., Merz, S.: Verifying safety properties with the TLA$^+$ proof system. In: Giesl, J., Hähnle, R. (eds.) IJCAR 2010. LNCS, vol. 6173, pp. 142–148. Springer, Heidelberg (2010)

[CHQ15] Chevrou, F., Hurault, A., Quéinnec, P.: Automated verification of asynchronous communicating systems with TLA+. In: Electronic Communications of the EASST (PostProceedings of the 15th International Workshop on Automated Verification of Critical Systems), vol. 72, pp. 1–15 (2015)

[KS11] Kshemkalyani, A.D., Singhal, M.: Distributed Computing: Principles, Algorithms, and Systems. Cambridge University Press, Cambridge (2011)

[Lam78] Lamport, L.: Time, clocks and the ordering of events in a distributed system. Commun. ACM **21**(7), 558–565 (1978)

[Lam02] Lamport, L.: Specifying Systems. Addison Wesley, Reading (2002)

[Tel00] Tel, G.: Introduction to Distributed Algorithms, 2nd edn. Cambridge University Press, Cambridge (2000)

A Super Industrial Application of PSGraph

Yuhui Lin[1]([⊠]), Gudmund Grov[1], Colin O'Halloran[2], and Priiya G.[2]

[1] Heriot-Watt University, Edinburgh, UK
{Y.Lin,G.Grov}@hw.ac.uk
[2] D-RisQ Software Systems, Malvern, UK
{coh,priiya.g}@drisq.com

Abstract. The ClawZ toolset has been successful in verifying that Ada code is correctly generated from Simulink models in an industrial setting, using the Z notation. D-RisQ is now extending this technique to new domains of the C programming language, which requires changes to their highly complex proof technique. In this paper, we present initial results in the technology transfer of the graphical PSGraph language to support this extension, and show feasibility of PSGraph for industrial use with strong maintainability requirements.

1 Introduction

The ClawZ toolset is used to verify that automatically generated code from (a sub-set of) Simulink[1] into (a verifiable subset of) Ada is correct [7]. This is achieved by encoding the semantics of the two representations in the Z notation. Correctness of the generation is then ensured by a formal proof of a refinement conjecture from the Z representation of Simulink into the Z representation of Ada. This proof is supported by a very powerful proof tactic for the ProofPower theorem prover called *Supertac* [7]. This tactic has been developed over a number of years and has a very high degree of automation as it is tailor-made for these types of conjectures.

Whilst Ada is used extensively for avionics software, other sectors, such as automotive, normally use the C programming language. The TargetLink code generator from dSPACE[2] is able to generate C code from a Simulink model. However, Supertac is configured for Ada and will, as will be shown in the next section, not be able to verify correctness of C generation.

A side-effect of the automation achieved by Supertac is that the code base has become highly complex and large. In fact, it consists of almost 50 K lines of dense ML code[3]. Adapting it from Ada to C is therefore a non-trivial problem.

Tactic languages, such as the one used to encode Supertac, are often difficult to analyse and debug as the error may manifest itself a different place from

---

This work has been supported by EPSRC grants EP/J001058, EP/K503915, EP/M018407 and EP/N014758. The second author is supported by a SICSA Industrial Fellowship.

[1] See www.mathworks.com.

[2] See www.dspace.com.

[3] As far as we know, this is the largest proof tactic ever made in terms of code size.

© Springer International Publishing Switzerland 2016
M. Butler et al. (Eds.): ABZ 2016, LNCS 9675, pp. 319–325, 2016.
DOI: 10.1007/978-3-319-33600-8_28

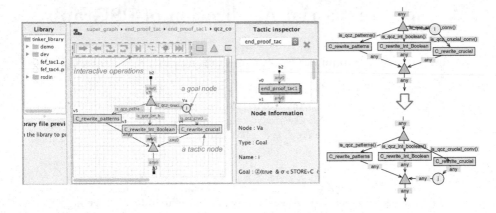

Fig. 1. The GUI of tinker (left) & an illustrative evaluation step (right)

where the problem actually lies. This is further complicated by: (1) the *non-deterministic nature* of tactics, meaning multiple branches can be generated; and (2) that they tend to be *untyped* in the sense that subgoals cannot be differentiated. This non-deterministic nature makes it more difficult to debug than most software, and often the only method to find the mistakes is to insert "writeln" statements in the code to print information. Needless to say, this can be a hard task for 50 K LoC. As proofs are getting larger and more commonplace, proof maintenance is going to be a problem. Improved debugging features for ML would be beneficial and a step in the right direction, but this alone will not be sufficient: the non-deterministic nature of tactics and the non-trivial flow of sub-goals will require their own solutions.

To ease maintenance, debugging and general understanding of tactics, we have previously developed the *PSGraph* language [1] and the supporting *Tinker* tool [2,5], as can be seen in Fig. 1 (left). Here, proof tactics are encoded as directed hierarchical graphs, where the boxes contain tactics or nested graphs, and are composed by labelled wires. The labels are called *goal types* and are predicates that describe the expected properties of sub-goals. Each sub-goal becomes a special *goal node* in the graph, which "lives" on a wire. Evaluation is handled by applying a tactic to a goal node that is placed on one of its input wires. The resulting sub-goals are sent to the out wires of the tactic node. To add a goal node to a wire, the goal type must be satisfied. Figure 1 (top right) illustrates a single evaluation step where tactic C_rewrite_crucial, which is discussed in Sect. 3, is applied to a goal labelled i:

$$\ldots (* \ ?\vdash *)_{\overline{Z}}^{\Gamma}(true \wedge \sigma \in STORE_C \wedge true)$$
$$\wedge \ IntVal_C \ (\ IntOf_C \ (IntVal_C(\ mk\_signed\_int_C \ (IntOf_C \ fa_v + 1) \ ))$$
$$==_{CZ} \ mk\_signed\_int_C \ (IntOf_C \ fa_v + 1)) \neq IntVal_C \ 0^{\Gamma}$$

The tactic produces a goal labelled j on its output wire (bottom right of Fig. 1)

$$... \ (* \ ?\vdash *)_{\overline{z}}^{\Gamma}(true \ \wedge \ \sigma \in STORE_C \ \wedge \ true)$$
$$\wedge \ IntVal_C \ (mk\_signed\_int_C \ (IntOf_C \ fa_v \ + \ 1)$$
$$==_{CZ} \ mk\_signed\_int_C \ (IntOf_C \ fa_v \ + \ 1)) \neq IntVal_C \ 0^\lnot$$

The goal types introduce a notion of 'types' to the tactic language, which improves upon the static properties for composition and evaluation found in the tactic language currently used to encode Supertac[4]. As can be seen in Fig. 1 (left), Tinker provides support for tactic developers to step through the proofs using 'interactive operations' to e.g. 'step over' or 'step into' a tactic. A novel breakpoint feature has recently been introduced [5], where sub-goals are evaluated automatically until a sub-goal reaches the breakpoint. At this point evaluation will stop, and the user can guide the evaluation step-by-step from this point onwards. This is particularly useful for large and complex tactics such as Supertac. We will return to how we have exploited these special Tinker features for this work in Sect. 3. We believe that this support for inspection and adaptations through simple graph visualization is novel.

We will show our initial encoding of Supertac in PSGraph in Sect. 2. We will then show how this is used to analyse Supertac using PSGraph and adapt this to support C code in Sect. 3. We conclude and detail our next steps in Sect. 4.

2 PSGraph Encoding of Supertac for Ada

PSGraph handles modularity and complexity through *hierarchies*, represented by boxes in the graph that contains subgraphs. The architecture of Supertac consists of four subtactics executed in sequence, where the first of these has been decomposed in a hierarchical node. This is done by unfolding and breaking it, and its sub-tactics, down, and then recompositing the sub-tactics in PSGraph with proper goaltypes. In the future we plan to decompose the remaining three. Figure 2 shows the top-level of our Supertac encoding:

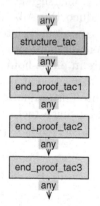

Fig. 2. Supertac

structure_tac is used to classify the conjectures and enhancing them with meta-information. In addition, it unpacks structure surrounding conjectures, such as quantifiers.

end_proof_tac1 gets rid of Simulink and Ada vocabulary, to make it easier to reason about.

end_proof_tac2 deals with mathematical statements, and gets rid of high-level concepts (e.g. functions) to reduce it to set theoretical primitives that are easier to reason about.

end_proof_tac3 handles case statements and is mainly used as a brute-force strategy if end_proof_tac2 has not been able to discharge the conjecture.

---

[4] Recently, several *typed* tactic language, e.g. Mtac [9], have been developed. They will have comparable static properties to PSGraph, albeit they do not have the dynamic inspection features Tinker provides.

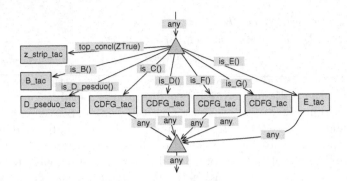

Fig. 3. The structure_tac sub-tactic of Supertac

Figure 3 shows the tactic nested by the structure_tac node in PSGraph. As a proof of concept, this encoding has been successfully applied to discharge VCs generated from a *Nose-Gear Velocity* case study [8] in ClawZ.

3   Adapting Supertac to C-Code

The main difference between conjectures generated for Ada and C is that Ada specific semantics in the refinement conjecture has now been replaced by C constructs[5]. The remaining parts, e.g. the encoding of Simulink using Z Notation, are unchanged. The key challenge is thus to replace the parts of Supertac that reduce the Ada vocabulary with pure set theory into parts that reduce the C vocabulary into set theory (and develop these). This is non-trivial, because semantically variables in Ada can be treated simply when aliasing restrictions are in place, however the semantics of a variable in C come with a side condition stating that it is disjoint from other C objects. PSGraph enables users to step through evaluation and see how the goals evolve. This is a very useful when adapting a tactic in this way. To illustrate, the following VC has been generated from a simple C program:

$$\lfloor (* \ ? \vdash \ *)^{\Gamma}_Z \forall \ [fa_v : VALUE_C; \ \sigma : STORE_C \ | \ \langle (fa_v, fa_l) \rangle \ AllocatedIn \ \sigma] \ \bullet$$
$$(Test01_v! \ \hat{=} \ IntVal_C \ (mk\_signed\_int_C \ (IntOf_C \ fa_v + 1)), fa_v \ \hat{=}$$
$$fa_v, \sigma \ \hat{=} \ \sigma) \in Test01_{post}^{\urcorner}$$

With the debugging support of Tinker to interactively step through our PSGraph version of Supertac, we were able to pinpoint where in end_proof_tac1 it was assumed that the target programming language constructs need to be eliminated. First we split this sub-tactic into end_proof_tac1_1 and end_proof_

---

[5] A subset of C called C♭ is used. It has been designed with safety critical applications in mind. C♭'s formalisation in ProofPower is based on work by Norrish for the HOL system [6]. The formalisation is comparable to Frama-C, however it has not been designed to act as a framework for other analysis tools.

tacl_2. The C specific parts will need to be eliminated between these two sub-tactics. To illustrate, after end_proof_tacl_1 our VC looks as follows:

$$(* \ \ 3 \ *)^{\Gamma}_{Z}\sigma \in STORE_C^{\neg}$$
$$(* \ \ 2 \ *)^{\Gamma}_{Z}\langle\langle fa_v, \ fa_l\rangle\rangle \ AllocatedIn \ \sigma^{\neg}$$
$$(* \ \ 1 \ *)^{\Gamma}_{Z}clawz\_hint1\,"Supertac:VC\_Origin:Empty\_Block\_List"^{\neg}$$

$$(* \ ?\vdash \ *)^{\Gamma}_{Z}(true \wedge \sigma \in STORE_C \wedge true)$$
$$\wedge \ IntVal_C \ (IntOf_C \ (IntVal_C \ (mk\_signed\_int_C \ (IntOf_C \ fa_v + 1))))$$
$$==_{CZ} \ mk\_signed\_int_C \ (IntOf_C \ fa_v + 1)) \neq IntVal_C \ 0^{\neg}$$

where assumption *(* 1 *)* has been inserted by structure_tac to guide the rest of the proof. The intuition behind this VC is that the C value of comparing $IntOf_C(IntVal_C(mk\_signed\_int_C(IntOf_C fa_v + 1)))$ with $mk\_signed\_int_C(IntOf_C fa_v + 1)$, using the C version of integer comparison operator $==_{CZ}$, is not equal to the C value of 0. When the VC contains C specific parts, then these have to be reduced to set theory before end_proof_tacl_2 executes. As can be seen in Fig. 4 (left), we introduce a new nested tactic called qcz_conversion, and insert it between the two tactics discussed above. VCs containing C vocabulary are identified by the is_qcz_conv predicate, found on the wire leading to the qcz_conversion tactic.

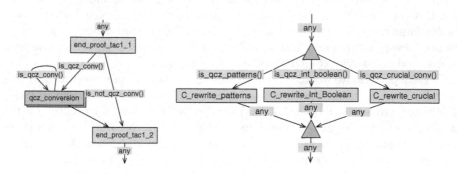

Fig. 4. New end_proof_tacl (left) & code conversion types (right)

The nested graph of this tactic is shown in Fig. 4 (right). Depending on certain properties of the goal, identified by the wire labels of the PSGraph, one of three tactics may be applied.

Each of them applies some *conversions*, which are rewrite rules proven from the semantics of the language formalisation, to simplify the goal: C_rewrite_patterns does general simplifications of C constructs; C_rewrite_Int_Boolean deals with special simplifications related to C booleans (represented as integer in C); while C_rewrite_crucial does the crucial steps in eliminating C vocabulary. To illustrate, the following are some of the *crucial* conversions of C_rewrite_crucial:

$$\forall\ x\ :\ \mathbb{U}\ \bullet\ IntOf_C\ (IntVal_C\ x) = x$$
$$\forall\ x\ :\ \mathbb{U};\ y\ :\ \mathbb{U}\ \bullet\ IntVal_C\ x = IntVal_C\ y \Leftrightarrow x = y$$

These changes were sufficient to complete the running example, and has been successfully applied to fully automate six handcrafted VCs.

4   Conclusion and Future Work

In this paper we have reported on a successful technology transfer project, where the PSGraph language has been used to start the adaptation of an industrial proof strategy to a new domain – using the state-based Z notation. Through an example, we have illustrated how it has been used to pinpoint and support the development of this adapted proof strategy.

Proofs are becoming more commonplace and increasingly complex. Like with software maintenance, *proof maintainence* is going to be a problem, as people who did the proof originally will have moved on or retired. For example, the substantial proof effort on the sel4 kernel [4] will be around for decades, and even small changes to the underlying kernel is likely to break many proofs. We have shown the advantages of PSGraph for the maintainence of Supertac; this experience has given us confidence that it has potential to play a supporting role in *proof engineering* [3] other large proof-based developments.

So far we have only decomposed structure_tac. In the medium term we aim to complete this work by decomposing the remaining tactics and we are hoping that PSGraph can be used to remove some of the "clutter" that has been the result of updating Supertac to new applications. Particular challenges will be to discover new goal types and find suitable graphical representations of some domain specific and non-trivial combinators used in Supertac.

In the long term we would like to re-implement the overall structure of Supertac from scratch, using the existing Supertac components as building blocks. Ideally, we will support both Ada and C. This will enable us to reflect on the intuition and develop a more conscious strategy, where the overall proof plan is clear. For example, we should separate reasoning about the denotational semantics given to expressions from the operational semantics of statements, as these will be tackled in different ways. We believe that this will give a much cleaner proof strategy that would be easier to analyse and adapt for future applications and changes to the modelling language or the target programming language.

References

1. Grov, G., Kissinger, A., Lin, Y.: A graphical language for proof strategies. In: McMillan, K., Middeldorp, A., Voronkov, A. (eds.) LPAR-19 2013. LNCS, vol. 8312, pp. 324–339. Springer, Heidelberg (2013)
2. Grov, G., Kissinger, A., Lin, Y.: Tinker, tailor, solver, proof. In: UITP 2014, vol. 167 of ENTCS, pp. 23–34. Open Publishing Association (2014)
3. Klein, G.: Proof engineering considered essential. In: Jones, C., Pihlajasaari, P., Sun, J. (eds.) FM 2014. LNCS, vol. 8442, pp. 16–21. Springer, Heidelberg (2014)

4. Klein, G., Elphinstone, K., Heiser, G., Andronick, J., Cock, D., Derrin, P., Elkaduwe, D., Engelhardt, K., Kolanski, R., Norrish, M., et al.: seL4: formal verification of an OS kernel. In: SOSP, pp. 207–220. ACM (2009)
5. Lin, Y., Bras, P.L., Grov, G.: Developing & debugging proof strategies by tinkering. In: TACAS (2016). to appear
6. Norrish, M.: C formalised in HOL. Ph.D. thesis, University of Cambridge (1999)
7. O'Halloran, C.: Automated verification of code automatically generated from Simulink. ASE **20**(2), 237–264 (2013)
8. O'Halloran, C.: Nose-gear velocity-a challenge problem for software safety. In: Australian System Safety Conference (ASSC 2014), Held in Melbourne 28–30, May 2014 (2014)
9. Ziliani, B., Dreyer, D., Krishnaswami, N.R., Nanevski, A., Vafeiadis, V.: Mtac: a monad for typed tactic programming in Coq. J. Funct. Program. **25**, e12 (2015)

Articles Contributing to the Hemodialysis Machine Case Study

The Hemodialysis Machine Case Study

Atif Mashkoor[✉]

Software Competence Center Hagenberg GmbH, Hagenberg, Austria
atif.mashkoor@scch.at

1 Introduction

This documents presents a description of a case study concerning the control of a hemodialysis (HD) machine. It provides an overview of the requirements and the design of an HD machine including a sketch of the machine's functionality, related safety conditions, and a top-level system architectural description. This case study is supposed to stimulate research and pedagogical activities related to the use of formal methods in a medical application domain with challenging requirements concerning safety and hardware/software interaction.

Kidney diseases are becoming endemic in recent times. There are several contributing factors such as changed life style as well as increase in hypertension and diabetes. Kidneys are the organs that are responsible for cleaning human blood by removing excess fluid, minerals and wastes. They also aid regulating blood pressure, electrolyte balance, and red blood cell production. When kidneys fail, toxic waste products are accumulated in the human body, causing a rise in blood pressure and a decline in the production of red blood cells. When kidneys fail completely, an artificial system is required that can substitute their functionality.

Hemodialysis (HD) is a treatment for kidney failure that uses a machine to send the patient's blood through a filter, called a dialyzer, for extracorporeal removal of waste products. The blood is transported to and from the patient's body through a surgically created vein during this process. The blood then travels through a tube that takes it to the dialyzer. Inside the dialyzer, the blood flows through thin fibers that filter out wastes and extra fluid. The machine returns the filtered blood to the body through a different tube. A vascular access lets large amounts of blood flow continuously during HD treatments to filter as much blood as possible per treatment. A specific amount of blood (approx. a pint) is conducted through the machine every minute.

This document is structured as follows: Sect. 2 presents a high-level architectural description of the HD machine describing its main components. Section 3 gives an overview of different types of dialysis therapies and how a therapy is conducted. Finally, Sect. 4 lists several general and software safety requirements related to the correct operation of the HD machine.

Acronyms. Table 1 contains a list of acronyms used in this document.

© Springer International Publishing Switzerland 2016
M. Butler et al. (Eds.): ABZ 2016, LNCS 9675, pp. 329–343, 2016.
DOI: 10.1007/978-3-319-33600-8_29

Table 1. The list of acronyms

AP	Arterial Pressure
BEP	Blood-Side Entry Pressure
BP	Blood Pump
DF	Dialyzing Fluid
EBC	Extracorporeal Blood Circuit
HD	Hemodialysis
SAD	Safety Air Detector
TMP	Trans Membrane Pressure
UF	Ultra Filtration
UFP	Ultra Filtration Pump
UI	User Interface
VP	Venous Pressure
VRD	Venous Red Detector

2 System Architecture

As shown in Fig. 2, the HD machine under consideration is composed of the following components:

Extracorporeal Blood Circuit (EBC)
  The EBC consists of the machine's peristaltic pumps, which are used to transport the blood to and from the dialyzer. Blood is pumped through a disposable system that is mainly composed of tubing, connectors, and (arterial/venous) drip chambers connected to the dialyzer. Peristaltic pumps withdraw blood from the patients vascular access into the dialyzer. A syringe pump pumps heparin into the bloodlines in a quantity and time set by the user to avoid the coagulation of the blood in the disposable circuit and the dialyzer filter. The Safety Air Detector (SAD) component is a combined air and red detector. It detects air bubbles and blood in the EBC while transporting blood back to the patient's body. The functionality of the EBC is depicted in Fig. 1.

Dialyzer
  The capillary dialyzer houses semipermeable hollow fibers encased in a plastic canister. The dialyzer is used to correct the concentration of water-soluble substances in the patients blood before delivering it back to the patient. The blood is separated from the Dialyzing Fluid (DF) by a semi-permeable membrane that permits bidirectional diffusive transport and Ultrafiltration (UF). The process also allows diffusion of substances from the DF into the blood.

Balance Chamber
  The balance chamber is a closed system that consists of two chambers, each with a flexible membrane, allowing it to fill the chambers from one side while an identical volume is emptied to the other side. Therefore the outlet fluid volume is equal to the input fluid volume. Each membrane has a magnetic sensor

which reads the membrane position and controls the opening and the closing of each sub-compartment. The control of the DF volume is also carried out by the balance chamber. The difference between the used DF and the fresh DF is the UF volume, which is removed from the blood side of the dialyzer. The removal of the UF volume is carried out by the Ultra Filtration Pump (UFP).

DF Preparation

In bicarbonate dialysis, which is the most common procedure, concentrate preparation consists of mixing the heated and degassed water with bicarbonate concentrate and acid concentrate. The accuracy of the DF concentration is controlled by the conductivity sensors. If the concentration is incorrect, the dialyzer will be bypassed.

DF Water Preparation

Purified water coming from the reverse osmosis system has to be degassed and tempered to a predetermined temperature, which is set by the user (usually 37°C), before the concentrate is prepared. A degassing chamber and a heater are integral to the system.

Bypass

Bypass mode occurs when the DF conductivity or temperature goes beyond permissible limits. This is used for safety reasons. This feature allows that the dialyzer is separated from the DF flow without interrupting the DF preparation until temperature and conductivity are back within the acceptable limits. The DF path is then directed to the waste without passing the dialyzer.

User Interface (UI)

The UI is a display panel that provides communication between the HD machine and the user. On the display, it is possible to visualize all the dialysis parameters and the relevant information about the procedure and the alarm conditions. By touching the icons on the screen, the user can input all the parameters for the treatment such as the dialysis time, the UF volume and the heparin pump flow.

Control System

The control system is divided into two parts: The top-level control system connects the interface with the user and transmits data to and from other modules. The low-level control system controls and monitors the HD machine and its functions and also communicates with the top-level control system. Both systems operate independent of each other.

3   Therapy

During a therapy, an HD machine pumps blood through a vascular access from the patient into the dialyzer. Inside the dialyzer, metabolic waste products are separated from the blood. The dialyzer operates as a filter that is divided into two parts by a semipermeable membrane. On one side, the patient's blood is flowing and on the other side, the DF.

The DF, a chemical substance that is used in HD to draw fluids and toxins out of the bloodstream and to supply electrolytes and other chemicals to

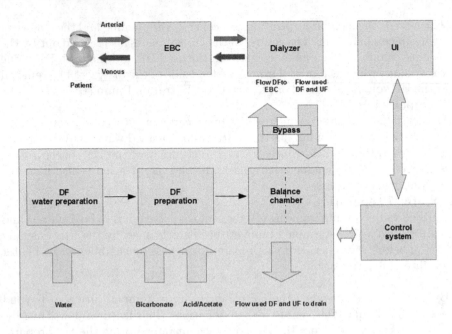

Fig. 1. Schematic view of EBC

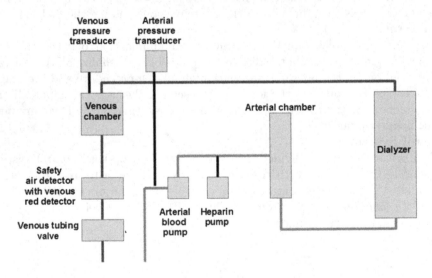

Fig. 2. Architecture of HD machine

the bloodstream, is prepared by the HD machine for the therapy. It consists of prepared water that contains certain quantities of electrolyte and bicarbonate, depending on the individual patient's requirements. The concentrations of electrolyte and bicarbonate in the DF are adjusted in such a way that certain substances can be removed from the blood through convection, diffusion and osmosis, while other substances are added at the same time. This is achieved mainly by diffusive clearance through the semipermeable membrane of the dialyzer. The DF transports the metabolic waste products from the dialyzer into the discharge line. The cleaned blood is then recycled back to the patient.

During therapy, the HD machine performs extracorporeal monitoring of blood circulation, pumps blood and DF in separate circulation systems through the dialyzer and monitors the composition and volume balance of the DF. The heparin pump, which is also part of the HD machine, is used to add anticoagulants to the blood so as to prevent the formation of blood clots.

In addition to cleaning metabolic waste from the blood, the HD machine removes water from the blood which would be excreted through the kidney in healthy humans.

3.1 Types of Therapy

Double-Needle Therapy. If arterial and venous vascular access are different, the therapy method is called double-needle therapy. The double-needle procedure is the standard technique in HD. Blood is extracted from the patient through the arterial vascular access. The blood pump continuously pumps the blood through the arterial tube system to the dialyzer. There, the exchange of metabolic waste products between the blood and the DF proceeds through the semipermeable membrane of the dialyzer. After that, the blood is taken back through the venous tube system, the bubble trap and a second vascular access to the vein.

Single-Needle Therapy. If arterial and venous vascular access are identical, the therapy method will be called single-needle therapy. The single-needle therapy is applied when patients experience problems with the predominantly used double-needle dialysis. In the single-needle procedure, only one needle is applied to the patient. The arterial and venous ends of the tube system are connected via a Y-connector. This procedure allows reducing the number of punctures by half compared to double-needle dialysis, thus preserving the patient's shunt. The single-needle clamp procedure allows ending a running double-needle dialysis in case of problems (e.g., at the shunt). The single-needle therapy is out of scope of this case study.

3.2 Phases of Therapy

A therapy session follows the following sequence of phases: (1) preparation, (2) initiation, and (3) ending.

Therapy Preparation

1. **Automated Self Test**

   First the HD machine automatically checks all control functions relevant to the safety of the machine. If the test is successfully completed, signal lamps on the monitor change to green.

2. **Connecting the Concentrate**

   On successful completion of the test, the requested concentrates, e.g., acetate/acid, are connected to the HD machine.

3. **Setting the Rinsing Parameters**

   The rinsing parameters are entered in the machine. The parameters must be within the defined ranges as shown in Table 2.

Table 2. Rinsing parameters

Text	Range	Description
Filling BP rate	50–600 mL/min	The rate with which the blood side is filled or rinsed
Filling BP volume	0–6000 mL	The BP stops after it has rinsed the blood side using the set volume
Rinsing BP rate	50–300 mL/min	BP rate for rinsing program
DF flow	50–300 mL/min	DF flow rate for rinsing program
Rinsing time	0–59 mins	Duration of adjusted rinsing program
UF rate for rinsing	0–3000 mL/h	When rinsing with a physiological saline solution
UF volume for rinsing	0–2950 mL	When rinsing with a physiological saline solution
Blood flow for connecting patient	50–600 mL/min	

4. **Inserting, Rinsing and Testing the Tubing System**

   (a) Standard arterial/venous (A/V) tubing is inserted. Both arterial and venous tubes must be properly connected. All other components of the machine as depicted in Fig. 1 should be in place.

   (b) Saline bag levels in the blood line chambers for preparation are set.

   (c) Bloodlines are inserted.

   (d) Rinsing and testing the tubing system.

   – The BP speed is increased to 150 mL/min to complete priming. Priming is the process by which a pump is filled with fluid and made able to operate.

   – Once the saline solution has filled the venous line past the drip chamber, the drip chamber and the vent line are completely filled.

- When the BP stops, the arterial and venous patient ends are connected for recirculation and the BP is restarted. The membrane inside the pods should be moving slightly when the BP is running. The faster the BP speed is, the more movement should be detectable.

5. **Preparing the Heparin Pump**

   (a) The heparin syringe is inserted.

   (b) The heparin line is vented.

6. **Setting Treatment Parameters**

   (a) The DF parameters are set. They are described along with their ranges in Table 3.

Table 3. DF parameters

Text	Range	Description
Conductivity	12.5–16.0 mS/cm	The rate with which the blood side is filled or rinsed
Bicarbonate/Acetate		Selection of the concentrate
Bicarbonate conductivity	2–4 mS/cm	
DF temperature	33–40°C	
Rinsing time	0–59 mins	
DF flow	300–800 mL/min	

   (b) The UF parameters are set. They are described along with their ranges in Table 4.

Table 4. UF parameters

Text	Range
UF volume	100–20000 mL
Therapy time	10 min–10 hrs
Min UF rate	0–500 mL/h
Max UF rate	0–4000 mL/h

   (c) The pressure limits are set. They are described along with their ranges in Table 5.

   Limits Window for Arterial Entry Pressure Control. The arterial entry pressure AP (pressure between a patient and a BP) is monitored by an automatically set limits window. A maximum lower arterial limit is set in the machine (max. -400 mmHg). The automatically set lower limit

Table 5. Pressure parameters

Text	Range	Description
Limit delta Min/Max AP	10–100 mmHg	Limits window for arterial entry pressure AP. Distance to min and max AP
Actual TMP/max TMP	300–700 mmHg	
Limits TMP	ON/OFF	Monitoring the TMP at the dialyzer
Low/high	2–99 %	Limits window for TMP in % of actual value
Extended TMP limit range	ON/OFF	
DF flow	300–800 mL/m	

cannot fall below this value. The size of the arterial limits window is defined through the respective distance (delta) between the actual value and the lower and upper limits. The total of the two distances to the actual value gives the width of the arterial limits window. When the actual AP is changed slowly, the limits window is continuously adapted to the actual value, but only within the absolute limits set in the machine.

Limits Window for Trans-Membrane Pressure Control. The trans-membrane pressure (TMP) of the dialyzer is controlled by an automatically set limits window. The size of the limits window is entered as a percentage of the actual value. The limits window is therefore independent of the dialyzer in use. When the limits window is switched off, the control of the dialyzer-dependent maximum TMP is still active. Activating the bypass icon or changing the blood flow causes the limits window to be re-centered.

(d) The heparin parameters are set as described along with their ranges in Table 6.

7. **Rinsing the dialyzer**
   When the DF is prepared, the machine asks to connect the dialyzer. The (arterial drip) chamber in front of the dialyzer entry (BEP) is filled nearly half full and the venous drip chamber up to approx. 1 cm from the upper edge. Once the dialyzer has been filled, the BP stops running. It is ensured that the blood tubing system and the dialyzer are filled and rinsed with physiological saline solution. It is also ensured that the level in the venous chamber is correct.

Therapy Initiation. After completion of the preparation work, the icon for connecting the patient on the UI is enabled. The signal lamps on the monitor change to yellow.

Table 6. Heparin parameters

Text	Range	Description
Heparin stop time	00:00–10:00 hrs:mins	The heparin pump is switched off by the set time prior to the end of the therapy
Heparin bolus volume	0.1–10.0 mL	Bolus volume for a bolus administration during dialysis
Heparin profile/rate	0.1–10.0 mL/h	Continuous heparin rate over the entire duration of heparin administration
Treatment without heparin	deactivated/activated	Switching on/off the heparin monitoring function
Syringe type	10/20/30 mL	A list of permissible syringe types

1. **Connecting the patient and starting therapy**
   - The patient is connected arterially.
   - The BP is started by pressing the START/STOP button on the UI.
   - The blood flow is set.
   - The blood tubing system is filled with blood. The BP stops automatically when blood is detected on the VRD in the SAD.
   - The patient is connected venously.
   - The blood pump is started and the prescribed blood flow is set.
   - The machine is taken out of bypass mode. The HD machine switches to main flow and bicarbonate running. The signal lamps on the UI switch to green.
2. **During therapy**
   (a) Monitoring the blood-side pressure limits

   Venous Return Flow Pressure. The venous return flow pressure (VP) is monitored by an automatically set limits window. The limits window is set 10 s after the last activation of the BP and is identified by markings on the bar showing the venous return flow pressure. The width and thresholds of the limits window are set in the HD machine. The venous lower limit value is automatically adjusted during treatment. This means that the distance between the lower limit and the actual pressure decreases. This compensates for the hematocrit increase generally caused by UF. The adjustment is carried out every 5 min and adds up to 2.5 mmHg at a time. The minimum distance of 22.5 mmHg is, however, always maintained. The venous lower pressure limit during HD is checked. An optimal interval is approximately 35 mmHg between the lower pressure limit and the current value. By changing the speed of the BP for a brief period, it is possible to reposition the limits window.

Arterial Entry Pressure. The arterial entry pressure is automatically monitored within set limits. The limits window is set 10 s after the last activation of the BP. These limits are active in the initiation phase and during final circulation.

Blood-side Entry Pressure at the Dialyzer. The BEP must be connected during preparation. If the BEP transducer protector is used, the BEP at the dialyzer is controlled by its upper limit. The BEP monitoring function warns or signals a possible blockage of the dialyzer due to a kinked tube or increased clotting within the dialyzer. The BEP measurement allows the user to monitor the formation of a secondary membrane layer in the dialyzer. A possible filter clotting might be avoided. The limits can only be set via the alarm limits screen at the beginning of the therapy.

(b) Treatment at minimum UF rate

Treatment at minimum UF rate can be activated to achieve, for instance, an immediate lowering of the set UF rate in case of falling blood pressure and unstable circulation.

(c) Heparin bolus

There is a risk of blood loss due to blood clotting in case of insufficient anticoagulation. Complete heparin bolus as required.

(d) Arterial bolus (saline)

Using the function "arterial bolus" a defined volume of sodium chloride is infused from a Sodium Chloride (NaCl) bag. The required bolus volume is entered on the UI. The BP stops automatically and a safety message appears on the UI. A bag with physiological saline solution to arterial infusion connector is connected to the machine. The arterial bolus is infused. The values can be monitored in the settings window. Once the set quantity has been infused or the arterial bolus has been terminated by an alarm, a window appears to confirm bolus termination.

(e) Interrupting dialysis (bypass)

The therapy can be interrupted by switching to bypass mode. The signal lamps on the UI switch to yellow. The bypass mode can also be terminated likewise.

(f) Completion of treatment

On completion of the treatment, an acoustic signal can be heard, the message "treatment time completed" is displayed, and the signal lamps on the monitor switch to yellow. The UF rate is set to 50 mL/h. The BP is still running.

Therapy Ending

1. Reinfusion

The physical arterial connection is removed from the patient. The arterial line to the infusion Y on the saline line containing the physiological saline solution is connected. The arterial disconnection on the UI is confirmed. The BP starts the reinfusion. The reinfusion screen appears on the UI. The HD machine monitors the reinfusion volume and reinfuses until the red detector (VRD) detects the physiological saline solution. The BP stops. To continue reinfusion, start the BP by pressing the START/STOP button on the UI. The BP stops automatically after 400 mL have been reinfused or when a reinfusion time of 5 min has elapsed. The procedure can be repeated and the HD machine will carry out reinfusion of another 400 mL, or reinfusion for 5 min. The venous patient connection is disconnected.

2. **Emptying the dialyzer**
   The dialyzer is emptied by pressing the respective button on the UI.
3. **Emptying the cartridge after dialyzer drain**
   The cartridge is emptied automatically by the machine.
4. **Overview of the therapy carried out**
   An overview of the therapy is shown with the actual values, e.g., the treated blood volume, the UF volume, the heparin volume, and the time lapsed.

4 Safety Requirements

4.1 General Requirements

S-1 Arterial and venous connectors of the EBC are connected to the patient simultaneously.

S-2 As an intervention, in case of a drop of the patient's blood pressure, an infusion is applied in order to stabilize the cardiovascular circulation. Flow saline at the set volume and the set rate from the saline bag/bottle with closed arterial access through the dialyzer to their venous connection.

S-3 To prevent coagulation of blood, an anti-coagulation pump doses an anti-coagulant into the bloodline between the BP and the dialyzer. Anti-coagulant is flown at the set rate or the set volume from the syringe into the EBC during treatment.

S-4 When the patient is connected to the EBC, the saline solution in the EBC is replaced with blood. The saline solution can be discarded to a bag or a bucket connected to the venous connector of the EBC. The saline solution and the blood are exchanged using the BP. The BP is stopped when either the VRD detects blood or the BP has transported a predefined volume.

S-5 The patient cannot be connected to the machine outside the initiation phase, e.g., during the preparation phase.

S-6 The BP cannot be used for infusion outside the initiation phase, e.g., during saline infusion.

S-7 When the blood flow stops because of BP failure during loss of main power, blood clotting could cause blood loss. The blood should be returned to the patient manually.

S-8 There is a risk to the patient due to hemolysis if the blood flow rate setting is too high for the selected fistula needle (AP too low)! The blood flow rate should be adjusted, taking into consideration the AP.

S-9 There is a risk to the patient due to reduced dialysis effectiveness if the actual blood flow rate is lower than the displayed flow rate if the AP is highly negative. The blood flow rate setting should be corrected and treatment time should be extended.

S-10 There is a risk to the patient due to reduced dialysis effectiveness if the blood flow rate is too low. It should be ensured that the blood flow rate is optimal.

S-11 Once "empty dialyzer" has been confirmed, the BP cannot be started anymore.

4.2   Software Requirements

Arterial Bolus

R-1 During arterial bolus application, the software shall monitor the infusion of saline into the patient and if the infused volume exceeds 0.4 l, then the software shall stop the blood flow and execute an alarm signal. The blood volume can be detected by measuring the pump rotations with the speed sensor of the BP.

Blood Pump

R-2 During initiation, the software shall monitor the blood flow in the EBC and if no flow is detected for more than 120 s, then the software shall stop the BP and execute an alarm signal.

R-3 During initiation, if the machine is not in bypass, then the software shall monitor the blood flow in the EBC and if the actual blood flow is less than 70 % of the set blood flow, then the software shall execute an alarm signal.

R-4 During initiation, the software shall monitor the rotation direction of the BP and if the software detects that the BP rotates backwards, then the software shall stop the BP and execute an alarm signal.

Blood-Side Entry Pressure

R-5 During initiation, if the software detects that the pressure at the VP transducer exceeds the upper pressure limit, then the software shall stop the BP and execute an alarm signal.

R-6 During initiation, if the software detects that the pressure at the VP transducer falls below the lower pressure limit, then the software shall stop the BP and execute an alarm signal.

R-7 During initiation, if the software detects that the pressure at the AP transducer exceeds the upper pressure limit, then the software shall stop the BP and execute an alarm signal.

R-8 During initiation, if the software detects that the pressure at the AP transducer falls below the lower pressure limit, then the software shall stop the BP and execute an alarm signal.

R-9 While connecting the patient, if the software detects that the pressure at the VP transducer exceeds +450 mmHg for more than 3 s, then the software shall stop the BP and execute an alarm signal.

R-10 While connecting the patient, if the software detects that the pressure at the VP transducer falls below the defined lower pressure limit for more than 3 s, then the software shall stop the BP and execute an alarm signal.

R-11 While connecting the patient, if the software detects that the pressure at the AP transducer falls below the lower pressure limit for more than 1 s, then the software shall stop the BP and execute an alarm signal.

R-12 During reinfusion, if the software detects that the pressure at the VP transducer exceeds +350 mmHg for more than 3 s, then the software shall stop the BP and execute an alarm signal.

R-13 During reinfusion, if the software detects that the pressure at the AP transducer falls below −350 mmHg for more than 1 s, then the software shall stop the BP and execute an alarm signal.

Connecting the Patient

R-14 While connecting the patient, the software shall monitor the blood flow in the EBC and if no flow is detected, then the software shall stop the BP and execute an alarm signal. The blood flow can be detected by measuring the pump rotations with the speed sensor of the BP.

R-15 While connecting the patient, the software shall monitor the filling blood volume of the EBC and if the filling blood volume exceeds 400 mL, then the software shall stop the BP and execute an alarm signal. The blood volume can be detected by measuring the pump rotations with the speed sensor of the BP.

R-16 While connecting the patient, the software shall use a timeout of 310 s after the first start of the BP. After this timeout, the software shall change to the initiation phase.

R-17 While connecting the patient, the software shall monitor the blood flow direction and if the reverse direction is detected, then the software shall stop the BP and execute an alarm signal. The blood flow direction can be detected by the direction sensor of the BP.

Flow Bicarbonate Concentrate into Mixing Chamber

R-18 During preparation of the DF in the bicarbonate mode, if acid concentrate is provided instead of bicarbonate concentrate, then the software shall detect the mix-up of concentrates and disconnect the dialyzer from the DF and execute an alarm signal.

R-19 During preparation of the DF in bicarbonate mode, if acetate concentrate is provided instead of bicarbonate concentrate, then the software shall detect the mix-up of concentrates and disconnect the dialyzer from the DF and execute an alarm signal.

Heat and Degas DF Water

R-20 If the machine is in the preparation phase and performs priming or rinsing or if the machine is in the initiation phase and if the temperature exceeds the maximum temperature of 41°C, then the software shall disconnect the dialyzer from the DF and execute an alarm signal.

R-21 If the machine is in the initiation phase and if the temperature falls below the minimum temperature of 33°C, then the software shall disconnect the dialyzer from the DF and execute an alarm signal.

Heparin

R-22 If anticoagulant delivery is running, then the software shall monitor the anticoagulant flow direction and if the reverse direction is detected, then the software shall stop the blood flow and the anticoagulant flow, and execute an alarm signal.

Safety Air Detector

R-23 If the machine is in the preparation phase and performing rinsing of the EBC or if the machine is connecting the patient or if the machine is in the initiation phase or if the machine is in the reinfusion process, then the software shall monitor the flow through the SAD sensor and if the flow through the SAD sensor exceeds 1200 mL/min, then the software shall stop the BP and execute an alarm signal.

R-24 If the flow through the SAD sensor is in the range of 0 to 200 mL/min, then the software shall use an air volume of 0.2 mL as limit for air detection by the SAD sensor.

R-25 If the flow through the SAD sensor is in the range of 200 to 400 mL/min, then the software shall use an air volume of 0.3 mL as limit for air detection by the SAD sensor.

R-26 If the flow through the SAD sensor is greater than 400 mL/min, then the software shall use an air volume of 0.5 mL as limit for air detection by the SAD sensor.

R-27 The software shall update the air volume detected by the SAD sensor every 1 mS.

R-28 If the machine is in the preparation phase and performing rinsing of the EBC and if the software detects that the air volume exceeds the air volume limit depending on the actual flow through the venous blood line, then the software shall stop the blood flow and execute an alarm signal.

R-29 During the application of arterial bolus, if the software detects that the air volume exceeds the air volume limit depending on the actual flow through the venous blood line, then the software shall stop the blood flow and execute an alarm signal.

R-30 While connecting the patient, if the software detects that the air volume exceeds the air volume limit depending on the actual flow through the venous blood line, then the software shall stop the blood flow and execute an alarm signal.

R-31 During the initiation phase, if the software detects that the air volume exceeds the air volume limit depending on the actual flow through the venous blood line, then the software shall stop the blood flow and execute an alarm signal.

R-32 During the reinfusion process, if the software detects that the air volume exceeds the air volume limit depending on the actual flow through the venous blood line, then the software shall stop the blood flow and execute an alarm signal.

Ultrafiltration

R-33 The software shall monitor the net fluid removal rate in the balance chamber and if the net fluid removal rate exceeds a safe upper limit, then the software shall stop flow from and to the dialyzer and execute an alarm signal.

R-34 If the machine is in the initiation phase and net fluid removal is enabled, then the software shall monitor the rotation direction of the UFP and if backward rotation of the UFP is detected, then the software shall put the machine in bypass and execute an alarm signal. The backward delivered volume shall not exceed 400 mL.

R-35 If the machine is in the initiation phase and net fluid removal is enabled, then the software shall monitor the net fluid removal volume and if the net fluid removal volume exceeds (UF set volume + 200 mL), then the software shall put the machine in bypass and execute an alarm signal. When the alarm is acknowledged by the user, then the software shall increase the UF set volume by 200 mL.

R-36 If the machine is in the initiation phase and net fluid removal is enabled and if the bypass valve is opened, then the software shall stop the DF flow to and from the dialyzer and execute an alarm signal.

5 Final Remarks

We have documented the requirements and design of an HD machine to provide an open example of model-based engineering of safety-critical active medical devices. We are now looking for solution models formulated in various rigorous methods and encompassing related high-level safety and efficacy properties whose correctness can be effectively asserted using conventional or unconventional verification and validation methods.

How to Assure Correctness
and Safety of Medical Software:
The Hemodialysis Machine Case Study

Paolo Arcaini[1]([✉]), Silvia Bonfanti[2],
Angelo Gargantini[2], and Elvinia Riccobene[3]

[1] Faculty of Mathematics and Physics,
Charles University in Prague, Prague, Czech Republic
arcaini@d3s.mff.cuni.cz
[2] Department of Economics and Technology Management,
Information Technology and Production,
Università degli Studi di Bergamo, Bergamo, Italy
{silvia.bonfanti,angelo.gargantini}@unibg.it
[3] Dipartimento di Informatica, Università degli Studi di Milano, Milano, Italy
elvinia.riccobene@unimi.it

Abstract. Medical devices are nowadays more and more software dependent, and software malfunctioning can lead to injuries or death for patients. Several standards have been proposed for the development and the validation of medical devices, but they establish general guidelines on the use of common software engineering activities without any indication regarding methods and techniques to assure safety and reliability.

This paper takes advantage of the Hemodialysis machine case study to present a formal development process supporting most of the engineering activities required by the standards, and provides rigorous approaches for system validation and verification. The process is based on the Abstract State Machine formal method and its model refinement principle.

1 Introduction

In medical treatments depending on the use of a medical device (e.g., a hemodialysis machine), patient safety depends upon the correct operation of the device hardware/software. For this reason, validation and verification of medical devices are mandatory, and methods and techniques to assure medical software safety and reliability are highly demanded. Along this research line, the Hemodialysis machine case study [20] (HMCS) has been proposed, within the ABZ 2016 conference, as an example of medical device that should be formally validated and verified to assure safety and hardware-software correct interaction.

The research reported in this paper has been partly supported by the Grant Agency of the Czech Republic project 14-11384S, and by the Austrian Ministry for Transport, Innovation and Technology, the Federal Ministry of Science, Research and Economy, and the Province of Upper Austria in the frame of the COMET center SCCH.

© Springer International Publishing Switzerland 2016
M. Butler et al. (Eds.): ABZ 2016, LNCS 9675, pp. 344–359, 2016.
DOI: 10.1007/978-3-319-33600-8_30

Although several standards for the validation of medical devices have been proposed – as ISO 13485 [1], ISO 14971 [4], IEC 60601-1 [2], EU Directive 2007/47/EC [16] –, they mainly consider physical aspects and electrical components of a device rather than its software. The main references concerning regulation of medical software development are the standard IEC 62304 [3] (International Electrotechnical Commission) and the "General Principles of Software Validation" [21] established by the FDA (Food and Drug Administration). Both documents establish general guidelines on the use of common software engineering activities, but they do not provide any indication regarding life cycle models, or methods and techniques to assure safety and reliability. These qualities could be assured by adopting rigorous approaches of software development, based on the use of formal methods, that allow the designers to specify what the software is intended to do by means of mathematical models, and demonstrate that the use of the software fulfills those intentions by means of precise validation and verification techniques [3, 19].

Among the different existing formal methods, Abstract State Machines [13] have been already successfully used in the context of medical software for the rigorous development of an optometric measurement device software [5].

The ASM-based design process [10] has an incremental life cycle model. It is based on model refinement, includes the main software engineering activities, and is supported by techniques for model validation and verification at any desired level of detail. The process can guide the development of software and embedded systems seamlessly from requirements capture to their implementation, and this has been shown by numerous and successful case studies [13]. Although the ASM method has a rigorous mathematical foundation, practitioners need no special training to use the method since ASMs can be correctly understood as pseudo-code or virtual machines working over abstract data structures.

In this paper, we apply the ASM-based process for modeling, validating, and verifying the HMCS. At the same time, we present a software development process compliant with two current standards for medical software. It provides a life cycle model for IEC62304 and embeds most of the software engineering activities required by this standard. Moreover, it reflects the principles established by the FDA, especially those regarding the integration of validation and verification activities into the development process.

Section 2 briefly introduces the ASM-based development process. Section 3 presents the specification of the HMCS given in terms of a chain of refined models. Section 4 presents some validation activities we have applied to the developed models, and Sect. 5 reports the verification of the requirements. Section 6 recalls the normative for medical software and shows how the ASM-based development process captures the existing regulations. Section 7 concludes the paper.

As at the moment of writing this paper no other solutions for the HMCS are available, we do not report any related work. For an example of application of different formal methods to a safety-critical system, we remind the reader to the solutions proposed for the Landing Gear System case study [12].

2   ASM-Based Development Process

ASMs allow an iterative design process, shown in Fig. 1, based on model refinement. Validation and verification (V&V) are fully integrated into the process, and are possible at any level of abstraction. Tools supporting the process are part of the ASMETA (ASM mETAmodeling) framework[1] [11].

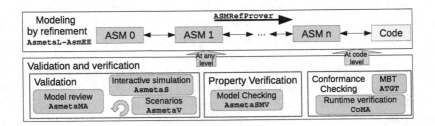

Fig. 1. ASM-based development process

Modeling requirements starts by developing a high-level model called *ground model* (ASM 0 in Fig. 1). It is specified by reasoning on the informal requirements (generally given as a text in natural language) and using terms of the application domain, possibly with the involvement of all stakeholders. The ground model should *correctly* reflect the intended requirements and should be *consistent*, i.e., without possible ambiguities of initial requirements. It does not need to be *complete*, i.e., it may not specify some given requirements. The ground model and the other ASM models can be edited in AsmEE by using the concrete syntax AsmetaL [18].

Starting from the ground model, through a sequence of *refined* models, further functional requirements can be specified and a complete architecture of the system can be given. The refinement process permits to tackle the complexity of the system, and allows to bridge, in a seamless manner, specification to code.

Each refinement step should be proved to be correct: the refinement correctness proof can be done by hand or, for a particular kind of refinement called *stuttering refinement*, using the tool ASMRefProver.

At each level of refinement, already at the level of the ground model, different *validation* and *verification* (V&V) activities can be applied.

Model validation helps to ensure that the specification really reflects the intended requirements, and to detect faults and inconsistencies as early as possible with limited effort. ASM model validation is possible by means of the model simulator AsmetaS [18] and by the validator AsmetaV [15] that allows to build and execute *scenarios* of expected system behaviors. A further validation technique is *model review* (a form of static analysis) to determine if a model has sufficient *quality* attributes (as minimality, completeness, consistency). Automatic ASM model review is possible by means of the AsmetaMA tool [7].

---

[1] http://asmeta.sourceforge.net/.

Model verification requires the use of more expensive and accurate methods. Formal verification of ASMs is possible by means of the model checker AsmetaSMV [6], and both *Computation Tree Logic* (CTL) and *Linear Temporal Logic* (LTL) formulas are supported.

If a system implementation is available, either derived from the model (as last low-level refinement step) or externally provided, also *conformance checking* is possible. Both *model-based testing* (MBT) and *runtime verification* can be applied to check if the implementation conforms to its specification [9]. We support conformance checking w.r.t. Java code. The tool ATGT [17] can be used to automatically generate tests from ASM models[2] and, therefore, to check the conformance *offline*; CoMA [8], instead, can be used to perform runtime verification, i.e., to check the conformance *online*.

3 Hemodialysis Device: Modeling

In modeling the hemodialysis device[3] we have proceeded through refinement. The complete model has been developed in five levels, as shown in Fig. 2 where some data of the five models are reported. Each refined model introduces, w.r.t. the previous model, some additional details to the machine behaviour. The ground model abstractly describes the transitions between the three hemodialysis phases: preparation, initiation, and ending. In the first three refinement steps, we refine the three phases singularly. In the last refinement step, to comply with the given requirements, we introduce the modeling of the checks and the error handling performed by the device. For each step of refinement, we prove the refinement correctness using the SMT-based tool AsmRefProver.

Ground model	1st refinement	2nd refinement	3rd refinement	4th refinement
Machine phases	Preparation Phase	Initiation Phase	Ending Phase	Checks and errors
size: 0 m, 1 c, 0 d, 5 rd, 11 r, 0 p	*size*: 15 m, 9 c, 2 d, 58 rd, 179 r, 0 p	*size*: 32 m, 22 c, 3 d, 89 rd, 303 r, 2 p	*size*: 39 m, 25 c, 3 d, 103 rd, 363 r, 3 p	*size*: 60 m, 39 c, 12 d, 154 rd, 608 r, 41 p

Fig. 2. Refinement steps - *size*: m = #monitored functions, c = #controlled, d = #derived, rd = #rule declarations, r = #rules, p = #properties

3.1 Ground Model

The ground model simply describes the transition between the phases that constitute a hemodialysis treatment, without any additional detail. Code 1 shows the ground model written using the AsmetaL syntax. The machine, at each step, checks the current **phase** of the treatment and executes the corresponding rule. Rule r_preparation performs all the activities necessary to prepare the hemodialysis device for the treatment (modeled in Sect. 3.2). Rule r_initiation

---

[2] Note that sequences generated by ATGT could be used to test programs written in any programming language.

[3] Models are available at http://fmse.di.unimi.it/sw/ABZ2016caseStudy.zip.

specifies the hemodialysis therapy in which the patient is connected to the device and the blood cleaning is performed (modeled in Sect. 3.3). Rule r_ending models the ending of the therapy, in which the patient is disconnected from the device and the device is cleaned for subsequent treatments (modeled in Sect. 3.4).

```
asm Hemodialysis_GM                      rule r_run_dialysis =
signature:                                 switch(phase)
  enum domain Phases =                         case PREPARATION: r_preparation[]
  {PREPARATION | INITIATION | ENDING}          case INITIATION: r_initiation[]
  controlled phase: Phases                     case ENDING: r_ending[]
                                             endswitch
definitions:                             main rule r_Main =
  rule r_preparation = phase := INITIATION   r_run_dialysis[]
  rule r_initiation = phase := ENDING      default init s0:
  rule r_ending = skip                       function phase = PREPARATION
```

Code 1. Ground model

3.2 First Refinement: Preparation Phase

The first refinement extends the ground model by refining the preparation phase specified in rule r_preparation (see Code 2). In this phase, different activities are performed to prepare the device for the therapy, as the preparation of the heparin or the rinsing of the dialyzer. Function modePreparation specifies which activity (i.e., which rule) must be executed in the current state.

```
asm Hemodialysis_ref1                          rule r_set_treatment_param =
signature:                                       switch(treatmentParam)
  enum domain ModePreparation = { ... |              case BLOOD_CONDUCTIVITY:
  RINSE_DIALYZER | SET_TREAT_PARAM}                    r_set_blood_conductivity[]
  enum domain TreatmentParam = { ... |               case ACTIVATION_H:
  BLOOD_CONDUCTIVITY | ACTIVATION_H}                   r_set_h_activation[]
  controlled modePreparation: ModePreparation         ...
  controlled treatmentParam: TreatmentParam        endswitch
  ...
definitions:                                   rule r_preparation =
                                                 switch(modePreparation)
  rule r_insert_param($low_lim in Integer,           case AUTO_TEST:
                 $up_lim in Integer,                   r_auto_test[]
                 $next_param in TreatmentParam,     case CONNECT_CONCENTRATE:
                 $mon_param in Integer,               r_connect_concentrate[]
                 $contr_param in Integer) =        case SET_RINSING_PARAM:
    if $mon_param <= $up_lim and                      r_set_rinsing_param[]
              $mon_param >= $low_lim then          case TUBING_SYSTEM:
      par                                             r_tubing_system[]
        $contr_param := $mon_param                 case PREPARE_HEPARIN:
        treatmentParam := $next_param                 r_prepare_heparin[]
      endpar                                       case SET_TREAT_PARAM:
    endif                                             r_set_treatment_param[]
                                                   case RINSE_DIALYZER:
  rule r_set_h_activation =                           r_rinse_dialyzer[]
    r_insert_param[SYRINGE_TYPE,                  endswitch
               activation_h, activation_h_contr]  ...
```

Code 2. First refinement

For the lack of space, we here only report the setting of the treatment parameters, that is modeled by rule r_set_treatment_param. The machine sets one parameter at a time, using function `treatmentParam` to keep track of the parameter that has to be updated. Each parameter is updated by rule `r_insert_param`, shown in Code 2. For each parameter, there exist one monitored function and one controlled function. The model takes the value of the monitored function decided by the environment (in this case, by the user), and, if the specified value is allowed, saves it in the controlled function. When a correct value is acquired, the machine updates function `treatmentParam` with the name of the next parameter that must be set.

3.3   Second Refinement: Initiation Phase

This step refines the initiation phase (see Code 3), which is divided into two steps identified by function `initPhaseMode`: `r_initiate_patient` and `r_run_therapy`. During all the initiation phase, some checks on the device are performed by rule `r_check_initiation_phase`; such rule is left abstract in this refinement step, and it will be modeled in the fourth refinement (see Sect. 3.5).

```
asm Hemodialysis_ref2

signature:
  enum domain ModelInitiation =
  {CONNECT_PATIENT | THERAPY_RUNNING}
  enum domain TherapyPhase =
  {START_HEPARIN | ... | THERAPY_END}
  controlled initPhaseMode: ModelInitiation
  controlled therapyPhase: TherapyPhase
  ...

definitions:

  rule r_initiate_patient =
    par
      r_check_patient[]
      r_connect_patient[]
    endpar
```

```
rule r_run_therapy =
  switch(therapyPhase)
    case START_HEPARIN:
      r_start_heparin[]
    case THERAPY_EXEC:
      r_therapy_exec[]
    case THERAPY_END:
      r_therapy_end[]
  endswitch

rule r_initiation =
  par
    r_check_initiation_phase[]
    switch(initPhaseMode)
      case CONNECT_PATIENT:
        r_initiate_patient[]
      case THERAPY_RUNNING:
        r_run_therapy[]
    endswitch
  endpar
```

Code 3. Second refinement

`r_initiate_patient` is the first activity to perform. It is divided into two parallel actions, `r_check_patient` and `r_connect_patient`. Rule `r_check_patient`, that is responsible for doing some additional checks on the patient, is also left abstract in this refinement step (as rule `r_check_initiation_phase`), and it will be modeled in the fourth refinement (see Sect. 3.5). Rule `r_connect_patient` (shown in Code 4) implements the procedures described in section "Connecting the patient and starting therapy" of the case study document [20]; it executes a series of activities related to the patient connection, specified by function `patientPhase`. At the beginning, the patient is connected arterially (by rule `r_conn_arterially`). Then, the blood pump is started (by rule `r_start_bp`), the blood flow is set (by rule `r_set_blood_flow`), and the blood tubing system is filled with blood (by rule `r_fill_tubing`). Then, the patient is connected venously (by rule `r_conn_venously`). Finally, the blood pump is started and the blood flow is set again; the connection procedure is terminated (by rule r_end_-connection).

```
asm Hemodialysis_ref2                              rule r_therapy_exec =
signature:                                            par
  enum domain PatientPhase =                            r_pump_heparin[]
  {CONN_ARTERIALLY | ... | END_CONN}                    r_monitor_ap_vp_limits[]
  controlled patientPhase: ModeInitiation               r_therapy_min_UF[]
  ...                                                    r_update_blood_flow[]
definitions:                                            r_check_therapy_run[]
rule r_connect_patient =                                r_arterial_bolus[]
  switch(patientPhase)                                  r_interrupt_dialysis[]
    case CONN_ARTERIALLY: r_conn_arterially[]           r_therapy_completion[]
    case START_BP: r_start_bp[]                       endpar
    case BLOOD_FLOW: r_set_blood_flow[]
    case FILL_TUBING: r_fill_tubing[]
    case CONN_VENOUSLY: r_conn_venously[]
    case END_CONN: r_end_connection[]
  endswitch
```

Code 4. Second refinement – Rules r_connect_patient and r_therapy_exec

r_run_therapy is executed after r_initiate_patient and it is composed of a set of activities, as shown in Code 3. Function therapyPhase specifies which activity must be executed in a given moment. At the beginning, rule r_start_-heparin activates the heparin, if required by the doctor. Then, the therapy is executed (rule r_therapy_exec) and terminated (rule r_theraphy_end).

Rule r_therapy_exec is shown in Code 4. A set of parallel activities are performed as specified in section "During therapy" of the case study document [20]: the monitoring of the blood pressure limits (rule r_monitor_ap_vp_limits), the activation of the treatment at the minimum ultra filtration rate (rule r_therapy_min_UF), the update of the blood flow (rule r_update_blood_flow), the infusion of sodium chloride (rule r_arterial_bolus), the pumping of the heparin in the blood (rule r_pump_heparin). Rule r_check_therapy_run does some checks during the blood cleaning: it is left abstract at this level of refinement and it will be modeled in the fourth refinement (see Sect. 3.5). Activities r_interrupt_dialysis and r_therapy_completion can terminate the therapy (by updating therapyPhase to THERAPY_END): r_interrupt_dialysis models the premature interruption of the therapy, whereas r_therapy_completion models the normal therapy conclusion.

3.4   Third Refinement: Ending Phase

The third refinement details the behaviour of the ending phase (see Code 5) consisting of a set of activities that are performed at the end of the treatment. Function endingPhaseMode specifies the current activity. At the beginning, the reinfusion of the blood takes place by means of rule r_reinfusion. Then, the dialyzer is emptied (by rule r_empty_dialyzer), and afterwards the cartridge is emptied (by rule r_empty_cartridge). At the end, rule r_therapy_overview gives an overview of the executed therapy (e.g., how much blood has been treated).

```
asm Hemodialysis_ref3                          definitions:
                                                  rule r_ending =
signature:                                           switch(endingPhaseMode)
  enum domain ModeEnding =                              case REINFUSION:
  {REINFUSION | ... | THERAPY_OVERVIEW}                   r_reinfusion[]
  controlled endingPhaseMode: ModeEnding                 case DRAIN_DIALYZER:
  ...                                                     r_empty_dialyzer[]
                                                        case EMPTY_CARTRIDGE:
                                                          r_empty_cartridge[]
                                                        case THERAPY_OVERVIEW:
                                                          r_therapy_overview[]
                                                     endswitch
```

Code 5. Third refinement

3.5 Fourth Refinement: Handling of Checks and Errors

In the last refinement step, we model the checking activities and the error handling performed by the device. During the whole hemodialysis process, the device checks some parameters acquired using sensors. Every parameter is checked and, when an error occurs, the system raises an alarm and an error signal. The alarms are reset just after the device operator presses the button. The errors, instead, are reset when the user fixes them.

For example, we here consider the error due to the high temperature of the dialysing fluid (see Code 6). The machines checks the temperature only if an error related to high temperature has not been raised yet (rule r_check_temp_high in Code 6). When the temperature exceeds the threshold value, an error and an alarm signal are generated. If the alarm is running, the device operator can decide to reset it by setting reset_alarm function to true in rule r_turnOff_alarm. Once the alarm has been reset (but the error is still flagged) and the temperature has returned to an acceptable value, the model resets the error (rule r_error_temp_high).

```
asm Hemodialysis_ref4                          rule r_turnOff_alarm =
signature:                                       if reset_alarm then
  enum domain ErrorAlarmType =                      par
  {TEMP_HIGH | UF_BYPASS |...}                         forall $a in ErrorAlarmType
  controlled error: ErrorAlarmType −> Boolean                      with alarm($a) do
  controlled alarm: ErrorAlarmType −> Boolean            alarm($a) := false
  ...                                                  ...
definitions:                                        endpar
  rule r_check_temp_high =                       endif
    if not(error(TEMP_HIGH)) then
      if current_temp > 41 then                 rule r_error_temp_high =
        par                                       if error(TEMP_HIGH) and
          error(TEMP_HIGH) := true                   not(alarm(TEMP_HIGH)) then
          alarm(TEMP_HIGH) := true                   if current_temp <= 41 then
        endpar                                          error(TEMP_HIGH) := false
      endif                                         endif
    endif                                         endif
```

Code 6. Fourth refinement – Error of high temperature of the dialyzing fluid

4   Hemodialysis Device: Validation

We here describe the validation activities we performed on the produced models.

Interactive Simulation and Scenario-Based Validation. As first validation activity, we performed *interactive simulation* by means of the simulator
AsmetaS [18] that allowed us to observe some particular system executions. Interactive simulation consists in providing inputs (i.e., values of monitored functions)
to the machine and observing the computed state. The simulator, at each step,
performs *consistent updates checking* to check that all the updates are consistent
(two updates are inconsistent if they update the same location to two different
values at the same time [13]), and *invariant checking*.

As interactive simulation is a tedious and time-consuming activity, after we
gained enough confidence that the system roughly captured the intended requirements, we performed a more powerful form of simulation, called *scenario-based
validation* [15], that permits to automatize the simulation activity. In scenario-
based validation the designer provides a set of scenarios specifying the expected
behaviour of the models (using the textual notation Avalla [15]). These scenarios are used for validation by instrumenting the simulator AsmetaS. During
simulation, AsmetaV captures any check violation and, if none occurs, it finishes
with a *PASS* verdict. Avalla provides constructs to express execution scenarios
in an algorithmic way, as interaction sequences consisting of actions committed
by the user to set the environment (i.e., the values of monitored/shared functions), to check the machine state, to ask for the execution of certain transition
rules, and to enforce the machine itself to make one step (or a sequence of steps
by command step until) as reaction of the actor actions.

While developing our models, we wrote several scenarios for exercising the
models under different operating conditions. Soon we discovered that such scenarios had several common parts, since they had to perform the same actions
and same checks in different parts of their evolution. Therefore, we extended
the validator with the possibility to define some *blocks of actions* that can be
reused in different components: a block is a named sequence of Avalla commands delimited by keywords begin and end. A command block can be defined
in any Avalla scenario and can be called by means of the command execblock
in other parts of the same scenario or in other scenarios.

Code 7 shows an example of scenario (over the last refined machine) repro-
ducing the situation in which an error is detected and resolved. The scenario
calls some blocks (defined in completeTherapy) containing actions related to
the therapy that are used in several different scenarios (e.g., the initial checks
and the operations performed during the preparation phase). Code 8 shows the
definition of some blocks.

```
scenario errorPressureTherapy
load Hemodialysis_ref4.asm

execblock completeTherapy.initiationCheck;
step
execblock completeTherapy.preparationPhase;
step
execblock completeTherapy.patientConn;
step
check therapyPhase = THERAPY_EXEC;
check heparin_running = true;

set passedSec(10) := true;
set current_ap := 100;
set vp_limit_low := 50;
set vp_limit_up := 150; ...
step
check therapyPhase = THERAPY_END;

step
execblock completeTherapy.therapyEnd;
```

```
scenario completeTherapy
load Hemodialysis_ref4.asm

begin initiationCheck
    check phase = PREPARATION and
    prepPhaseMode = AUTO_TEST and
    rinsingParam = FILLING_BP_RATE;
    ...
    set auto_test_end := true;
end

begin preparationPhase
    check prepPhaseMode =
            CONNECT_CONCENTRATE;
    set conn_concentrate := true;
    step
    ...
end
```

Code 7. Example of scenario using blocks   **Code 8.** Example of scenario blocks

Static Analysis. Although interactive simulation and scenario-based validation can discover many faults in the models, some of them may pass undetected. Moreover, even if the models correctly capture the requirements, they may still be improved from the stylistic point of view. We therefore applied *model review*, whose aim is to determine if a model has some particular *qualities* that should help in developing, maintaining, and enhancing it. The `AsmetaMA` tool [7] (based on `AsmetaSMV`) performs *automatic* review of ASMs. Common vulnerabilities and defects that can be introduced during ASM modeling are checked as violations of suitable *meta-properties* (*MPs*, defined in [7] as CTL formulae). The violation of a meta-property means that a quality attribute (*minimality, completeness, consistency*) is not guaranteed, and it may indicate the presence of an actual fault (i.e., the ASM is indeed faulty), or only of a *stylistic defect* (i.e., the ASM could be written in a better way). For example, the presence of an inconsistent update (meta-property MP1) is the sign of a real fault in the model; the presence of functions that are never read nor updated (meta-property MP7), instead, may simply indicate that the model is not minimal, but not that it is faulty.

In our last step of refinement, we found several violations of meta-property MP1, i.e., inconsistent updates. These were due to a wrong mechanism in the handling of errors. Different kinds of errors can be detected by the device (see Sect. 3.5), and, in all these cases, the state of the device is set to the BYPASS mode. In the faulty model, when a given error is resolved, the error handler puts the device back to the normal mode MAIN_FLOW; however, if another error is simultaneously detected, another error handler puts the device in the BYPASS mode: in this way, an inconsistent update occurs, since two rules try to update the same function to two different values. We have restructured the model so that error handlers only detect errors and mark them as resolved after the error handling (but do not modify the device state); the real mode of the device is now determined by a derived function, whose value is BYPASS if at least one not resolved error exists, otherwise its value is determined by the phase of the therapy. Note that we were not able to detect such a problem by means of simulation, since in simulations we considered errors occurring singularly or simultaneously, but we did not consider errors occurring while others are being resolved.

We also found some minimality violations related to some domain values that were not useful. For example, we discovered that value RED of domain SignalLamps was useless: indeed, in the manual of the device [14], we found that the lamp could be red in particular situations, but this case was not considered in the given requirements [20]. We may decide to remove the value from the domain or keep it for further improvements of the model.

5   Hemodialysis Device: Verification of Requirements

We verified the 11 general requirements (S-1 – S-11) and the 36 software requirements (R-1 – R-36) specified in the case study description [20]. We specified requirements as LTL formulas and we used the model checker AsmetaSMV [6] for their verification. A formula may capture more than one requirement.

As an example of requirement formalization, let us consider the general requirement S-11: "Once *empty dialyzer* has been confirmed, the blood pump cannot be started anymore". Such requirement has been specified as follows:

g(empty_dialyzer implies g(bp_status_der = STOP))

Along the validation activity, each requirement has been proved as soon as possible in the chain of refinements, i.e., in the first model describing all the elements involved in the requirement. Requirement S-1, regarding the connection of arterial and venous connectors, has been added in the second refinement, that models the initiation phase in which the patient is physically connected to the device. Also requirement S-4, regarding the exchange between saline solution and blood during the connection of the patient, has been specified in the second refinement. Requirement S-11, regarding the stopping of the blood pump at the end of the therapy, has been added in the third refinement that models the ending phase. All the other general requirements and all the software requirements are related to errors and alarms, and, therefore, have been specified in the last refinement step that models error handling.

Although we specified all of the requirements, some of them are trivial since already captured by the semantics of the transition rules. In particular, all the properties (from R-1 to R-36) having form $G(\varphi \text{ implies } X(\rho))$, where φ and ρ are predicates over the ASM state, are directly captured by the model structure of nested conditional rules.

We discovered that some requirements specified in [20] are not correct. For example, we specified the following property for S-1:

g(art_connected_contr iff ven_connected_contr)

checking that "arterial and venous connectors of the EBC are connected to the patient simultaneously" [20]. However, the property is false since the patient is *before* connected to the arterial connector and *then* to the venous connector.

We had problems in verifying some requirements, since these were ambiguous. Requirement S-5, for example, states that "the patient cannot be connected to the machine outside the initiation phase, e.g., during the preparation phase". We were not sure whether to interpret "be connected" as the status of the patient who is attached to the machine, or as the atomic action in which the doctor connects the patient to the machine. Following the former interpretation, the temporal property would be

g((art_connected_contr or ven_connected_contr) implies phase = INITIATION)

However, the property is false since the patient can be attached to the machine also outside the INITIATION. Following the former interpretation, we wrote the two following properties

g((not art_connected_contr and x(art_connected_contr)) implies
                                                phase = INITIATION)
g((not ven_connected_contr and x(ven_connected_contr)) implies
                                                phase = INITIATION)

that are both true. It could be the case that this interpretation is indeed the correct one; however, this is a clear example of an ambiguous requirement that would need a clarification from the stakeholders.

For keeping verification time reasonable, we applied some abstractions to our models. For example, in the preparation phase introduced in the first refinement (see Sect. 3.2), the model retrieves values for the parameters from the environment (by means of monitored functions), checks their validity (i.e., whether they are in allowed ranges), and stores them in some controlled functions. For model checking, we abstract from that mechanism: for each parameter, we simply have a monitored function saying whether its new value is allowed or not.

6   ASM Process and Normatives for Medical Software

We want here to relate the ASM-based design process to the current normative for developing medical software. The aim is to evaluate how far we are from having a formal process compliant with the standards. The two main normative references for development and validation of medical software are the standard IEC 62304 [3] and the "General Principles of Software Validation" [21] established by the FDA.

ASM Process and IEC 62304 Standard. The standard IEC 62304 classifies medical software in three classes on the basis of the potential injuries caused by software malfunctions, and defines the life cycle activities (points 5.1–5.8 of Sect. 5 in [3], also shown in Fig. 3) that have to be performed and appropriately documented when developing medical software. Each activity is split into tasks that are mandatory or not, depending on the class of software. The standard does

5.1 Software development planning	5.2 Software requirements analysis	5.3 Software architectural design	5.4 Software detailed design	5.5 Software unit implementation and verification	5.6 Software integration and integration testing	5.7 Software system testing	5.8 Software release

Fig. 3. IEC 62304 development process

not prescribe a specific life cycle model, nor it gives indications on methods and techniques to apply. Users are responsible to map their model to the standard.

Step (5.1) essentially consists in defining a life cycle model, planning procedures and deliverables, establishing how to achieve traceability among system requirements, software requirements, software test and risks control. By using the ASMs, we follow an iterative life cycle model based on refinement. Procedures are modeling, validation, verification, and conformance checking. Deliverables are given in terms of a sequence of refined models, each model equipped with validation and verification results. Traceability is given by the conformance relation between abstract and refined models, at each refinement step. We do not consider risk management, although ASM tools can be used to predict possible risks by reasoning on models and checking incorrect behaviors or potential faults. Risk management was also not part of the case study description in [20].

Step (5.2) consists in defining, documenting and verifying software requirements. This is a continuous activity along the ASM process till the desired level of refinement, possibly to code level. For the HMCS, this is what is reported in Sects. 3, 4, 5 (there is no implementation in this case).

Steps (5.3 - 5.4) regard the definition of *design* from *specification*. In the ASM process, these steps are performed in a seamless manner along the chain of refined models. Already at ground level, system structure is captured, even if not completely, by the model signature. Design decisions and architectural choices are added at a certain level of the refinement. For example, for the hemodialysis machine, patient and hardware checks and error handling are dealt at level four.

Steps (5.5 - 5.7) regard software implementation and testing, at unit and integration levels. Using ASMs, a first code prototype could be obtained as last refinement step and conformance checking is possible. Otherwise, techniques for model-based testing and runtime verification can be used having models available. These techniques have not been shown for the HMCS since not requested. However, we refer to [5] as an example of application.

Step (5.8) includes the demonstration, by a device manufacturer, that software has been validated and verified. It is intrinsic to the ASM process.

ASM Process Compliant with the FDA Principles. FDA accepts the standard IEC 62304 for all levels of concern and pushes for an integration of software life cycle management and risk management activities. The organization establishes some general principles [21], reported in Fig. 4, as guidelines for software validation, and promotes the use of formal approaches.

1. A documented software requirements specification should provide a baseline for both V&V.
2. Developers should use a mixture of methods and techniques to prevent and to detect software errors.
3. Software V&V must be planned early and conducted throughout the software life cycle.
4. Software V&V should take place within the environment of an established software life cycle.
5. Software V&V process should be defined and controlled through the use of a plan.
6. Software V&V process should be executed through the use of procedures.
7. Software V&V should be re-established upon any (software) change.
8. Validation coverage should be based on software complexity and safety risk.
9. V&V activities should be conducted using the quality assurance precept of "independence of review".
10. Device manufacturer has flexibility in choosing how to apply these V&V principles, but retains ultimate responsibility for demonstrating that the software has been validated.

Fig. 4. General Principle for Software Validation (FDA)

We here discuss how the ASM process realizes these FDA guideline principles.

(1) In our process, requirements are specified and documented through a chain of models providing a *rigorous baseline for both validation and verification*.

(2) A continuous *defect prevention* is supported. At each level, faults and unsafe situations can be checked. Safety properties are proved on models, while software testing for conformance verification of the implementation is possible.

(3)–(6) The ASM process allows *preparation for software validation and verification* as early as possible, since V&V can start at ground level. These activities are *part of* the process, can be *planned* at different abstract levels, are *documented*, and supported by precise *procedures*, i.e., methods and techniques.

(7) In case of an implementation local change not affecting the model, our process requires to re-run conformance checking only; in case of a change affecting the specification at a certain level, it requires to re-prove refinement correctness, and to re-execute V&V from the concerned level down to the implementation.

(8) Regarding *validation coverage*, by simulation and testing, we can collect the coverage in terms of rules or code covered. This can be used by the designer to estimate if the validation activity is commensurate with the risk associated with the use of the software for the specified intended use.

(9) Since V&V are performed by exploiting mathematical-based techniques, they facilitate *independent evaluation* of software quality assurance.

(10) The ASM process allows a *device manufacturer to demonstrate that the software has been validated and verified*, both when an implementation is obtained as last model refinement step, and when it is an external code that has been checked for conformance.

7   Conclusions

We have presented the specification, validation, and property verification of a hemodialysis device by using ASMs and model refinement. By taking advantage of the case study, we have related our ASM-based design process with the current normative for developing medical software. The main advantages offered to the standards are: (i) an iterative software life cycle; (ii) models as a rigorous means for safety properties assurance; (iii) validation and verification performed continuously along the software life cycle, and always aimed at defect prevention; (iv) software quality evaluation performed in an objective and repeatable manner; and (v) demonstration that software has been validated and verified.

References

1. ISO 13485: medical devices - quality management systems - requirements for regulatory purposes (2003)
2. IEC 60601–1:2005 medical electrical equipment part 1: General requirements for basic safety and essential performance (2005)
3. IEC 62304 - medical device software - software lifecycle processes (2006)
4. ISO 14971: medical devices - application of risk management to medical devices (2007)
5. Arcaini, P., Bonfanti, S., Gargantini, A., Mashkoor, A., Riccobene, E.: Formal validation and verification of a medical software critical component. In: ACM/IEEE International Conference on Formal Methods and Models for Codesign (MEMOCODE), pp. 80–89. IEEE, Sept 2015
6. Arcaini, P., Gargantini, A., Riccobene, E.: AsmetaSMV: a way to link high-level ASM models to low-level NuSMV specifications. In: Frappier, M., Glässer, U., Khurshid, S., Laleau, R., Reeves, S. (eds.) ABZ 2010. LNCS, vol. 5977, pp. 61–74. Springer, Heidelberg (2010)
7. Arcaini, P., Gargantini, A., Riccobene, E.: Automatic review of abstract state machines by meta property verification. In: Proceedings of the Second NASA Formal Methods Symposium (NFM 2010), pp. 4–13. NASA (2010)
8. Arcaini, P., Gargantini, A., Riccobene, E.: CoMA: Conformance monitoring of Java programs by abstract state machines. In: Khurshid, S., Sen, K. (eds.) RV 2011. LNCS, vol. 7186, pp. 223–238. Springer, Heidelberg (2012)
9. Arcaini, P., Gargantini, A., Riccobene, E.: Offline model-based testing and runtime monitoring of the sensor voting module. In: Boniol, F., Wiels, V., Ait Ameur, Y., Schewe, K.-D. (eds.) ABZ 2014. CCIS, vol. 433, pp. 95–109. Springer, Heidelberg (2014)
10. Arcaini, P., Gargantini, A., Riccobene, E.: Rigorous development process of a safety-critical system: from ASM models to Java code. Int. J. Softw. Tools Tech. Transf., 1–23 (2015)

11. Arcaini, P., Gargantini, A., Riccobene, E., Scandurra, P.: A model-driven process for engineering a toolset for a formal method. Softw. Pract. Experience **41**, 155–166 (2011)

12. Boniol, F., Wiels, V.: The landing gear system case study. In: Boniol, F., Wiels, V., Ameur, Y.A., Schewe, K.-D. (eds.) ABZ 2014. Communications in Computer and Information Science. Springer International Publishing, Switzerland (2014)

13. Börger, E., Stärk, R.: Abstract State Machines: A Method for High-Level System Design and Analysis. Springer, Heidelberg (2003)

14. BRAUN. Dialog $+^{®}$ Dialysis Machine - Instructions for Use: Software Version 9.1x

15. Carioni, A., Gargantini, A., Riccobene, E., Scandurra, P.: A scenario-based validation language for ASMs. In: Börger, E., Butler, M., Bowen, J.P., Boca, P. (eds.) ABZ 2008. LNCS, vol. 5238, pp. 71–84. Springer, Heidelberg (2008)

16. EU. Directive 2007/47/EC of the European Parliament and of the Council. Official Journal of the European Union, September 2007

17. Gargantini, A., Riccobene, E., Rinzivillo, S.: Using spin to generate testsfrom ASM specifications. In: Börger, E., Gargantini, A., Riccobene, E. (eds.) ASM 2003. LNCS, vol. 2589, pp. 263–277. Springer, Heidelberg (2003)

18. Gargantini, A., Riccobene, E., Scandurra, P.: A metamodel-based language and a simulation engine for abstract state machines. J. UCS **14**(12), 1949–1983 (2008)

19. Jetley, R., Purushothaman Iyer, S., Jones, P.L.: A formal methods approach to medical device review. Computer **39**(4), 61–67 (2006)

20. Mashkoor, A.: The hemodialysis machine case study. In: Butler, M., Schewe, K.-D., Mashkoor, A., Biro, M. (eds.) ABZ 2016. LNCS, vol. 9675, pp. 329–343. Springer, Heidelberg (2016)

21. U.S. Food and Drug Administration (FDA). General principles of software validation; final guidance for industry and FDA staff, version 2.0, January 2002

Validating the Requirements and Design of a Hemodialysis Machine Using iUML-B, BMotion Studio, and Co-Simulation

Thai Son Hoang[1(✉)], Colin Snook[1(✉)], Lukas Ladenberger[2], and Michael Butler[1]

[1] ECS, University of Southampton, Southampton, UK
{t.s.hoang,cfs,mjb}@ecs.soton.ac.uk
[2] University of Dusseldorf, Dusseldorf, Germany
ladenberger@cs.uni-dusseldorf.de

Abstract. We present a formal specification of a hemodialysis machine (HD machine) using Event-B. We model the HD machine using iUML-B state-machines and class diagrams and build a corresponding BMotion Studio visualisation. We focus on validation using (i) diagrams to aid the modelling of the sequential properties of the requirements, and (ii) ProB-based animation and visualisation tools to explore the system's behaviour. Some of the safety properties involve dynamic behaviour which is difficult to verify in Event-B. For these properties we use co-simulation tools to validate against a continuous model of the physical behaviour.

Keywords: Hemodialysis machine · Event-B · Validation · Visualisation · iUML-B · BMotion Studio · Co-Simulation

1 Introduction

This paper describes our approach to formally model the requirements and design of a *hemodialysis machine* (HD machine) [8]. The HD machine is used by patients with kidney failure to remove waste products from their blood. We identify how we deal with the safety requirements that are defined for the HD machine [8].

We use Event-B [1], a formal method for system development, and structure our model using refinements to deal with complexity. Since the HD machine's requirements involve extensive sequencing and user interactions as well as dynamic interaction with the HD machine, we focus on validation using (i) diagrams to aid the modelling of the sequential properties of the requirements, and (ii) ProB-based animation, visualisation and simulation tools to explore the behaviour of our models. Where appropriate we use the proof capabilities of Event-B to verify safety constraints.

Our contribution is to demonstrate how different kinds of safety requirements can be verified or validated using the tools available. To do this we provide (i) a model of the HD machine using iUML-B [10–12] state-machines and class

© Springer International Publishing Switzerland 2016
M. Butler et al. (Eds.): ABZ 2016, LNCS 9675, pp. 360–375, 2016.
DOI: 10.1007/978-3-319-33600-8_31

diagrams, (ii) a BMotion Studio visualisation for the developed Event-B model, (iii) a co-simulation of the closed-loop parts of the controller with a continuous domain model of the environment. The graphical model and visualisation enable us to analyse and validate the behaviour of the HD machine.

The rest of the paper is structured as follows. Section 2 gives some background on the methods and tools that we use. The main content of the paper is in Sect. 3 describing the development of the HD machine and validation of its requirements and design using iUML-B, BMotion Studio, and co-simulation. Finally, we summarise and conclude in Sect. 4. For more information and resources, we refer the reader to our website: http://stups.hhu.de/ProB/ABZ16. The website contains the Event-B model and the BMotion Studio visualisation of the HD machine.

2 Background

Event-B. Event-B [1] is a formal method for system development. Main features of Event-B include the use of *refinement* to introduce system details gradually into the formal model. An Event-B model contains two parts: *contexts* and *machines.* Contexts contain *carrier sets*, *constants*, and *axioms* constraining the carrier sets and constants. Machines contain *variables*, *invariants* constraining the variables, and *events.* An event comprises a guard denoting its enabled-condition and an action describing how the variables are modified when the event is executed. A machine in Event-B corresponds to a transition system where *variables* represent the states and *events* specify the transitions. More information about Event-B can be found in [5]. Event-B is supported by the *Rodin Platform* (Rodin) [2], an extensible toolkit which includes facilities for modelling, verifying about the consistency of models using theorem proving and model checking techniques, and validating models with simulation-based approaches.

iUML-B. iUML-B provides a diagrammatic modelling notation for Event-B in the form of state-machines and class diagrams. The diagrammatic models are contained within an Event-B machine and generate or contribute to parts of it. For example a state-machine will automatically generate the Event-B data elements (sets, constants, axioms, variables, and invariants) to implement the states while Event-B events are expected to already exist to represent the transitions. Transitions contribute further guards and actions representing their state change, to the events that they elaborate. A choice of two alternative translation encodings are supported by the iUML-B tools. State-machines are typically refined by adding nested state-machines to states. Class diagrams provide a way to visually model data relationships. For the HD machine we use state-machine diagrams extensively to model the sequential processes and exploit both Event-B encodings. We used class diagrams to model environmental interfaces but do not show this here.

BMotion Studio. In this paper we have used the new version[1] of BMotion Studio [6] to create a *domain specific visualisation* (DSV) of our Event-B model of the

---

[1] Originally BMotion Studio was developed as a separate plug-in for Rodin [7].

HD machine. BMotion Studio comes with a graphical environment including a visual editor that provides various *graphical elements* to create a visualisation of the model. A graphical element is based on Scalable Vector Graphics (SVG) and HTML, two markup languages which support widgets like shapes, images, labels, tables and lists. Moreover, *observers* are used to link the model with the visualisation. For instance, the tool provides a *formula observer* that binds a formula (e.g. an expression or a variable) to a graphical element and allows the tool to compute a visualisation for any given state by changing the properties of the graphical element (e.g. the colour or position) according to the evaluation of the formula in the respective state. Finally, *event handlers* can be attached to the visualisation to provide interactive functionalities, such as an *execute event handler* that binds an Event-B event to a graphical element and executes the event when the user clicks on the graphical element.

Co-Simulation. The Rodin tools and plug-ins are aimed at modelling discrete state-changing events; they are not so good at validating continuous behaviour. Despite this we often need to model the requirements for a system that periodically controls some continuous dynamic behaviour. In order to validate such models a *MultiSim* plug-in [9] was developed by Savicks et al. The plugin allows an Event-B model and a continuous model to be simulated synchronously. Typically the Event-B part will model a cyclic control system that monitors process variables from the continuous model and calculates a controlled output variable. The Event-B model is simulated programmatically using ProB and the continuous model is a *Functional Mock-up Unit* (FMU) [4] which has been exported from a continuous domain modelling tool such as Dymola [3]. The plug-in controls the coordination and communication between the co-operating simulations.

3 Development

In this section, we give some highlights of our formal development of the HD machine. The requirements and design of the HD machine are given in [8] and we will not repeat those requirements in this paper. We suggest the readers to study this section together with the requirements document [8] and the formal model available from the web site http://stups.hhu.de/ProB/ABZ16. We first give an overview of the development strategy that we have applied for this formal model in Sect. 3.1. Subsequently, we highlight the key important modelling decisions using iUML-B (Sect. 3.2), how we use BMotion Studio to validate our model (Sect. 3.3). For dynamic properties that cannot be expressed in Event-B, we show how co-simulation helps us to validate such properties (Sect. 3.4).

3.1 Development Strategy

The hemodialysis process is highly sequential with several sub-processes. In the design described in [8], the HD machine's control system contains two parts: a top-level and a low-level control system, working independently. The top-level control system manages the communication with the users, and transmits data from/to other modules. The low-level control system manages the HD machine while interacting with the top-level control system. Our formal model reflects this design of the control system: the top-level one manages the overall hemodialysis process, interacting with the users, while the low-level one controls the sub-processes by monitoring and regulating the behaviour of the HD machine.

We omitted requirements related to *Ultra Filtration* (UF), the *Safety Air Detector* (SAD), the temperature of the HD machine, loss of main power, and explit real-time modelling. These can easily be incorporated via refinements using similar modelling techniques.

Refinement strategy in Event-B is often influenced by the correctness proofs. In this case, we do not find any verification difficulty, hence our refinement strategy follows the abstraction levels of the sequential steps of the system. This strategy also fits the nested state-machine architecture in iUML-B.

m0: Models the main phases of the hemodialysis process for the top-level control system, i.e., *PREPARATION*, *INITIATION*, and *ENDING*

m1: Models the *sub-processes* within each main phases for the top-level control system.

m2: Models the low-level control's *automatic testing of control functions*.

m3: Models the actual (physical) result of testing the HD machine's control functions.

m4: Model the *message passing communication* between the low-level control system and the HD machine for testing control functions.

m5: Models the set of *signals*.

m6: Models the signal for indication of control function testing result.

m7: Models the connection of concentrate to the HD machine.

m8: Models the *setting of rinsing parameters*.

m9: Models the sequence of connecting patient

m10: Models the physical connection of the patient (arterially and venously).

m11: Models the three pressure monitors and the system normal/abnormal states.

m12: Models various abnormal blood-side pressures.

m13: Models the blood pump, actual blood flow and abnormal situations when monitoring the blood flow.

m14: Models arterial bolus.

m15: Models heparin bolus.

3.2   Modelling Using iUML-B

Modelling Sequential Processes. The hemodialysis process contains three main phases: *preparation, initiation, ending*. Each main phase is composed of several sequential steps. Using iUML-B state-machines, it is straight-forward to model such sequential processes/sub-processes. Furthermore, the notion of nested state-machines (which can be naturally introduced via refinement) fits perfectly for refining the processes further into smaller sequential steps.

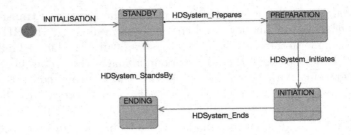

Fig. 1. State-machine *CS_TopLevel* in **m0**

Figures 1 and 2 illustrate how we model the sequential processes in **m0** and **m1**. Figure 1 shows the main phases of the hemodialysis process (with the additional *STANDBY* state). In **m1**, we introduce nested state-machines for states *PREPARATION*, *INITIATION*, *ENDING* to model the sequential sub-steps of each main phase. Figure 2 gives an example of the nested state-machine for the *PREPARATION* state.

The incoming/outgoing transitions of the super-state *PREPARATION* in **m0**, i.e., HDSystem_Prepares and HDSystem_Initiates, are respectively refined to HDSystem_StartsTestingCF and HDSystem_StartsConnectingPatient in **m1**. Using the encoding where each iUML-B state is represented by a constant from an enumerated carrier set, HDSystem_Prepares and HDSystem_StartsTestingCF are straightforwardly translated into Event-B as follows. Here, variable *CS_TopLevel* indicates the current state of the top-level state-machine, and the current state of the nested state-machine in state *PREPARATION* is represented by variable *Preparation_sm*.

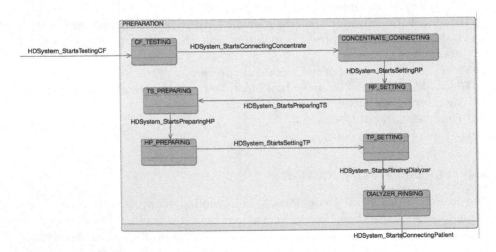

Fig. 2. Nested state-machine for state *PREPARATION* in **m1**

HDSystem_Prepares :
when
  $CS\_TopLevel = STANDBY$
then
  $CS\_TopLevel := PREPARATION$
end

HDSystem_StartsTestingCF :
when
  $CS\_TopLevel = STANDBY$
then
  $CS\_TopLevel := PREPARATION$
  $Preparation\_sm := CF\_TESTING$
end

Top-level vs. Low-level Control Systems. The top-level control system, which maintains the sequential hemodialysis process and its sub-steps, is modelled in a single state-machine. The low-level, responsible for direct control of the HD machine to perform certain tasks, is modelled using several state-machines each corresponding to a particular task. Figure 3 illustrates the state-machine for the low-level control system performing testing of *Control Functions* (CF).

The CS_LowLevel_StartsTestingCF and CS_LowLevel_StandsBy transitions are guarded by $Preparation\_sm = CF\_TESTING$ and $CS\_TopLevel = STANDBY$, respectively, to ensure that they can only be carried out in the correct phases as specified in the top-level control state-machine $CS\_TopLevel$.

The top-level control system can only move from state $CF\_TESTING$ to the next when the CF testing is successful. Using the alternative Event-B encoding for state-machine $CS\_LowLevel\_CFTesting$, where each state is represented by a Boolean variable, this is modelled by a guard on the transition elaborating HDSystem_StartsConnectingConcentrate in state-machine $CS\_TopLevel$ (Fig. 2) stating that $CS_LL_CF_TESTED_OK = $ TRUE.

A Pattern for Controlling Physical Equipments. A common pattern that we used in modelling the HD machine is to formalise how the controller interacts with the physical equipment in the environment. The pattern involves two refinement levels. At the abstract level, the controller and the physical equipments can have access to the states of the other components. In the refinement, this direct access is refined by message passing communication. This pattern is similar to those in [1] and we applied them to iUML-B state-machines. We show below how we incorporate the physical testing of CF into the formal model. The low-level control system for CF testing is illustrated in Fig. 3. In **m3**, a variable *HDMachine_CFTestedOK* is introduced to denote the result of the HD machine's CF test and a new event HDMachine_CFTests is allowed to

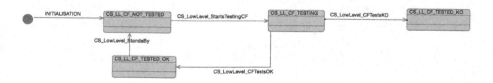

Fig. 3. Low-level control system for CF testing

set this variable non-deterministically representing the result of the test. The guard of the event ensures that the physical tests are carried out only when the low-level controller is in the testing state, i.e. $CS\_LowLevel\_CFTesting$. Transitions CS_LowLevel_CFTestsOK and CS_LowLevel_CFTestsKO of state-machine $CS\_LowLevel\_CFTesting$ are directed by the actual result of the test: they are guarded to select a pass/fail response according to $HDMachine\_CFTestedOK$, i.e., $HDMachine\_CFTestedOK = \mathrm{TRUE}$ or $HDMachine\_CFTestedOK = \mathrm{FALSE}$.

In the refinement **m4**, we introduce the communication between the controller and the HD machine. Two new variables $CS\_2\_HD\_StartsCFTesting$ and $HD\_2\_CS\_CFTestingFinished$ are introduced to model the message exchange. Flag $CS\_2\_HD\_StartsCFTesting$ is set in event CS_LowLevel_StartsTestingCF and unset in HDMachine_CFTests. Invariant

$$CS\_2\_HD\_StartsCFTesting = \mathrm{TRUE} \Rightarrow \mathsf{CS\_LowLevel\_StartsTestingCF} = \mathrm{TRUE}$$

allows us to refine HDMachine_CFTests's guard to $CS\_2\_HD\_StartsCFTesting = \mathrm{TRUE}$. Similarly, $HD\_2\_CS\_CFTestingFinished$ is set in HDMachine_CFTests and unset in CS_LowLevel_CFTestsOK and CS_LowLevel_CFTestsKO, while the guards of CS_LowLevel_CFTestsOK and CS_LowLevel_CFTestsKO are strengthened by adding the condition $HD\_2\_CS\_CFTestingFinished = \mathrm{TRUE}$.

Safety Properties as State-Machine Invariants. An important feature of iUML-B is state invariants. They can be used to express safety properties that must hold in a certain state. Consider the state-machine $CS\_TopLevel$ in **m2**. We wish to ensure that when the system is in the $INITIATION$ phase, the CF should have been successfully tested. We add an invariant CS_LowLevel_CFTestsOK = TRUE to state $INITIATION$. The translation of the state-invariant in Event-B is, as expected, i.e.,

$$CS\_TopLevel = INITIATION \Rightarrow \mathsf{CS\_LowLevel\_CFTestsOK} = \mathrm{TRUE} .$$

To prove the above invariant, an invariant is added to the $PREPARATION$ state stating that $Preparation\_sm \neq CF\_TESTING \Rightarrow$ CS_LowLevel_CFTestsOK = TRUE.

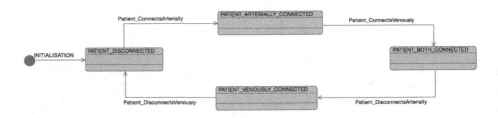

Fig. 4. Patient connections to the HD machine

Animation/Model Checking to Validate Requirements. Consider **m10**, we introduce a state-machine to model the physical connections/disconnections of the patient to the HD machine arterially and venousely (Fig. 4). The patient is connected to the machine in the first step of the *INITIATION* phase and disconnected from the machine in the first step of the *ENDING* phase. Requirements **S-1** and **S-5** from [8] are directly related to the connections status of the patient and are as follows.

S-1. Arterial and venous connectors of the EBC are connected to the patient simultaneously.
S-5. The patient cannot be connected to the machine outside the initiation phase, e.g., during the preparation phase.

Initially, we model **S-1** as an invariant

$$PATIENT\_CONNECTIONS \neq PATIENT\_DISCONNECTED \Rightarrow$$
$$PATIENT\_CONNECTIONS = PATIENT\_BOTH\_CONNECTED .$$

and **S-5** as state-invariants for states *PREPARATION* and *ENDING* specifying that *PATIENT_CONNECTIONS = PATIENT_DISCONNECTED*. Attempts to prove these invariants lead to failure. We use the ProB model checker to find counter-example traces and iUML-B state-machine animation to visualise the obtained traces. The visualisation helps us to identify the cause of the problems and how to fix them. In this case, the requirements are clearly too strong and contradict other requirements. During *PREPARATION*, the patient is connected first arterially and then venously contradicting **S-1**. The patient is still connected both arterially and venously when the reinfusion step starts, i.e., outside the initiation phase contradicting **S-2**. We therefore weaken the requirements **S-1** and **S-5** as follows.

S-1' Arterial and venous connectors of the EBC are *both* connected to the patient *during therapy.*
S-5' The patient cannot be connected to the machine outside the *initiation and reinfusion phases.*

Modelling Abnormal Behaviours. During hemodialysis treatment, an important part of the HD machine is to monitor the various aspects of the patient and the machine, and raise the alarm when abnormal behaviours are detected. This includes low/high blood pressures, incorrect blood flow directions, etc. We have developed a pattern for modelling such behaviour. An abstract state-machine for the low-level control system is added in **m11** (Fig. 5a). When we introduced the pressure monitor in **m12**, various abnormal conditions can be detected. The events modelling such detection are a refinement of the abstract event CS_LowLevel_Abnormal (Fig. 5b). Note that we still keep the abstract event CS_LowLevel_Abnormal to be able to detect more abnormal behaviours in future refinements.

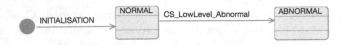

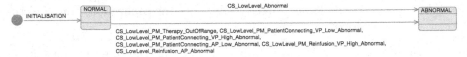

(a) State-machine *CS_LowlLevel_Overall* in **m11**

(b) Nested state-machine *CS_LowlLevel_Overall* in **m12**

Fig. 5. Modelling abnormal behaviours

3.3   Validating Using BMotion Studio

We use BMotion Studio to create a DSV of our iUML-B/Event-B model of the HD machine. The DSV consists of two views: a view of the user interface (UI) and a view of the environment of the HD machine. The description of the DSV is supported by listings in which observers and event handlers are described using JavaScript. However, the visual editor of BMotion Studio also provides a graphical user interface for creating observers and event handlers.

Visualising the UI Display Panel. Figure 6a demonstrates the DSV of the UI display panel. Each dialysis parameter is represented using simple graphical elements to display its description, unity and current value. In addition, for

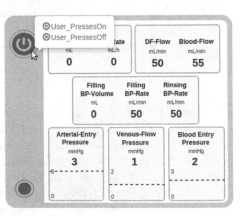

```
1  bms.observe("formula", {
2    selector: "#bloodFlow",
3    formulas: ["bloodFlow"],
4    trigger: function(e, v) {
5      e.text(v[0]);
6  }});
7
8  bms.executeEvent({
9    selector: "#bt_power",
10   events: [
11   {name: "User_PressesOn"},
12   {name: "User_PressesOff"}
13   ]});
```

(a) UI display panel visualisation

(b) Blood flow observer and on/off button execute event handler

Fig. 6. Domain specific visualisation of UI display panel (Color figure online)

pressure parameters, the width and thresholds of the limits window are shown with the current value being represented by a horizontal dashed line.

Each dialysis parameter binds a *formula observer* that observes the respective state variable of the parameter. For instance, Fig. 6b shows the formula observer for the blood flow parameter. Line 1 and 2 state that we register a new formula observer on the graphical element that matches the selector "#bloodFlow" (The prefix "#" is used to match a graphical element by its ID.[2]) Line 3 states that the observer should observe the variable *bloodFlow* during the animation. In lines 4 to 6 we define a trigger function that is called whenever a state change occurs. The reference to the matched graphical element (e) and the state values of the observed formulas (v) are passed as arguments to the trigger function. In line 5 we define the action which is made on the label whenever a state change occurs: the observer sets the text content of the label to the current value of the state variable *bloodFlow* (val[0]). We have defined the observers for the other dialysis parameters in a similar fashion.

The visualisation of the UI display panel also contains graphical elements and observers for the automated self test signal lamp (see lower left side of Fig. 6a), which is represented by a circle. The corresponding observer is responsible for indicating whether the automated self test has been successfully completed (change the signal lamp to green) or not (change the signal lamp to red) based on the observed formula signal_status(CF_TESTING_SIGNAL).

We have used the *execute event handler* feature of BMotion Studio to add interactive components to our visualisation. As an example, Fig. 6b (lines 8 to 13) shows the execute event handler for the HD machine on/off button (#bt_power) that wires the events User_PressesOn and User_PressesOff. In case of hovering the graphical element with the mouse a tooltip with the available events will be shown as demonstrated in Fig. 6a.

Visualising the Environment of the HD Machine. The DSV of the HD machine provides a second view that visualises the HD machine's environment as shown in Fig. 7. The objective of this view is to show how the different parts of the system are connected together. For instance, it contains graphical elements and observers that represent the dialysis pressure parameters (arterial-, venous-, and blood entry pressure) and their connection to the environment.

The visualisation is subdivided into SVG groups, where each group represents a different refinement level. Furthermore, each group binds a *refinement observer* that is responsible for showing or hiding the group depending on whether the observed refinement is part of the running animation or not. For instance, the group for refinement **m11** that contains the dialysis pressure parameter graphical elements, binds a refinement observer that observes refinement **m11**. Whenever **m11** is part of the running animation the observer sets the *visibility* attribute of the group to the value "visible" otherwise to "hidden". We have also created similar refinement observers for the UI display panel view. This helped us to focus on relevant parts of the system while validating a specific refinement level.

---

[2] See jQuery selector API: http://api.jquery.com/category/selectors/.

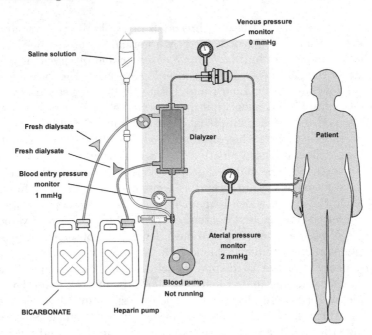

Fig. 7. Domain specific visualisation of the environment of the HD machine (Figure CC BY 3.0 YassineMrabet (https://commons.wikimedia.org/wiki/File:Hemodialysis-en. svg))

Application of the DSV. The benefits of more effective validation of the HD machine's Event-B model justify the extra effort required to develop a DSV. The visualisation helped us reach a common understanding about the model and to identify faulty behaviour and errors during development. This is particularly valuable when the formal model becomes complex in later refinement levels. Animation tools with only textual representation of the state are insufficient for validation purposes. In the case of the HD machine, requirements such as **S-2–S-4**, **R-5–R-13** are modelled by the enabled-ness of iUML-B transitions (ultimately events). Such properties are cumbersome to formulate as a proof obligation in Event-B but can be readily demonstrated via BMotion Studio. Hence in many cases we use BMotion Studio to validate whether the requirements have been adequately taken into account. BMotion Studio also helped us to discover problems with our model during its development, especially mistakes leading to liveness problems, where the HD machine cannot make any progress.

The DSV also enables domain experts to validate our formal model in terms of user interactions. This can be compared with prototyping techniques. Indeed, we plan to show our formal model and DSV to medical scientists and physicians at the university medical centre of Duesseldorf.

BMotion Studio provides a feature for sharing a DSV online. This is useful for demonstration purposes as well as for sharing the visualisation between the persons involved in developing the model. We refer the reader to the website stated in Sect. 1 which contains an online live visualisation of the HD machine.

3.4   Validating Using Co-Simulation

Safety requirements **S-8**, **S-9** and **S-10** concern adjustments to the *Blood Flow Rate* (BF). **S-8** requires the demanded BF to be lowered if *Arterial Pressure* (AP) is low. We conclude that there is an inverse relationship between BF and AP. **S-8** also states that the AP to BF relationship is affected by the fistula needle type. **S-9** indicates that low AP can result in reduced BF. Hence the achieved BF should be monitored and treatment time adjusted accordingly. **S-10** requires that BF should be optimal (presumably after consideration of **S-8** and **S-9**). We assume that this means as close to the user selected BF as possible and that a stable closed loop control of the blood pump is needed. In order to validate the specification of a suitable control system we use the continuous domain modelling tool Dymola to create a model of the environment which we co-simulate with the Event-B model of the control system (Fig. 9), which is extracted from the formal model developed in Sect. 3.2.

Figure 8 shows the continuous domain Dymola model of the physical interaction between BF and AP. This detail of the environment being controlled was not given in the specification. We have invented an example behaviour, based on typical pump suction properties, for the purposes of illustration. In order to

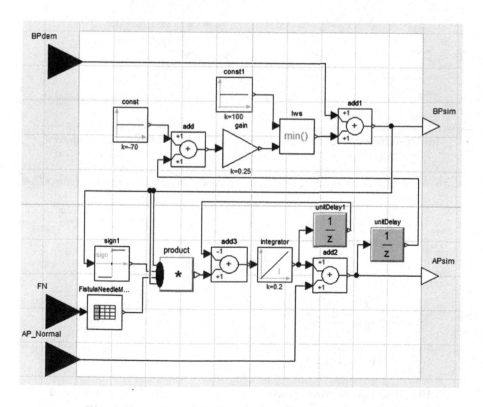

Fig. 8. Dymola model of interaction between BF and AP

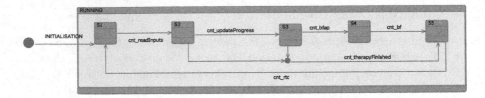

Fig. 9. iUML-B model of BF and AP control cycle

validate this model we also developed a Dymola model of the control system. Once the environment model behaved as desired it was exported as a FMU which allows it to be run as a simulation outside of the Dymola tool. We then imported the FMU into the Rodin co-simulation tool, linked its I/O with our Event-B model of the control system and co-simulated the combined models.

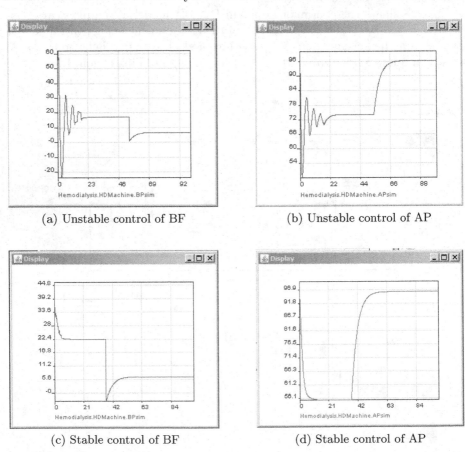

(a) Unstable control of BF   (b) Unstable control of AP

(c) Stable control of BF   (d) Stable control of AP

Fig. 10. Co-simulation plots showing unstable and stable control of BF and AP

The transition *cnt_readinputs* obtains new values provided by the Environment FMU simulation. The transition *cnt_updateProgress* subtracts the achieved BF for the cycle period from the total blood volume required to be processed. If the total has been processed, *cnt_therapyFinished* sets the demanded BF to 0. Otherwise, *cnt_bfap* calculates the demanded BF which is the user configured BF adjusted for AP (i.e. in accordance with S8). This adjustment is implemented as a simple linear interpolation function from (0,0) to (70,initial BF) which is limited outside this domain. Transition *cnt_bf* adjusts the output commanded BF to adjust for the achieved BF using a proportional error control. This final adjustment corresponds to control of the *Blood Pump* (BP) speed to achieve the desired BF except that we abstracted from BP units for simplicity. We chose to model AP in % of some nominal initial AP and BF in ml/min with a control cycle period of 100 ms. The initial BF is set 30 ml/min in the following analysis.

Our initial co-simulation results (Fig. 10) showed that the AP was correctly controlled to a steady value of 72 % by lowering BF to below 20 ml/min. However, the initial response is very unstable. It is interesting to note that we did not see this instability when we first tested the environment model in Dymola only (i.e. using a Dymola model of the control in place of the Event-B model). We believe this is because we did not accurately model the discrete periodic cycle and therefore the response rate of the Dymola version of the control was fast enough to mask the problem. This demonstrates the advantage of testing the actual Event-B model which is inherently discrete. To improve stability we decreased the gain of the proportional control. This improved stability but resulted in a degraded AP of approx. 55 %. This is due to a larger residual offset error which is an inherent problem of proportional controllers. A possible solution would be to introduce an integral term to the controller which would remove the residual offset.

4   Conclusion

The HD machine is predominantly a sequential process of user interactions with few safety properties that can be expressed as constraints on state. During the therapy stage the machine controls the dynamic properties of AP and BF. At first sight it appeared that this case study would not illustrate the strengths of our modelling tools very well since Event-B verifies the preservation of invariant properties over discrete state-changing events. However, the case study gave us an opportunity to focus on the validation tools that we use to develop useful models. Proofs may result in a correct model but we need user validation to ensure the usefulness of our models. For this case study, we therefore used the validation tools to drive a manual assessment of the model. iUML-B state-machine modelling tools map readily to the process steps of the requirements and their animation enables us to 'see' the sequential flows of the model. BMotion Studio visualisation tools link the process to a more realistic representation of the HD machine which allows us to disassociate ourselves from the model giving a stronger validation. For validation of the dynamic control of AP versus BF we use a continuous domain model of the controlled parameters to co-simulate with our iUML-B/Event-B models to provide a strong validation of the stability and effectiveness of the modelled control scheme.

The summary of the requirements (from [8]) that have been modelled and verified/validated within our development is as follows.

- *Invariant Proofs*: **S-1', S-5', S-6, S-11**
- *Simulation Validation*: **S-2, S-3, S-4, R-1–R-15, R-17–R-19, R-22**
- *Co-simulation*: **S-8, S-9, S-10**

Many requirements are validated using simulation/animation techniques. One exception is requirement **R-16**: "while connecting the patient, the software shall use a timeout of 310 s after the first start of the BP. After this timeout, the software shall change to the initiation phase". An attempt to model this requirement leads to an invalid iUML-B state-machine. We found that the requirement is inconsistent: while connecting patient, the system is *already in the initiation phase*. It is not clear to us what the intended meaning of the requirement is.

In the future, we plan to address the remaining requirements using similar techniques. We will continue to develop the BF control using co-simulation to improve its accuracy without degrading stability and responsiveness. We plan to investigate ways to provide validation records that might be used as evidence in a safety case. For example, BMotion Studio could be enhanced to provide and replay traces of animations.

References

1. Abrial, J.-R.: Modeling in Event-B: System and Software Engineering. Cambridge University Press, New York (2010)
2. Abrial, J.-R., Butler, M., Hallerstede, S., Hoang, T.S., Mehta, F., Voisin, L.: Rodin: An open toolset for modelling and reasoning in Event-B. Softw. Tools Technol. Transf. **12**(6), 447–466 (2010)
3. Dassault Systemes. Catia Systems Engineering Dymola. http://www.3ds.com/products-services/catia/products/dymola (Accessed on January 2016)
4. FMI Steering Committee. Functional Mock-up Interface. https://www.fmi-standard.org (Accessed on January 2016)
5. Hoang, T.S.: An introduction to the Event-B modelling method. Industrial Deployment of System Engineering Methods, pp. 211–236. Springer, Heidelberg (2013)
6. Ladenberger, L.: BMotion Studio for ProB project website. http://stups.hhu.de/ProB/w/BMotion_Studio,
7. Ladenberger, L., Bendisposto, J., Leuschel, M.: Visualising event-B Models with B-Motion studio. In: Alpuente, M., Cook, B., Joubert, C. (eds.) FMICS 2009. LNCS, vol. 5825, pp. 202–204. Springer, Heidelberg (2009)
8. Mashkoor, A.: The hemodialysis machine case study (2015). http://www.cdcc.faw.jku.at/ABZ2016/HD-CaseStudy.pdf
9. Savicks, V., Butler, M., Colley, J.: Co-simulating Event-B, continuous models via FMI. In: Proceedings of the 2014 Summer Simulation Multiconference, SummerSim 2014, pp. 37:1–37:8. Society for Computer Simulation International, San Diego (2014)

10. Savicks, V., Snook, C.: A framework for diagrammatic modelling extensions in Rodin. In: Rodin Workshop 2012, Fontainbleau (2012)
11. Snook, C.: Modelling control process and control mode with synchronising orthogonal state machines. In: B2011, Limerick (2011)
12. Snook, C.: iUML-B statemachines. In: Proceedings of the Rodin Workshop, Toulouse, France (2014). http://eprints.soton.ac.uk/365301/

Hemodialysis Machine in Hybrid Event-B

Richard Banach[✉]

School of Computer Science, University of Manchester,
Oxford Road, Manchester M13 9PL, UK
banach@cs.man.ac.uk

Abstract. The hemodialysis machine case study is examined in Hybrid Event-B (an extension of Event-B that includes provision for continuously varying behaviour as well as the usual discrete changes of state). A broadly component based strategy is adopted, using the multi-machine and coordination facilities of Hybrid Event-B. Since, like most medical procedures, hemodialysis is under overall human control, it is largely a sequential process, with some branching to deal with exceptional circumstances. This makes for a relatively uncomplicated modelling framework, provided a model of the operator is included in order to capture the handling of exceptions.

1 Introduction

This paper reports on a treatment of the hemodialysis machine case study using Hybrid Event-B. Hybrid Event-B [6] is an extension of the well known Event-B framework, in which continuously varying state evolution, along with the usual discrete changes of state, is admitted. Although tool support for Hybrid Event-B is lacking at the time of writing, theoretical case studies like the present one are of great benefit in exercising the formalism, to develop the most appropriate modelling metaphors and to confirm that the formalism lives up to its expectations in terms of expressivity in the face of a wide variety of applications challenges.

The hemodialysis machine case study requirements are in [16]. As almost always happens in such situations, the requirements document itself does not completely document all aspects of the situation, and supplementary information is helpful to assist in framing the development. Additional details can be found in, e.g. [2,12,13,15,17]. While such references are certainly helpful in putting the problem in context, they actually say little about the detailed operation of hemodialysis machines, perhaps because such details are commercially sensitive. Nevertheless, they reveal that hemodialysis is overwhelmingly a sequential process, with branches corresponding to a number of exceptional circumstances that may arise along the way. What also becomes clear is that the most critical component of the process is the human operator, since the operator is relied on to steer the hemodialysis procedure through its various phases, and to initiate handling of any exceptions — in the most benign cases, operators are the patients themselves when hemodialysis is self-administered (though an assistant

© Springer International Publishing Switzerland 2016
M. Butler et al. (Eds.): ABZ 2016, LNCS 9675, pp. 376–393, 2016.
DOI: 10.1007/978-3-319-33600-8_32

is normally present).[1] The sequential aspect, and the ultimate reliance on human expertise, are in fact common to most medical procedures, which take place in a series of steps, perhaps branching on the basis of the outcomes of earlier steps, and always under ultimate hands-on human control. Thus, the case study we are faced with is simplified in being essentially sequential, but needs to have the human operator as an agent in the system, since vital parts of the management of the procedure are under direct operator control.

A feature of the requirements in [16] is that they do not go into many details regarding the physical/mechanical processes involved in the operation of the machine. At a low enough level of description, it becomes clear that there are a number of subsystems in the machine, and we can infer that they enjoy a certain amount of autonomy, but without details, it is difficult to do more than guess at some interactions with the control system. The same goes for the behaviour of the operator, which is so vital. To truly model the operation of hemodialysis, we would need to understand how experienced hemodialysis nurses go about their work. All of this rather underutilises the capacity of Hybrid Event-B to model such sytems faithfully, and specifically, to write the more incisive invariants that would strengthen confidence in the design.

The rest of the paper is as follows. In Sect. 2 we overview the hemodialysis machine and its operation. In Sect. 3 we give a sketch of multi-machine Hybrid Event-B as used here. In Sect. 4 we give some generalities about the Hybrid Event-B development of the case study, and in Sect. 5 we discuss it in more detail. Section 6 discusses the role of invariants (and hence the prospects for verification), in a system like the present one, where so much depends on the operator. Section 7 concludes.

2   The Hemodialysis Machine and Its Operation

In Fig. 1 we reproduce the internal structure of a hemodialysis machine from [16]. The patient's direct connection to the hemodialysis machine is via the extracorporeal blood circuit (EBC), shown in more detail in Fig. 2. This monitors blood pressure on entry, infuses Heparin to prevent clotting inside the machine, and pumps the blood into the dialyser. On exit from the dialyser the blood pressure is monitored again, and passes a sensor that detects blood arrival (for filling of the EBC at procedure startup) and blood departure (for emptying the EBC of blood

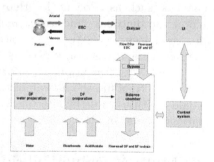

Fig. 1. Architectural overview of the hemodialysis machine, reproduced from [16].

at procedure end). The sensor also monitors the presence of air in the blood returned to the patient to prevent dangerous amounts from being infused into

---

[1] In the most extreme cases, the operator or assistant undertakes a manual reinfusion of the patient when a power failure causes the equipment to halt.

378   R. Banach

the patient's bloodstream. Although not shown in Fig. 1, the EBC is also directly connected to the control system.

The other main component of the hemodialysis machine prepares the dialysing fluid. This is filtered, degassed and heated water, with bicarbonate and acetate suitably added. The fluid passes through one side of the balance chamber, then into the dialyser, upon return from which it passes through the other side of the balance chamber, and then to the drain. If the properties of the dialysing fluid being supplied stray outside permitted ranges at any time during

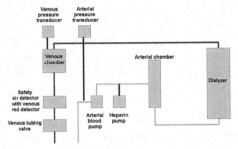

Fig. 2. Overview of the extracorporeal blood circuit and dialyser, reproduced from [16].

treatment, a bypass circuit isolates the dialyser from the dialysing fluid flow to prevent harm to the patient's blood.

Both the balance chamber and dialyser are rigid structures, with a flexible membrane separating the two sides of the balance chamber. Therefore, the rate of flow of fresh dialysing fluid entering the balance chamber can be checked against the rate of flow of waste dialysing fluid exiting it (up to the latitude permitted by the flexible membrane, which has a limited freedom of movement). This being so, the rigidity of the dialyser structure then implies that the blood flow leaving the patient is strictly related to the blood flow returned to the patient (modulo the latitude permitted by the flexible membrane), this, despite the osmotic and ultrafiltration processes taking place in the dialyser which can exchange both salts and liquid between blood and dialysing fluid, the latter also altering the volume of both. This conclusion (and variations that allow for other changes, such as allowing for Heparin) is based on the assumption that both blood and dialysing fluid are *incompressible fluids*, which is true to a very high degree of accuracy.[2] Were it the case that the requirements in [16] covered such mechanical and treatment aspects of the hemodialysis machine in more detail, the insights just mentioned could be reexpressed as suitable Hybrid Event-B invariants (modulo the leeway of the flexible membrane). As noted above, this would capture a worthwhile safety property, one that was important from the patient's point of view.

3   Multi-Machine Hybrid Event-B

In this section we give a brief outline of multi-machine Hybrid Event-B for purposes of orientation. In Sect. 5 we cover part of the development in more detail, amplifying what we say here.

---

[2] In [16] it is stated that the volumes in and out of the balance chamber are equal, but this is only true modulo the additional considerations stated here.

A large Hybrid Event-B system is organised syntactically in a PROJECT file. This names a number of MACHINEs and INTERFACEs. A machine is a hybrid extension of an Event-B machine (of which more shortly), while an interface declares a set of variables, their initialisations, and the invariants that involve them, so that several machines can access these variables. The project file also declares synchronisations, each of which is a set of events drawn from different machines, which must execute simultaneously at runtime. At runtime, all the machines in the project must execute concurrently, so projects must be designed in such a way that this makes sense.

Individual Hybrid Event-B machines are inspired by their Event-B predecessors. Event-B machines define discrete events that define individual changes of state, along with the supporting definitions needed to do this, and, crucially, the invariants that are claimed to hold when these events execute, thus embodying the safety properties of the system. The usual interpretation is that occurrences of the discrete events take place at isolated time points (though that is outside the actual formalism of Event-B). In a hybrid system it is natural to have discrete changes of state embodying change of 'mode', after each of which there is a period of continuously varying state change. Hybrid Event-B thus takes the individual discrete events of Event-B, naming them 'mode events', and interleaves them with periods of continuously varying state change defined in 'pliant events'. The latter may be defined in various ways: direct (continuous) assignment, (ordinary) differential equations, and implicit definition via a (time dependent) predicate expression.

Figure 3 shows a schematic Hybrid Event-B machine. In this, the VARIABLES, INVARIANTS, *INITIALISATION* and *MoEv* are standard, as in Event-B (the latter defining discrete state change via the before-after predicate *BApred*), while TIME, CLOCK, PLIANT and the pliant event *PliEv* are new for Hybrid Event-B. The first three of these are declarations (PLIANT declaring those variables that are to be permitted to evolve continuously). The continuous behaviour is defined in *PliEv*. Its 'STATUS pliant' proclaims it as such, and its contents describe what is needed to do this. Thus, there are two

```
MACHINE  HyEvBMch        ... ...              ... ...
TIME t                    MoEv                  PliEv
CLOCK clk                  STATUS ordinary       STATUS pliant
PLIANT x, y                ANY i?, l, o!         INIT iv(x, y, t, clk)
VARIABLES u                WHERE                 WHERE grd(u)
INVARIANTS                   grd(x, y, u, i?, l, t, clk)    ANY i?, l, o!
  x, y, u ∈ ℝ, ℝ, ℕ       THEN                  COMPLY
EVENTS                       x, y, u, clk, o! :|    BDApred(x, y, u,
 INITIALISATION               BApred(x, y, u, i?, l, o!,       i?, l, o!, t, clk)
  STATUS ordinary             t, clk, x', y', u', clk')   SOLVE
  WHEN                    END                     𝒟 x =
    t = 0                ... ...                    φ(x, y, u, i?, l, o!, t, clk)
  THEN                                             y, o! :=
    clk, x, y, u := 1, x₀, y₀, u₀                    E(x, u, i?, l, t, clk)
  END                                            END
... ...                                        END
```

Fig. 3. A schematic Hybrid Event-B machine.

guards, *grd* concerning mode variables only, and *iv* concerning all aspects connected with initial values in continuous behaviour (when needed). The COMPLY clause is used to specify generic properties of the required behaviour (e.g. bounds for the permitted range of pliant variables, or other nondeterministic aspects), whereas the SOLVE clause is used to specify more precise behaviour. Thus the SOLVE clause contains an ordinary differential equation for the pliant variable x, i.e. $\mathcal{D}\,x = \phi(x, y, u, i?, l, o!, t, clk)$, and a direct (continuous) assignment for the pliant variable y and output $o!$, i.e. $y, o! := E(x, u, i?, l, t, clk)$.

At runtime, mode events and pliant events are required to execute alternately, in line with the intuition sketched above. With care, a semantics may be constructed that makes for a very clean interaction between the discrete and contiuous parts, and proof obligation schemas may be designed to statically ensure that the runtime behaviour is as required.

Observing that refinement in Event-B is of mode events to mode events (exclusively), in Hybrid Event-B this is preserved and extended to refinement of pliant events to pliant events (exclusively). Time is required to progress at the same rate in both abstract and concrete models. Synchronisations must be refined to synchronisations. We refer to [6] for a more detailed presentation of single machine Hybrid Event-B, and to [7] for details on the multi-machine version.

4   Hemodialysis Machine Development Generalities

Figure 4 shows an architectural overview of the Hybrid Event-B development of the hemodialysis machine. The rectangle in the middle represents the interface *Central_IF*. This sits at the centre of the development, and holds all of the shared variables needed. Four machines connect to it: *Operator*, modelling essential ele-

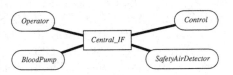

Fig. 4. Overview of the hemodialysis machine development architecture.

ments of the operator; *Control*, modelling the control system; *BloodPump*, modelling the blood pump; and *SafetyAirDetector*, modelling the safety air detector and venous red detector. Aside from the operator and controller, the BP and SAD/VRD systems were selected for more detailed modelling in separate machines, since [16] gives just enough detail about their working to make them interesting in their own right. For the remaining subsystems in the hemodialysis machine, even if a degree of independent working might be inferred, the relatively low level of detail given in [16] suggested that more precise modelling would be guesswork, so they were represented more superficially. The practical consequence of this is that whereas, in principle, a lot more use of the capabilities of Hybrid Event-B could have been made to model the physical aspects of the system more faithfully, and hence, where appropriate, to tie these model aspects to the software control events via suitable invariants, in practice, this was not done due to the lack of the relevant low level physical detail. In fact this is a common problem in requirements targeted at software development, since overwhelmingly, software modelling formalisms are not capable of expressing, or dealing with, nontrivial physical properties.

A further example of this is the user interface, regarding which, [16] is quite sketchy. So UI matters are represented using the environment, or recording information in suitable state variables, or simply by including the appropriate events — on the understanding that these representations could be elaborated as needed.

In the present development, it was decided to experiment with a simple component-based approach, to see how this aligned with the formal elements of Hybrid Event-B. Of course, there are many component based approaches these days, see e.g. [11,14,18]. The approach here was to determine the architecture of Fig. 4 at the outset, and to then construct a development plan around it. The main issue that this threw up was squaring the desire to do the development via refinement, with the sequential nature of the description in [16]. A sequential approach to the development would complete an earlier phase before embarking on the next one, but this could easily break refinement properties, as the end of the earlier phase might have to be modified to accommodate the start of the next one. This required care in designing the top level 'umbrella' model, and in the order of introducing the other components.

5   The Development in Detail

The details of the development may be accessed at [4]. The development proceeds level by level. **Level 00** introduces the architecture of Fig. 4. There is a variable $mode \in \{NORMAL, BYPASS, ALARM\}$, and $phase \in \{PREP, INIT, TREAT, END, ENDED\}$. The $Operator$ and $Control$ machines move the $phase$ variable through its various values, as in treatment. With an eye to refinement, a $Reset$ event is introduced to enable the operator to set all system variables to arbitrary desired values when an $ALARM$ occurs. In fact, the guard of the $Reset$ event is weakened to TRUE, to enable its use whenever subsequent modelling is weakened due to uncertainties arising from [16].

Level 01. Level 01 introduces the blood pump, and fleshes out the $BloodPump$ machine of Level 00. The Level 01 development is shown in Fig. 5, which is complete aside from a couple of small details.

We discuss Level 01 fairly fully, since it illustrates many details of multi-machine Hybrid Event-B in a small space. The PROJECT file $HemoDialysisSystemLevel\_01$ states that it is a refinement of the Level 00 project, and names the components, namely the $Operator$, $Control$ and $BloodPump$ machines, and the $Central\_IF$ interface, these all being at Level 01, except for $Control$, which is unchanged from Level 00.

The $Central\_IF$ interface refines its Level 00 counterpart. New variables relevant to the blood pump are introduced: $bpOnOff$, saying whether the blood pump is ON or OFF, and $bpMeasFlow$ and $bpMeasVol$. These are respectively

PROJECT *HemoDialysisSystem_Level_01*
REFINES *HemoDialysisSystem_Level_00*
  INTERFACE *Level_01_Central_IF*
  MACHINE *Level_01_Operator*
  MACHINE *Level_00_Control*
  MACHINE *Level_01_BloodPump*
END

INTERFACE *Level_01_Central_IF*
REFINES *Level_00_Central_IF*
VARIABLES
  mode, phase, bpOnOff
PLIANT
  bpMeasFlow, bpMeasVol
INVARIANTS
  $mode \in \{NORMAL, BYPASS, ALARM\}$
  $phase \in \{PREP, INIT, TREAT, END, ENDED\}$
  $bpOnOff \in \{ON, OFF\}$
  $bpMeasFlow, bpMeasVol \in \mathbb{R}, \mathbb{R}$
INITIALISATION
  BEGIN
    mode, phase, bpOnOff := *BYPASS, PREP, OFF*
    bpMeasFlow, bpMeasVol := 0, 0
  END
END

MACHINE *Level_00_Control*
CONNECTS *Level_01_Central_IF*
EVENTS
  PliTrue   ...   .
  RaiseALARM
    STATUS asynch
    BEGIN
      mode := *ALARM*
    END
  PrepComplete
    STATUS asynch
    WHEN $mode = BYPASS \wedge phase = PREP$
    THEN *phase* := *INIT*
    END
  InitModeToNORMAL
    STATUS asynch
    WHEN $mode = BYPASS \wedge phase = INIT$
    THEN *mode* := *NORMAL*
    END
  InitComplete
    STATUS asynch
    WHEN $mode = NORMAL \wedge phase = INIT$
    THEN *phase* := *TREAT*
    END
  TreatComplete
    STATUS asynch
    WHEN $mode = NORMAL \wedge phase = TREAT$
    THEN *phase* := *END*
    END
  EndComplete
    STATUS asynch
    WHEN $mode = NORMAL \wedge phase = END$
    THEN *phase* := *ENDED*
    END
END

MACHINE *Level_01_Operator*
REFINES *Level_00_Operator*
CONNECTS *Level_01_Central_IF*
EVENTS
  PliTrue
    STATUS pliant
    COMPLY *INVARIANTS*
    END
  Reset
    STATUS asynch
    BEGIN
      mode, phase, bpOnOff
      bpMeasFlow, bpMeasVol
      :| *INVARIANTS*
    END
  PrepStart
    STATUS asynch
    WHEN $mode = BYPASS \wedge phase = PREP$
    THEN skip
    END
  InitStart
    STATUS asynch
    WHEN $mode = BYPASS \wedge phase = INIT$
    THEN skip
    END
  TreatStart
    STATUS asynch
    WHEN $mode = NORMAL \wedge phase = TREAT$
    THEN skip
    END
  EndStart
    STATUS asynch
    WHEN $mode = NORMAL \wedge phase = END$
    THEN skip
    END
END

MACHINE *Level_01_BloodPump*
REFINES *Level_00_BloodPump*
CONNECTS *Level_01_Central_IF*
EVENTS
  BPStopped
    STATUS pliant
    REFINES *PliTrue*
    WHEN $bpOnOff = OFF$
    COMPLY CONST(*bpMeasVol*)
    SOLVE *bpMeasFlow* := 0
    END
  BPOn
    STATUS asynch
    WHEN $bpOnOff = OFF$
    THEN *bpOnOff* := *ON*
    END
  BPRunning
    STATUS pliant
    REFINES *PliTrue*
    WHEN $bpOnOff = ON$
    SOLVE $\mathcal{D} \, bpMeasVol = bpMeasFlow$
    END
  BPOff
    STATUS asynch
    WHEN $bpOnOff = ON$
    THEN *bpOnOff* := *OFF*
    END
END

Fig. 5. Hemodialysis machine development, Level 01. Introduction of the blood pump.

the measured flow rate produced by the blood pump, and the volume of blood that has passed through it. These are expected to vary continuously during operation since they are determined by the physical behaviour of the patient/machine system, so are declared pliant. The invariants now include the basic constraints upon these variables. At this level, these amount to type declarations.

There are further pump-related variables to be introduced later: the demanded pumping rate $bpRate$, and its lower and upper limits, $bpRateL$ and $bpRateU$. These are data that are entered by the operator in the preparation phase using mode events (because they will not change in between occurrences of these mode events). This is covered at Level 02. The only coupling between these later data and the pump's Level 01 variables is the constraint $bpMeasFlow \leq bpRate$, which is added in a refinement from Level 01 to Level 02 of the *BloodPump* machine.[3] The remainder of the Level 01 *Central_IF* interface declares the initialisations of the variables.

Next, we look at the *Operator* and *Control* machines. Whereas the *Control* machine is an exact replica of the Level 00 machine, the *Operator* is a refinement of its ancestor, since the *Reset* event has to now also encompass the new blood pump variables.

Both machines have a *PliTrue* pliant event. Since any Hybrid Event-B machine runs over an extended period of time, every such machine needs at least one pliant event to be able to do so. The *PliTrue* events simply stipulate that any behaviour conforming to the invariants is acceptable. These *PliTrue* events thus embody a loose specification of the pliant behaviour of the operator and controller.

At Level 01, the operator and controller merely steer the hemodialysis machine through the phases identified earlier: $PREP, INIT, TREAT, END,$ $ENDED$. In addition, the various modes are visible at this level: $NORMAL, BYPASS, ALARM$. For each of the phases, the operator initiates it with a *Start* mode event, and the controller terminates it with a *Complete* event, which changes the *phase* variable. Besides this, the controller can register an *ALARM*, so an event is required for this.

The observant reader will notice that, in fact, only the controller events change the phase variable, so the alternation between the operator who initiates a phase, and the controller which terminates it, is not enforced at this level of abstraction. However, at lower levels of abstraction, the controller events can have their guards strengthened to include conditions that witness the completion of intermediate activities that must be initiated by the operator. So, in order for the *Complete* events to be enabled, a collection of prerequisite events must have been triggered by the operator, enforcing the claimed sequentialisation.

All of these mode events have STATUS asynch. We explain this now. In discrete Event-B, occurrences of events, insofar as they are meant to model events in the real world, are assumed to occur at moments of time which are

---

[3] Obviously, the actual, measured flow rate, cannot exceed the rate that the pump is trying to achieve. To do so would contravene the laws of thermodynamics. However, in order to assert this, we must assume that the measured flow rate is *dependable*.

isolated. So 'real world' mode event execution is *lazy* (see, e.g., the many case studies in [1]).

However, because Hybrid Event-B is aimed at modelling physical behaviour, its mode events must execute *eagerly* — e.g. a ball falling onto a hard surface must bounce immediately; it does not have the option of waiting for a while. So when a mode event is enabled, it immediately preempts the ongoing pliant behaviour. Nevertheless, lazy execution of mode events is needed in modelling real systems that are capable of some spontaneous behaviour, or that respond to unpredictable stimuli from the environment. The metaphor that Hybrid Event-B provides in its semantics to handle this (see [6,7]) is to stipulate that *inputs* to mode events arrive lazily (unless more tightly constrained in the events' guards). The 'STATUS asynch' pattern used in Fig. 5 is a syntactic shorthand introduced to replace introducing an 'artificial' input (that would not be used in the body of the event) for each such mode event, and simply indicates that the relevant mode event is to execute lazily.

We turn to the *BloodPump* machine. At Level 00 it was specified very loosely, via a *PliTrue* pliant event alone. At Level 01 the operational details of the pump are included. There are mode events $BPOn$ and $BPOff$ that turn the pump on or off. Occurrences of these events interleave the pliant events $BPStopped$ and $BPRunning$. These refine the Level 00 *PliTrue* specification.

$BPRunning$ says that the derivative of the pumped blood measured volume is given by the measured blood flow rate $\mathcal{D}\,bpMeasVol = bpMeasFlow$. Note that there is no specification of the value of $bpMeasFlow$. The earlier yet-to-be-introduced constraint $bpMeasFlow \leq bpRate$, will only amount to a loose specification of the measured blood flow rate, since the actual value of $bpMeasFlow$ will depend on what takes place in the mechanical elements of the hemodialysis system, in particular on the amount of resistance they exert to pumping.

$BPStopped$ is the analogue of $BPRunning$ when the pump is not running. It stipulates that the measured flow rate $bpMeasFlow$ is zero (which it must be if no flow rate is demanded), and specifies that the measured volume $bpMeasVol$ remains CONSTant over the duration of the pliant event occurrence (although a differential equation could have been used). Note that this assumes that when it is switched off, the blood pump is locked in a state such that it cannot rotate backwards, which may happen at other times. If this not the case, a different specification for $BPStopped$ would be needed.

Note that at Level 01, the blood pump is completely decoupled from the rest of the system (aside from the *Reset* event). The integration with other dialysis activities comes later. This remark completes Level 01.

Level 02. We treat the remaining levels of the development more briefly. Level 02 is similar to Level 01 in that it introduces the safety air detector. As for the blood pump, this is introduced in a standalone way at this stage. Thus there are mode events to turn it on and off, and pliant events to model its behaviour when on and off respectively. These behave in a way analogous to the blood pump. There is a further mode event to raise an alarm when too much air has been detected. What makes the SAD interesting to model independently, is the way that its permitted accumulated air volume varies dynamically with the current blood flow rate. In fact, this feature is only needed in the alarm event just mentioned, so that its guard has a three way test, depending on the current blood flow rate, which may be low, medium or high, corresponding to a low, medium or high permitted volume. The way that the permitted volume depends on the blood flow rate means that, in principle, it is possible for the flow rate to be high say, and for the accumulated volume to be within bounds, but then for the flow rate to decrease, crossing the boundary to the medium regime, and for the same accumulated volume to suddenly be outside the new permitted bounds and thus to cause an alarm. Whatever the merits of this possibility, it has been modelled faithfully in order to test the modelling capabilities of Hybrid Event-B. The SAD also contains the venous red detector. The only feature of this is to execute a mode event when red appears or disappears.

Level 03. Level 03 deals with the first phase of dialysis, namely the preparation of the machine. In principle, this consists of rinsing and filling the machine, and of entering a large amount of data concerned with the parameters of the desired treatment regime. The preparation phase starts with the self test. If this fails, the alarm is sounded, and it is presumed that the operator takes over. Thus the success of the self test does not figure as a guard in subsequent steps (beyond the first).

Manipulation of various items of equipment is modelled by async mode events in the *Operator* machine, following the flow of [16]. Since the rinsing of the tubing is described in a little more detail, more detailed modelling of this is attempted. The main point of interest is that the blood pump is used, necessitating the coupling of the *Operator* machine, the *Control* machine, and the *BloodPump* machine. The required synchronisations are declared in the Level 02 PROJECT file thus:

SYNCH(*PrepRinsingTestingTubingStart*)   SYNCH(*PrepRinsingTestingTubingEnd*)
    Level_02_Operator.PrepRinsingTestingTubingStart_S   *Level_02_Operator.PrepRinsingTestingTubingEnd_S*
    Level_02_Control.PrepRinsingTestingTubingStart_S   *Level_02_Control.PrepRinsingTestingTubingEnd_S*
    Level_01_BloodPump.BPOn_S   *Level_01_BloodPump.BPOff_S*
    END   END

Both of these state that collections of events in different machines, each event referred to in the form *MachineName.EventName_S* (where the '*_S*' suffix is a decoration for additional readability), must execute together. In *PrepRinsingTestingTubingStart*, the *Operator* machine takes the initiative as the operator starts the tubing test process. Simultaneously, the *Control* machine records the change of state, and the *BloodPump* machine starts to operate. When the required amount of pumping has taken place, in *PrepRinsingTestingTubingEnd*, the *Control* machine is preempted by the pumped volume reaching a value that triggers the guard of the *PrepRinsingTestingTubingEnd_S* event, and the synchronisation then stops the pump and alerts the operator.

Aside from this, a lot of data entry is needed, modelled by straightforward state machine controlled input techniques. One data entry task is modelled faithfully (the entry of the filling blood pump rate), and the remainder are merely indicated.

Other mechanical manipulations of the equipment are indicated more briefly in [16], and are modelled more superficially. The last of them is the rinsing of the dialyzer. The completion of this causes the transition from the preparation to the initiation phase.

The preparation phase is characterised by a couple of safety requirements in [16], namely **R-18** and **R-19** (concentrate mixup), and **R-20** (temperature alarm in heating/degassing). These fall into the first of the requirement patterns discussed more extensively in the context of Level 04 below. The concentrate mixup alarm has been modelled by an alarm covering both cases, but since so little has been included in [16] about heating/degassing, the latter has not been modelled.

Level 04. The initiation phase is modelled in the Level 04 project. Arterial connection is assumed completed via operator events, and then the filling of the blood tubing is achieved by events synchronised between the operator, control and blood pump, just as discussed above. This time the pumping process is stopped via the venous red detector. Change of phase and change of mode are handled by existing events.

The initiation/connection phase is characterised by a large number of safety requirements in [16]. There are eleven general requirements **S-1** to **S-11**. Especially given the lack of mechanical details regarding the hemodialysis machine, these appear to be requirements on the behaviour of the operator, thus outside the remit of the developed system. Further, there are software requirements **R-1** to **R-36**. These appear to be amenable to incorporation into the developed system, but are hampered by the lack of detail mentioned. Thus their treatment must be, to some extent, hypothetical.

The requirements are of two types. The first concerns 'pointlike' conditions. A physical quantity, presumed monitored in a continuous way within the machine,[4] crosses some threshold, and some response is required (usually raising an alarm). Such requirements are easy to model. Thus, provided there are enough variables whose state is sufficient to define the period during which the condition is to be watched for, this state, plus the crossing of the threshold, constitute the guard of a 'raise alarm' mode event.

The second concerns 'extended' conditions. A physical quantity, again presumed monitored in a continuous way, crosses an undesired threshold and remains there too long (or until an extensive quantity has grown too large), at which point a response is required (again usually an alarm). A slightly different model is required here. Upon the threshold being crossed, a mode event as in the previous case starts a clock (or initialises the relevant extensive variable). Two further mode events complete the modelling. Since the physical quantity is assumed to vary in a *continuous* manner (a crucial assumption), it either stays

---

[4] In reality, such a variable will be monitored discretely, with a short sampling period. But in Hybrid Event-B terms it would correspond to a pliant variable, varying piecewise smoothly.

on the undesired side of the threshold or it recrosses the threshold in the opposite direction. In the latter case, if it happens soon enough, the recrossing will trigger a recrossing mode event and the clock can be disabled. In the former case the clock will reach a trigger value and an alarm can be raised. Of course, once the alarm has been raised, any subsequent recrossing is of no interest.

Regarding the initiation/connection phase, **R-2** to **R-4** provide examples of both kinds of requirement. They have been modelled faithfully by the patterns just described since the blood pump has been modelled in reasonable detail.[5] Of the rest, **R-5** to **R-11** and **R-14** to **R-17** also concern initiation/connection, involve the blood pump, and fit the patterns described. Of these, **R-11** is modelled (though not the switching off of the blood pump, since it is unclear from the various parts of [16] when exactly the pump is to be switched on or off during initiation/connection). Given the lack of precision around a number of details, and the evident similarity of this set of requirements to one or other of the two patterns described, the remainder of **R-5** to **R-11** and **R-14** to **R-17** were not modelled explicitly. Requirements **R-18** to **R-21** have been mentioned already.

There are some requirements that concern the SAD in the initiation/connection phase: **R-23** to **R-32**. Of these **R-24** to **R-26** are part of its operational definition, and have been modelled directly in the safety air detector machine. The remainder specify, in the various phases of the hemodialysis procedure, that the flow, or the accumulated air, remains within permitted margins (as discussed above). Notably, among the phases mentioned, the treatment phase appears to be absent. It seems strange that there is no objection to pumping the patient full of air during treatment, whereas it is prohibited during all other phases of the hemodialysis process. Accordingly, in the present model, monitoring of the accumulated air remains enabled throughout hemodialysis, which also simplifies the modelling a little. In this manner, all of **R-23** to **R-32** are covered.

Level 05. The treatment phase is modelled in the Level 05 project. In general, is not completely clear which of the very many parameters entered previously are active at the commencement of treatment, so this aspect was not reflected accurately in the modelling. Besides this, a number of activities take place during the treatment phase that raise questions regarding modelling. We discuss these one by one.

Venous return flow pressure. The description in [16] speaks of a dynamic limits window around the current value of the pressure, but leaves many issues uncertain. How is the dynamic limits window calculated from the current value? If it is based around the current value only, it is impossible for the current value to cross it, except when the machine's hard limits are crossed, so what is its purpose? If it is *not* based solely on the current value, how is it determined? What is supposed to happen when the current pressure crosses the boundary of either the dynamic limits window (assuming this is mathematically possible) or

---

[5] Since the blood pump rate is continuous, the rate *must* drop below 70 % of desired (**R-3**) *before* it can go into reverse (**R-4**), so one might conclude from the text of [16] that **R-4** is redundant.

the machine's hard limits window? Owing to the lack of clarity on these points —to which it is possible to imagine many possible resolutions— modelling of the venous return flow pressure was not attempted. However, it seems apparent that whatever the truth surrounding these matters, modelling based on the same kinds of idea used in the modelling of the blood pump and of the safety air detector, would be sufficient to handle this aspect of the machine.

Arterial entry pressure. The description in [16] is much briefer here, but suggests many of the same uncertainties as in the previous case. Again, due to the large degree of doubt, modelling was not attempted.

Blood-side entry pressure at the dialyser. Again, the description in [16] raises many questions about precisely what is monitored, what is under control, and what controller actions are needed in response to what measured stimuli from the equipment. Again, this was not modelled due to the large amount of uncertainty.

Treatment at minimum UF rate. This is modelled with a simple operator command to set the UF rate.

Heparin bolus. This is modelled by an operator command to start the bolus, and a controller response when it finishes. The present study does not model the Heparin pump explicitly, due to lack of detail, but one could presume that it might work like the blood pump. In connection with this, **R-22** could be modelled like **R-4**.

Arterial bolus. In the case of the arterial (saline) bolus, a little more detail is given. The modelling is like for the Heparin bolus, but the blood pump is used. So the start and stop events are synchronised with pump starting and stopping, the latter of which is conditional on the pumped volume.

Interrupting dialysis. Dialysis can be interrupted by going to bypass mode. This has not been modelled in order to illustrate a specific point. The *mode* variable is defined at Level 00, in order to help structure the whole development. To conform with (Hybrid) Event-B refinement practices, all the relevant events involving a change in *mode* should be defined at that level. Mostly they are apparent from a high level reading of [16]. However, here, there is another *mode* transition, buried deep in the detail. It could be accommodated easily enough by reworking from Level 00 onwards, to incorporate the needed transition. But doing this would be increasingly cumbersome the larger the development, so it was not done. This raises an interesting point regarding components and refinement, which we ponder now.

In defining a family of components, we can start by writing down the components in a 'hollow' form, writing down their coordination, and leave the internal details of their working to be refined in later. Alternatively, we can start by writing the components in an independent uncoordinated 'soup', then work bottom up, filling in their details, until their coordination needs to be addressed. The first option risks not accounting for all the needed coordinations early enough (as we have seen). The second option leaves all coordination in doubt until a late point, which may put high level requirements at risk. Further discussion of such situations, including a convenient remedy in the context of the Event-B refinement approach, appears at the end of this paper.

Treatment end. An acoustic sound is modelled. However, the lamp(s) are ignored. It is not clear how many lamps there are; nor whether they are real lamps or part of a synthetic UI display; nor, if they are real, what they do when they are not displaying the colour(s) indicated in [16] at the specific moments mentioned in the description.

From the above account, it is clear that only a portion of the whole treatment regime has been captured. Nevertheless, it seems apparent that if more detail were included in [16], it would not be difficult to model it using the patterns used at various places earlier.

Level 06. Level 06 models the therapy ending phase. The last stages of this, namely emptying the dialyser/cartridge and overview seem relatively automatic, in that they simply take place in sequence. Since few details about them are provided in [16], they are modelled with a single generic command from the operator. The preceding step, reinfusing the patient, is described in a little more detail, so is modelled using the operator, control and blood pump machines. The general approach resembles similar activities earlier in the hemodialysis process, that is to say, the operator starts a portion of the reinfusion, and this is stopped when one of the terminating conditions is reached, i.e. time elapsed, volume pumped, or red detected.

Some software requirements remain unaddressed: **R-33** to **R-36**. These concern the dialyser, and fluid removal, etc. Again, there is a dearth of needed detail in [16]. It is not at all clear what components are present in the dialyser, which of them connect to the software controller, how they operate, and how this processing sits in the rest of the hemodialysis procedure. So no serious attempt was made to model these requirements. Still, if we were to conjecture on the basis of what might be guessed from [16], we may suppose that the operation might be modelled in ways similar to subsystems we have covered in detail earlier.

6 Invariants, Verification

It is very clear when considering the hemodialysis machine system, that interaction with the environment is a vital and inescapable ingredient of operating the machine. In this case the environment comprises the patient, the operator, and includes unpredictable (and undesired) facets of machine operation. There is certainly a 'desirable, default' sequence of steps to hemodialysis treatment that one expects to be the norm. But it is clear that it is not within the power of the machine's controller to *enforce* this sequence during any particular occurrence of treatment. For example, it is *desirable* that the blood pumping rate remains close to the rate specified, but the system has to cope with the possibility that it doesn't (usually by raising an alarm, and throwing the responsibility back on the operator). There are a host of similar issues noted in [16].

The preceding impacts what can be achieved using verification. Normally, a system will be characterised by a number of properties linked together during operation. The interdependencies are usually expressible via invariants, and the B-Method gains its strength by insisting on the proof of the invariants. But when almost any variation in behaviour has to be anticipated, the prospect for writing invariants that are actually provable, diminishes severely. About all that remains, is the essentially sequential structure, insisting that one step precedes another.

In the case of simple sequential precedence, the invariants that arise, do so in the form 'if $u = A$ guards an event that achieves $v' = B$, then $v = B \Rightarrow u = A$ is an invariant'. This is no more than the action/reaction pattern noted by Abrial in [1]; obviously there are variations on this in less elementary cases. Such invariants can be extracted mechanically for sequential systems, so the role of the human designer is reduced in that there are many fewer informative invariants for him to infer, in a system in which variables are allowed to misbehave in arbitrary ways. The human may test his diligence by writing the invariants and checking whether he anticipates everything that the invariant generator is able to infer, but equally, he may just get bored by doing so. In line with this observation (and noting again that Hybrid Event-B is currently not machine processable), in the present modelling exercise, these routine invariants have not been written down.[6]

Thus, what can be achieved by verification of invariants (the mainstay of the B-Method) is severely limited in cases where so much comes from the environment. The methodology is much stronger in the case of closed systems, where the robustness has to come from internal design consistency, rather than ceding responsibility to an omniscient operator. And so, to regain some of this closed system robustness, we would have to depend on more detailed knowledge of the procedures engaged in by an experienced hemodialysis nurse, so that we could couple the *Controller* and *Operator* machines in a more exacting manner.

7   Conclusions

In the previous sections, we overviewed the hemodialysis case study, and after a brief review of Hybrid Event-B, described the Hybrid Event-B treatment of the case study. Our approach was structured round a pure refinement based strategy, based on an early recognition that the essential elements of the case study were sufficiently straightforward that the treatment of faults could be included in the refinement based strategy relatively straightforwardly, i.e. without the faults derailing the strategy to an unconvincing degree. This aspects is to be contrasted with the landing gear case study treated in [3,5], where (and especially in [5]) great benefit was derived from focusing first on the nominal development, and then incorporating the faulty regime, using retrenchment

---

[6] Aside from the routine nature of the invariants, is the fact that the *actual* degree of required sequentiality in hemodialysis is not clearly delineated in [16].

[8–10]. There, the various levels of fault tolerance would have made a flat treatment unhelpfully complex. Here though, the basic strategy of throwing responsibility back on the operator when things go wrong, made things considerably simpler, but undermined what could be accomplished using invariants, as discussed in the preceding section. Additionally, the earlier experience in modelling a complex application coming from the landing gear case study, made the process go a lot more smoothly.

As we noted in many places above, in the requirements [16], many issues regarding the dialysis treatment procedure were mentioned, without enough detail being given regarding how they might be implemented. Also, many aspects of the machine's working were alluded to without enough precision being given to make clear the degree of software involvement. In view of this, a fairly conservative approach was taken in the present development. What was modelled, was what related fairly directly to machine elements discussed with adequate precision in the text. Aspects that were less well described, involving features that were perhaps only hinted at, were typically not modelled, though often, it seemed that had they been better described, they would fit patterns already utilised elsewhere in the development. Since one of the main objectives of this study was to test the modelling adequacy of Hybrid Event-B in the face of challenging applications, the approach taken seemed reasonable.

On the other hand, the focus on a relatively narrow portion of the hemodialysis spectrum, loses the chance to define richer machine properties, and to connect them via suitable invariants to wider issues, whether in the machine or in the treatment regime. It is to be hoped that the introduction of more expressive formalisms such as Hybrid Event-B could contribute to more broadly based and more robust problem definitions.

In the author's opinion, a major, and worthwhile, outcome of the present work has been the opportunity to experiment with a simple component based approach to Hybrid Event-B development that was described earlier. Although it demands some forethought at the beginning, to decide on an appropriate level of modelling for the embryonic form of the various components that constitute the system-to-be, the clarity about the system architecture that it brings right at the outset, is very beneficial. Nevertheless, it demands some experience with refinement to judge how these components may be defined at the top level in order to avoid extensive rework later. Too much of the wrong sort of detail too early in the development can conflict with the rigours of refinement, when lower level detail emerges that is in conflict with more simplistic structures set down earlier. We saw an example of this in the discussion of bypass mode during the treatment phase. There we commented on the pros and cons of trying to define just enough of the coordination structure between the components early enough to provide a clear picture at a high level of abstraction, versus the 'soup of components' approach in which no attempt to define this especially early was made, at the cost of leaving coordination issues completely unaddressed till quite late on.

One convenient way of dealing with the situation is via retrenchment [8–10]. This, in its most benign form, would permit the introduction of 'new' events that modified 'old' variables, that preserved all invariants, but that were nonetheless *not* refinements of skip. Such events would evidently not upset anything that was deduced on the basis of the invariants —in our syntactic organisation of refinements of projects, all relevant invariants could be located by ascending the refinement hierarchy of interfaces until all needed variable declarations had been found— but they could upset conclusions made at a more abstract level based on model checking, since they can properly enlarge the reachability relation of the system at that level of abstraction. Postponing model checking till all such retrenchments had been completed would be an evident way of circumventing this difficulty. (Although clearly available here, the opportunity to take advantage of this technical possibility was not exploited in the present work, just to illustrate the point.) The approach described would provide the opportunity to address major coordination concerns between the components at a high level of abstration, while leaving the possibility of fine-tuning them later in the light of further detail that emerged at a lower level. With this proviso, the simple component based approach pursued here can be recommended as a useful methodological approach.

References

1. Abrial, J.R.: Modeling in Event-B: System and Software Engineering. Cambridge University Press, Cambridge (2010)
2. Ahmad, S.: Manual of Clinical Dialysis. Springer, US (2009)
3. Banach, R.: The landing gear case study in hybrid event-B. In: Boniol, F., Wiels, V., Ait Ameur, Y., Schewe, K.-D. (eds.) ABZ 2014. CCIS, vol. 433, pp. 126–141. Springer, Heidelberg (2014)
4. Banach, R.: Hemodialysis Case Study in Hybrid Event-B Web Site (2015). http://www.cs.man.ac.uk/banach/some.pubs/ABZ2016HemodialysisCaseStudy/
5. Banach, R.: The landing gear system in multi-machine hybrid event-B. Int. J. Softw. Tools Tech. Trans., pp. 1–24 (2015, to appear)
6. Banach, R., Butler, M., Qin, S., Verma, N., Zhu, H.: Core hybrid event-B I: single hybrid event-B machines. Sci. Comp. Program. **105**, 92–123 (2015)
7. Banach, R., Butler, M., Qin, S., Zhu, H.: Core Hybrid Event-B II: Multiple Cooperating Hybrid Event-B Machines (Submitted) (2015)
8. Banach, R., Jeske, C.: Retrenchment and refinement interworking: the tower theorems. Math. Struct. Comp. Sci. **25**(1), 135–202 (2015)
9. Banach, R., Jeske, C., Poppleton, M.: Composition mechanisms for retrenchment. J. Logic Algebraic Program. **75**, 209–229 (2008)
10. Banach, R., Poppleton, M., Jeske, C., Stepney, S.: Engineering and theoretical underpinnings of retrenchment. Sci. Comp. Program. **67**, 301–329 (2007)
11. Crnkovic, I., Larsson, M.: Building Reliable Component-based Software Systems. Artech House, Norwood (2002)
12. Daugirdas, J., Blake, P., Ing, T.: Handbook of Dialysis. Wolters Kluwer, New York (2007)
13. Harris, D., Elder, G., Kairaitis, G., Rangan, G.: Basic Clinical Dialysis. McGraw Hill, Sydney (2005)

14. Heineman, G., Councill, W.: Component-Based Software Engineering: Putting the Pieces Together. Addison Wesley, Boston (2001)
15. Kallenbach, J., Gutch, C., Stoner, M., Corea, A.: Review of Hemodialysis for Nurses and Dialysis Personnel. Elsevier Mosby, Philadelphia (2005)
16. Mashkoor, A.: The hemodialysis machine case study. In: Butler, M., Schewe, K.-D., Mashkoor, A., Biro, M. (eds.) ABZ 2016. LNCS, vol. 9675, pp. 329–343. Springer, Heidelberg (2016)
17. Nissenson, A., Fine, R.: Handbook of Dialysis Therapy. Saunders Elsevier, Philadelphia (2008)
18. Somaia, Z.: Component-Based Software Development. Lambert Academic Publishing, Germany (2014)

Modelling a Hemodialysis Machine Using Algebraic State-Transition Diagrams and B-like Methods

Thomas Fayolle[1,2,3], Marc Frappier[1](✉), Frédéric Gervais[2], and Régine Laleau[2]

[1] Université de Sherbrooke, Sherbrooke, QC, Canada
marc.frappier@usherbrooke.ca
[2] Université Paris-Est, LACL, Paris, France
[3] Ikos Consulting, Levallois-Perret, France

Abstract. This paper presents the specification of the hemodialysis case study, proposed by ABZ'16 conference. The specification was carried out by a coupling of Algebraic State-Transition Diagrams (ASTD) and B-like methods. ASTD are a graphical notation, based on automata and process algebra operators. They provide an easy-to-read specification of the dynamic behaviour of the system. The data model is specified using the Event-B language. The system is incrementally designed using extended refinement of both methods.

Keywords: ASTD · Event-B · Process algebra · StateCharts · Refinement

1 Introduction

This paper presents an answer to the case study proposed by ABZ'16 conference [1]. The case study proposes to specify a hemodialysis (HD) machine. The HD machine is specified using a combination of Event-B and ASTD. ASTD are a formal graphical langage that combines statecharts and process algebra operators. Event-B allows to capture the data model and the safety properties of the system whereas ASTD is used to specify the ordering of actions and to constraint the execution of the events.

The case study document is presented in two main parts. The first part presents the general behaviour of a hemodialysis machine. The second part presents safety requirements that the system has to verify. Our work mainly focuses on the specification of the general behaviour of the machine, safety requirement are dealt with at the last steps of the specification. On some points, the text of the case study is not very precise, even contradictory, and can be interpreted in several ways. In such cases, we made hypotheses. An archive containing the whole specification can be found on the following web site [2].

© Springer International Publishing Switzerland 2016
M. Butler et al. (Eds.): ABZ 2016, LNCS 9675, pp. 394–408, 2016.
DOI: 10.1007/978-3-319-33600-8_33

The paper is organised as follows: Sect. 2 explains the used languages. Section 3 briefly describes the way Event-B and ASTD are combined. The specification work is described in Sects. 4–6: Sect. 4 fully details the first two specification levels in order to illustrate the method, Sect. 5 describes interesting developments of the behavioural specification and Sect. 6 explains how we deal with failures and safety requirements. Section 7 concludes.

2   Background

2.1   B and Event-B

B [3] is a formal method that allows to formally and incrementally design software. The software is described in terms of machines. A machine contains state variables constrained by an invariant and possibly modified by operations. The B method requires to verify that modifications specified in operations do not violate the invariant. Refinement allows to design a software from its most abstract specification to implementation. B is supported by *Atelier B*[1] tool, which generates proof obligations and contains interactive provers. In our specification, an operation is described as a *precondition* substitution. This means that variables are modified according to the postconditions if the precondition is verified. Otherwise, the behaviour is not guaranteed.

Event-B [4] is the successor of B. It is used to design systems. A system is described in machines and contexts. The context contains the properties. A machine contains state variables that can be modified by events. Events are guarded and can only be executed when their guards are true. Like in B, properties on state variables are specified in terms of invariants. In this case study, Event-B developpement has been carried out using *Rodin*[2] tool.

2.2   Algebraic State-Transition Diagrams

Algebraic State-Transition Diagrams [5] combine Statecharts [6] and process algebra operators of EB[3] [7]. ASTD types are either hierarchical automatas or process algebra operators. The process algebra operators could be kleene closure, interleaving/synchronization, sequence, call or choice. Choice and interleaving/synchronization can be quantified on a set of variables. The automatas are hierarchical statecharts, where each state can be of any ASTD types. The complete operational semantics of ASTD are described in [8]. A system can be incrementally designed using trace refinement definition. The ASTD refinement has been defined in [9].

Elaborating the case study has conducted us to define an extension of the sequence operator, namely the quantified sequence. It takes an ASTD, a variable and a sequence of values for this variable. The ASTD is executed for each value in the sequence of values, in the order given in the sequence. The operator is detailed on an example in Sect. 5.2.

---

[1] http://www.atelierb.eu.

[2] http://www.event-b.org.

3   Our Method: Combining ASTD and B-like Methods

The hemodialysis case study has been specified by a combination of ASTD and
B-like methods. ASTD provide a graphical and formal notation of the behaviour
of the system whereas the data model is defined in Event-B. This data model
contains variables that describe the system (here the HD machine) and, for each
transition label of the ASTD specification, an event that precises the actions of
the transition on the variables. The ASTD specification constraints the execution
of the events. Both specifications are translated into classical B in order to verify
that an event is executed when its guard is true. The idea of the translation is the
following: the state of the ASTD is coded by B variables; a B operation is defined
for each transition label; its precondition checks that the operation is executable;
its postcondition updates the variables and calls the operation that specifies the
actions of the transition on the data. The B semantics requires to prove that the
precondition of an operation is true when the operation is called. To achieve this
proof, the designer has to define an invariant that links the variables of the data
model to the state of the ASTD. An example is given in Sect. 4.1. An overview
of the methodology can be seen on Fig. 1 and is explained with more details in
Sect. 1.2 of [10]. The method is partially supported by tools: ASTD and Event-B
are automatically translated into B and a graphical editor is being built.

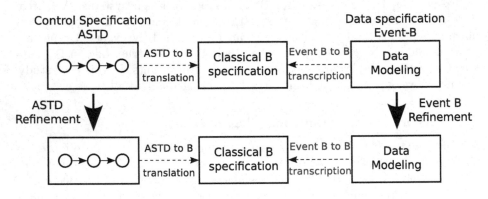

Fig. 1. Methodology of the specification

Both ASTD and Event-B notations include refinement notions. ASTD refinement is defined in [9] using properties on traces, deadlocks, etc... Event-B refinement is defined using properties on the state variables. But for some of the
refinement steps in this case study, Event-B and ASTD refinement definitions are
too restrictive. In those cases, we use a refinement definitions we wanted to add
in the methodology. Formal definition of these refinement is still an on-going
work.

4   The First Modelling Levels of the Hemodialysis Machine

4.1   Abstract Specification

At the most abstract level, the hemodialysis machine is described as a sequence of three main phases. At first, the HD machine has to be initialised. When it has been initialised, the blood can be filtered during the dialysis. Finally, an ending step is required to reinfuse the blood in the patient body. This is represented by an automata with 4 states and 3 transitions. We assume that after the completion of the treatment, the machine can be reused for an other dialysis. This is modelled by a kleene closure operator. The full specification can be seen on Fig. 2.

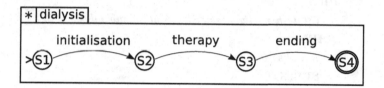

Fig. 2. First specification level

At this level, the data model of the specification is very simple: there is only one state variable that represents the state of the machine; each transition updates this state variable, named *state*, that can either be *OFF* before initialisation, *INITIALISED* after the initialisation phase and *HD_COMPLETE* after the dialysis. An Event-B event is associated to each transition of the ASTD specification. As an example, the event *initialisation_act* corresponding to the *initialisation* transition is guarded by the fact that the *state* variable is *OFF*. After the event, the new value of *state* is *INITIALISED*. The full event can be seen on Fig. 3.

$$\begin{aligned}
&\textbf{Event}\quad initialisation\_act \,\widehat{=} \\
&\quad \textbf{when} \\
&\qquad\quad \text{gu1} \;:\; state = OFF \\
&\quad \textbf{then} \\
&\qquad\quad \text{act1} \;:\; state := INITIALISED \\
&\quad \textbf{end}
\end{aligned}$$

Fig. 3. Event-B specification of *initialisation_act* event

In our methodology, this event is executed when the corresponding transition is executed. This means that when the precondition of the transition is true, the guard of the event has to be true. To generate this proof obligations, the Event-B specification and the ASTD specification are translated into classical B. The idea of the ASTD translation is the following: each transition is translated into a B operation; a variable codes the state of the ASTD; the precondition of the operation checks that this variable is in a state that allows the transition; the postcondition updates the variable to the state that is reached by the transition. In parallel, the B translation of the corresponding event is called. For this specification level, the state of the ASTD is coded with a variable named *State_hemodialysis*. The *initialisation* operation can be seen on Fig. 4. Note that the translation is automatically achieved which explains the two similar updates of variable *State_hemodialysis*.

$initialisation =$
PRE
$\quad(State\_hemodialysis = S1 \lor State\_hemodialysis = S4)$
THEN
$\quad initialisation\_act\|$
SELECT
$\quad State\_hemodialysis = S1$
THEN
$\quad State\_hemodialysis := S2$
WHEN
$\quad State\_hemodialysis = S4$
THEN
$\quad State\_hemodialysis := S2$
END
END

Fig. 4. B translation of *initialisation* transition

In B, when an operation is called, the proof that its precondition is true is required. In this example, we need to prove that the precondition of *intialisation_act* is true when the ASTD is in the state *S1* or in the state *S4*. In order to prove this, we write the following invariant:

$$State\_hemodialysis = S1 \lor State\_hemodialysis = S4 \Leftrightarrow state = OFF$$

This invariant is trivial and is automatically proved. Such invariants are added to link all the possible values of the *state* variable to the states of the ASTD.

4.2   First Refinement

Each phase of the hemodialysis is divided into two steps. In the initialisation phase, the tubbing system is filled with water, and then the water circulates to rinse the machine. During the therapy, the tubbing system is filled with blood and then the blood circulates to be filtered. Finally, during the ending phase, the tubbing system is filled with salted water and then the salted water circulates to reinfuse the blood and empty the dialysis machine.

This refinement level follows the existing definition of ASTD and Event-B refinement. Refinement patterns for ASTD have been introduced in [9]. One of these patterns allows to replace one abstract transition with new concrete transitions. Applying this pattern three times allows to replace the transition corresponding to each phase by a sequence of two transitions. The complete ASTD can be seen on Fig. 5.

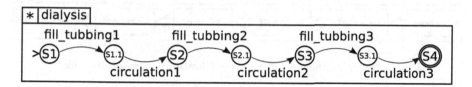

Fig. 5. First refinement of ASTD specification

The refinement of the data model follows the definition of Event-B refinement. A new variable named *detected_liquid* represents the liquid detected in the tubbing system. The possible values are *nothing*, *water* and *blood*. The three *circulation$_i$* events refine the three events from the abstract specification. The *fill_tubbing$_i$* events are new events and change the values of the detected liquid.

At the beginning of the hemodialysis, there is no liquid in the hemodialysis machine, and it is first filled with water. In the Event-B refinement machine, the *fill_tubbing1_act* event changes the value of the *detected_liquid* variable from *nothing* to *water*. *circulation1_act* event has the same actions as *initialisation_act* event and an additional guard that checks if the *detected_liquid* variable is set to water. The refinement proof is trivial and automatically proved.

Using the same mechanism as for the abstract level, ASTD and Event-B specifications are translated into B, in order to prove the horizontal consistency. For example, *circulation1_act* can only be executed if water has been detected and if the machine is in *OFF* state. The new invariant that we add says that if the ASTD is in state *S1.1*, water is detected and machine is in *OFF* state. Thus we prove that when *circulation1_act* is executed, its guard is true.

5   More Modelling Levels

In Sect. 4, the abstract specification and the first refinement were fully detailed, in order to illustrate the methodology. This section focuses on interesting behaviour development and will mainly detail the ASTD specification since the Event-B specification are rather trivial. A last specification level deals with failures and is detailed in Sect. 6.

5.1   Second Refinement

The aim of this specification level is to detail the behaviour of the system during each step of the previous specification. According to the specification document, each step is a sequence of at least five actions. Developing this system will make an unreadable flat automata of at least thirty transitions. To make it more readable, the sequence of transitions is replaced by a sequence of ASTD, defined elsewhere and called using the call operator of the ASTD language. The main ASTD specification is presented in Fig. 6.

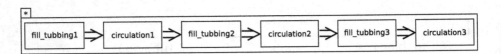

Fig. 6. Second refinement of ASTD specification: main specification

Let us detail the first two ASTD on Fig. 6. *fill_tubbing1* corresponds to the set of actions where the tubbing system is filled with water. A bag of saline solution is branched to the machine. The operator presses the start button which starts the pump. The machine keeps reading the VRD. When water is detected by the VRD (Venous Red Detector), the pump automatically stops. The specification says that "the venous and arterial patient ends are reconnected for recirculation", which assumes that the operator has been warned that the pump stopped.

This behaviour requires the collaboration of two entities: an operator and the HD machine. They are represented by two synchronised ASTD. The operator branches the saline solution. The HD machine starts the pump, reads the VRD until water is detected and stops the pump. Pushing the start button and signaling the end of *fill_tubbing1* step are synchronisation barriers. The specification of the *fill_tubbing1* is represented on Fig. 7.

circulation1 step corresponds to the step where the machine is rinsed. It consists in the following actions. The arterial and venous ends are connected one with another. The pump is started by pushing on START/STOP button. The case study document doesn't precise when it ends, so we assume that the machine is rinsed during a certain given time. After this given time, the pump stops and the signal lamp on the monitor changes to yellow.

Once again, this behaviour is executed by the operator and the HD machine. The operator branches the venous and arterial parts. The machine starts the pump, counts the given time and stops the pump. Pressing the START/STOP and the changing of color for the lamp are synchronization barriers. The specification of *circulation1* is represented on Fig. 8.

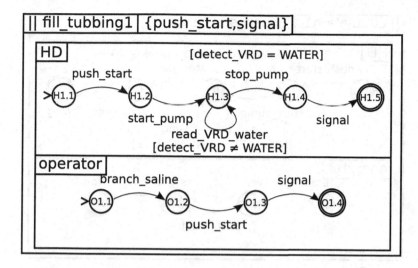

Fig. 7. Second refinement of ASTD specification: *fill_tubbing1* specification

fill_tubbing2 follows the same specification as *fill_tubbing1* except that the operator branches the arterial end to the patient and that the pump stops when the VRD detects blood, and not water. *circulation2* follows the specification of *circulation1* except that the operator branches the venous end to the patient and that the pump stops when a certain time is completed. In *fill_tubbing3* the operator branches saline to the arterial end, the pump stops when water is detected in the VRD. In *circulation3* the venous end is removed from the patient, and the pump is started to empty the machine. It stops when no liquid is detected by the VRD. In the specification document, a reinfusion part is described but is not modelled in our system. It could be added in further refinement.

The data model is an Event-B refinement of the specification of Sect. 4.1. Variables are added to express the fact that the start button has been pushed, that the pump starts, etc... The event that are associated to the final transition of each ASTD (the *signal_....* event) refines the event corresponding to each transition of the abstract model (*fill_tubbing*$_n$ and *circulation*$_n$). The other events are new events, which means they refine the **skip** substitution.

5.2 Factorising the ASTD Specification

We can observe that Figs. 7 and 8 are almost identical and only differ in three actions: (1) what is branched by the operator, (2) what is detected to end the pump and (3) what signal is given to the operator. The rest of the sequence remains unchanged. In Sect. 5.1 we only presented two of the six called ASTD, which means that the complete specification should take more place. To avoid such a cumbersome specification, we suggest to factorise the specification in order to make it more concise.

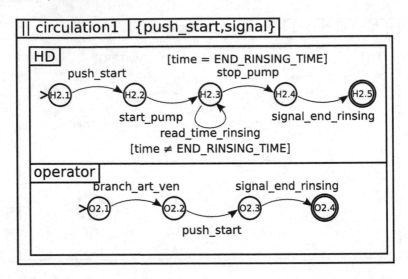

Fig. 8. Second refinement of ASTD specification: *ciruclation1* specification

The most abstract specification, described in Fig. 2, has three phases (initialisation, therapy and ending), represented by a variable p in an enumerated set $PHASES = \{INIT, THERAPY, ENDING\}$. Each of this phase has two steps (filling and circulation) represented by a variable s in an enumerated set $STEPS = \{FILL, CIRC\}$. The six ASTD corresponding to each phase are replaced by a unique ASTD that is parametrized by the phase and the step in which it is executed. The common operations (*push_start*, *start_pump* and *stop_pump*) remain unchanged, but the new ASTD has to model the specific behaviour of each phase. The transitions that differ between the steps are parametrized with the phase and the step. The ASTD that specifies the cycle can be seen on Fig. 9.

This ASTD is executed two times for each of the three phases. The quantified sequence is introduced to express the sequence of the three phases. The main specification can be seen on Fig. 10. It means that a sequence of two *cycle* ASTD is executed for p equal to $INIT$, then $THERAPY$ and then $ENDING$.

The specific behaviour corresponding to each phase is carried out by the data specification. As an example, the Event-B specification of the *branch_act* event is given in Fig. 11. This specifies what the operator has to branch depending on the phase and step. For example, in the initialisation phase and during the filling step ($p = INIT$ and $s = FILL$ - two first lines of act1 on Fig. 11), the operator does not change what is branched venously and has to branch water to the arterial end of the tubbing system.

This specification level is a rewriting of the specification of Sect. 5.1. Most of the transitions and events remain unchanged, only the three specific operations (*branch*, *signal* and *read*) are modified. In the ASTD specification they are replaced by a global transition which takes the specific phase and step as

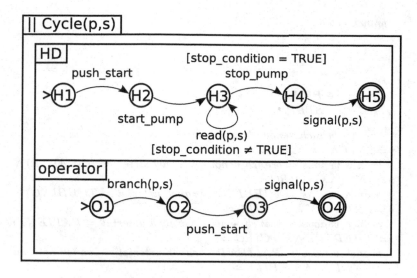

Fig. 9. *cycle* specification

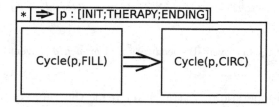

Fig. 10. Main specification

a parameter. For given parameters, the event corresponding to the transition should have the same behaviour as the specific abstract event. As an example, *branch_act(INIT, FILL)* (which means the event *branch* during the filling step of the initialization phase) has the same behaviour as the *branch_saline_act* event.

6 Introducing Failures and Dealing with Requirements

The requirement section of the case study document introduces many cases where failures can happen. The specification detailed in Sect. 5 is the normal mode specification, that is the specification where nothing goes wrong. The failures may happen when the pump is started, if some parameters take unsafe values. In our last level, the state where the pump is started is the state *H3*.

In [9], a refinement pattern allows to refine a loop transition by a kleene closure ASTD. This pattern is used to introduce failures in our system. In the specification of Fig. 9, when the pump is running (state *H3*), the machine keeps reading the value of some parameters, given by sensors, in order to detect when an ending condition holds. The idea to deal with failures is to replace the reading

Event $branch\_act \; \hat{=}$
    any
          p
          s
    where
          gu1 $: p \in PHASES$
          gu2 $: s \in STEPS$
    then
          act1 $: branch\_venous, branch\_arterial : |($

$(p = INIT \land s = FILL \Rightarrow$
$branch\_venous' = branch\_venous \land branch\_arterial' = WATER) \land$
$(p = INIT \land s = CIRC \Rightarrow$
$branch\_venous' = TUBBING \land branch\_arterial' = TUBBING) \land$
$(p = THERAPY \land s = FILL \Rightarrow$
$branch\_venous' = branch\_venous \land branch\_arterial' = PATIENT) \land$
$(p = THERAPY \land s = CIRC \Rightarrow$
$branch\_venous' = PATIENT \land branch\_arterial' = branch\_arterial) \land$
$(p = ENDING \land s = FILL \Rightarrow$
$branch\_venous' = branch\_venous \land branch\_arterial' = WATER) \land$
$(p = ENDING \land s = CIRC \Rightarrow$
$branch\_venous' = NOTHING \land branch\_arterial' = NOTHING))$

    end

Fig. 11. Event-B specification of $branch\_act$ event

transition by two operations. A transition reads the value of a list of sensors and a transition computes a value for a boolean *error* variable. The reading transition is guarded by the fact that the value of *error* and *ending_condition* are both false. If an error is detected, this specification would lead to a deadlock, which is solved by adding a transition on the main ASTD specification. This transition corresponds to the case where the therapy ends with an error. The new specification of a cycle can be seen on Fig. 12. The new main ASTD is on Fig. 13. Note that the *error* transition can be executed from any state of the kleene closure.

The idea of the data model in Event-B is the following: a variable is defined for each value that has to be monitored (*e.g.* blood flow in the EBC, presence of bicarbonate). The *read_act* event gives a value to each of these variables. This is defined using the *become element of* Event-B substitution. The *compute_act* event is parametrized by the phase and step of the dialysis, and computes the value of the *error* and *ending_condition* variables.

For example, requirements **R-5** to **R-8** deal with the blood pressure at the VP and AP transducers during initiation phase. In our specification, it corresponds to the circulation step of the therapy phase. The pressures at the VP and AP transducers are represented by two variables named *pressure_VP* and *pressure_AP*. The *read_act* event gives a value to these two variables. The part of *read_act* event that deals with the requirements **R-5** to **R-8** can be seen on Fig. 14. Symbole $:\in$ is the *become element of* substitution.

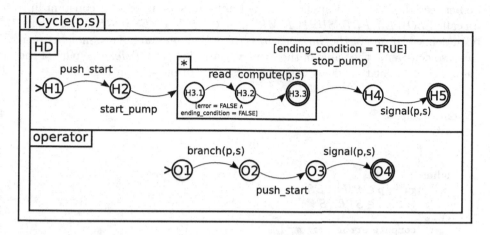

Fig. 12. Third refinement of ASTD specification: *cycle* specification

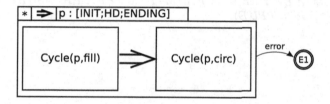

Fig. 13. Third refinement of ASTD specification: main specification

Event   *read_act* $\widehat{=}$
   when
        gu1 : *error* = *FALSE*
        gu2 : *ending_condition* = *FALSE*
   then
        ... : ...
        reading_pressure_AP : *pressure_AP* :$\in \mathbb{Z}$
        reading_pressure_VP : *pressure_VP* :$\in \mathbb{Z}$
        ... : ...
   end

Fig. 14. *read_act* event

In the case study document, it is said that the AP and VP pressures have to be between the lower and upper pressure limits. It is assumed that constant limits called $LOWER\_PRESSURE\_LIMIT$ and $UPPER\_PRESSURE\_LIMIT$ have been defined. If one of the variable $pressure\_VP$ or $pressure\_AP$ is not between this two constants, the value of the $error$ variable is set to $True$. The part of the $compute\_act$ event that deals with the pressure can be seen on Fig. 15.

Event $compute \;\widehat{=}$
    any
        p
        s
    where
        gu1 $: p \in PHASES$
        gu2 $: s \in STEPS$
    then
        compute_error $: error : |$
        $...$
      $((p = THERAPY \land s = CIRC \land$
       $((pressure\_AP \notin$
          $LOWER\_PRESSURE\_LIMIT .. UPPER\_PRESSURE\_LIMIT) \lor$
       $(pressure\_VP \notin$
          $LOWER\_PRESSURE\_LIMIT .. UPPER\_PRESSURE\_LIMIT)))$
       $\Rightarrow error' = TRUE)$
       $...$

    end

Fig. 15. *compute* event

7   Conclusion

This paper presents a specification of a hemodialysis machine using a combination of ASTD and Event-B. A formal graphical specification of a hemodialysis machine was obtained. This specification is based on a document that was written in natural language and therefore contained some imprecise descriptions, which is quite usual. The advantage of the graphical view provided by ASTD is that we can assume that medical experts can easily validate the specification and progressively correct the ambiguous textual document, which is more complicated, even impossible, with Event-B notations. On the other hand, system designers can benefit from the strengths of B-like notations, specially the different kinds of available tools such as provers, model-checkers or animators, to verify and validate specifications.

Moreover, our specification specifies the hemodialysis machine and its interaction with the operator. It means that we have a formal specification of what a nurse should do. It could be use to write a user manual.

Comparable results could have been obtained using iUML-B [11] for example. This language combines UML and B method and allows to graphically specify systems. It has graphical tools but ASTD contains more process algebra operators. The process algebra operators increase the expressiveness of the langage. On another side, $CSP\|B$ [12] or Circus [13] combine process algebra and state-based languages (resp. B and Z). The expressiveness is comparable, but ASTD use a graphical representation which is very usefull to make an easy-to-read specification.

Finally, the method we used allows to verify the overall consistency of the system. The events defined in the data model usually have guards. In Event-B semantics, it would mean that guards have to be checked before executing the event (if the specification is used as a user manual for example). Our method proves that if the actions are executed following the ordering defined in the ASTD, the guards of the events are true when they are to be executed. The counterpart is that the specification often generates many proof obligations (for example, proving the horizontal consistency of the second level generates about 70 -automatically proved- proof obligations). However, for this particular case study where the ordering of events is very complex whereas their actions on the state of the HD machine are quite simple, it is certainly easier to specify the system with a combination of ASTD and Event-B than with Event-B alone. Indeed, the main difficulty in Event-B is to define the guards of the events to ensure a correct ordering of these events.

References

1. Mashkoor, A.: The hemodialysis case study (2015)
2. http://www.lacl.fr/~tfayolle
3. Abrial, J.R.: The B-book: Assigning Programs to Meanings. Cambridge University Press, New York (1996)
4. Abrial, J.R.: The Event-B Book. Cambridge University Press, New York (2007)
5. Frappier, M., Gervais, F., Laleau, R., Fraikin, B., Saint-Denis, R.: Extending statecharts with process algebra operators. Inovation Syst. Softw. Eng. 4(3), 285–292 (2008)
6. Harel, D.: Statecharts: a visual formalism for complex systems. Sci. Comput. Program. 8, 231–274 (1987)
7. Frappier, M., St-Denis, R.: EB3: an entity-based black-box specification method for information systems. Softw. Syst. Model. 2, 134–149 (2003)
8. Frappier, M., Gervais, F., Laleau, R., Fraikin, B.: Algebraic State Transition Diagrams. Technical report, Université de Sherbrooke (2008). http://www.dmi.usherb.ca/~frappier/Papers/astd.pdf
9. Frappier, M., Gervais, F., Laleau, R., Milhau, J.: Refinement patterns for ASTDs. Formal Aspects Comput. 26, 919–941 (2014)
10. Fayolle, T.: Specifying a Train System Using ASTD and the B Method. Technical report (2014). http://www.lacl.fr/~tfayolle
11. Snook, C., Butler, M.: Uml-b: formal modeling and design aided by uml. ACM Trans. Softw. Eng. Methodol. 15, 92–122 (2006)

12. Schneider, S., Treharne, H.: Communicating B machines. In: Bert, D., Bowen, J.P., C. Henson, M., Robinson, K. (eds.) B 2002 and ZB 2002. LNCS, vol. 2272, pp. 416–435. Springer, Heidelberg (2002)
13. Woodcock, J., Cavalcanti, A.: A concurrent language for refinement. In: Butterfield, A., Pahl, C. (eds.) IWFM 2001: 5th Irish Workshop in Formal Methods. BCS Electronic Workshops in Computing, Dublin, Ireland (2001)

Modelling the Haemodialysis Machine
with *Circus*

Artur O. Gomes[✉] and Andrew Butterfield

School of Computer Science and Statistics, Lero,
The Irish Software Research Centre, Trinity College Dublin, Dublin, Ireland
{gomesa,butrfeld}@tcd.ie

Abstract. We present a formal model of aspects of the haemodialysis machine case study using the *Circus* specification notation. We focus on building a model in which each of the software requirements (**R-1–36**) are represented by a *Circus* action. All of these act in concert with actions that model the collection of sensor data and the progress through the various therapy phases and activities. We then present how we model check the system using FDR.

1 Introduction

This paper describes our experience in modelling the haemodialysis machine case study, that was issued for the ABZ 2016 conference [7]. We chose to do our modelling using *Circus*, a fusion of Z and CSP. We saw the case study as a way to assess how the ability to mix Z schemas with CSP-like processes would enable us to structure the model in a reasonably modular manner.

Our primary focus was on the software requirements (**R-1** through **R-36**) and our plan was to use a *Circus* process or action to model the behaviour implied by each of them. We also modelled some support services, such as clocks and sensor reading actions, as well as the control-flow prescribed for the various phases of a typical therapy session.

We make reference to the case-study document [7] using the shorthand [HMCS] or [HMCS, part X]. We present a quick overview of *Circus* in Sect. 2. We then give an overview of the approach and present some of the modelling infrastructure in Sect. 3, before describing how some of the software requirements were modelled as processes in Sect. 4, where we also discuss how they were then assembled to give the full model. An issue with *Circus* is the availability of tool support, and so we discuss in Sect. 5 how we translated our model by hand into machine-readable CSP (CSPm), so that we could use the FDR3 refinement-checker [3]. We talk about issues and inconsistencies spotting during the modelling process in Sect. 6, and then, in Sect. 7, we conclude.

This work was funded by CNPq (Brazilian National Council for Scientific and Technological Development) within the Science without Borders programme, Grant No. 201857/2014-6, and partially funded by Science Foundation Ireland grant 13/RC/2094.

© Springer International Publishing Switzerland 2016
M. Butler et al. (Eds.): ABZ 2016, LNCS 9675, pp. 409–424, 2016.
DOI: 10.1007/978-3-319-33600-8_34

2   Quick **Circus** Guide

Woodcock and Cavalcanti developed *Circus* [10,12], as a formalism which not only combines Z [13] and CSP [6], but also Dijkstra's guarded command language [2]. Its semantics is based on the Unifying Theories of Programming (UTP) [5] and it has a refinement calculus, developed by Oliveira [9] based on that of Morgan [8]. The thesis by Oliveria [9] is also the de-facto reference for *Circus*[1].

A *Circus* script can be considered as a series of "paragraphs", which can be either Z paragraphs or CSP process definitions, or a hybrid mix of CSP with "commands" to produce actions. The key feature here is that *Circus* uses Z-schemas to declare variables and state invariants, and then allows CSP actions to refer to and modify those variables.

We shall present a simplified version of *Circus* here, focussing on those CSP aspects as used in this paper. We shall simply model state-changing operations by variable assignment, as proxy for the Z schema parts. In the rest of the paper we make use of proper Z schemas.

For both CSP and Circus, a typical description consists of a series of definitions of the form

$$N(v_1, v_2, \ldots, v_n) \cong C$$

where N is a (process/action) name, the v_i are local parameters, and C is a process/action "construct" that may or may not refer to N and the v_i.

Basic building blocks include expressions over local state, and ways to describe events:

$$
\begin{aligned}
N &\in Name &&— \text{Process Names} \\
k &\in Const &&— \text{Concrete Values} \\
v &\in Var &&— \text{Local Variables} \\
e &\in Expr &&— \text{Expressions over Local State} \\
a &\in Event &&— \text{Atomic Events} \\
c &\in Chan &&— \text{Event Channel} \\
c.k &\in Event &&— \text{Channel Data Event}
\end{aligned}
$$

Events are observable, atomic (they either happen "in full" or not at all) but can be composite objects. So a common idiom is to describe events — atomic ! — of the form $c.k$ which is to be interpreted as the atomic event consisting of the transfer of a value k along a channel c. A process is an entity that is willing to perform some events, but not others, depending on its current state. We consider all processes as interacting with an environment, also considered as a process. A process willing to perform an event can be said to be "offering" that event. Whether or not the event actually occurs depends on both the willingness of the environment to perform it, and the synchronisation requirements between the process and its environment.

---

[1] More details and publications about *Circus* can be found at https://www.cs.york.ac.uk/circus/.

We shall first consider those constructs of *Circus* that are essentially the same as their CSP counterparts:

$$
\begin{array}{llll}
C & ::= & \textbf{Skip} & \text{— Termination} \\
  & | & a \longrightarrow C & \text{— Prefix} \\
  & | & c?v \longrightarrow C & \text{— Input} \\
  & | & c!e \longrightarrow C & \text{— Output} \\
  & | & C \,;\, C & \text{— Sequential Composition} \\
  & | & C \,[\!|\, cs \,|\!]\, C & \text{— Parallel Composition (CSP)} \\
  & | & C \,|||\, C & \text{— Parallel Interleaving} \\
  & | & C \,\Box\, C & \text{— External Choice} \\
  & | & e \,\&\, C & \text{— Guarded Process} \\
  & | & N(e_1, e_2, \ldots, e_n) & \text{— Process Call} \\
  & | & \mu X \bullet C & \text{— Recursion}
\end{array}
$$

Briefly, *Skip* terminates immediately; Prefix $a \longrightarrow C$ performs event a and then behaves like C; Input $c?v \longrightarrow C$ performs a channel event $c.k$ where k is a valid value for local variable v, and then behaves like $C[k/v]$; and Output $c!e \longrightarrow C$ performs event $c.k$, where k is the current valuation of e, and then behaves like C. We also have sequential composition where $C_1 \,;\, C_2$ behaves first like C_1 and then behaves like C_2. Another composition form is parallel, in which $C_1 \,[\!|\, cs \,|\!]\, C_2$ runs both commands in parallel, synchronising on events in cs and interleaving others. External choice ($C_1 \,\Box\, C_2$) allows the environment to choose between the events offered by C_1 and C_2, so determining which runs. The guarded command $e \,\&\, C$ behaves like C if e evaluates to *true*, otherwise behaves like the canonical deadlocked process *Stop*, which is a unit for external choice.

Now we look at those constructs which have been added to CSP to produce *Circus*, as well as CSP constructs that require modification in order to "play nice" with the extended semantics.

$$
\begin{array}{llll}
C & ::+ & v := e & \text{— Assignment} \\
  & | & \textbf{if } G [\!] \ldots [\!] G \textbf{ fi} & \text{— Guarded Choice} \\
  & | & C[\![us|cs|vs]\!]C & \text{— Parallel Composition (\textit{Circus})} \\
G & ::= & e \longrightarrow C & \text{— Guarded Command}
\end{array}
$$

We have assignment $v := e$, that updates a variable that can be local or global. We also have guarded commands (**if**...**fi**) in the "Dijkstra Style" [2], where a single guarded command is denoted by $e \longrightarrow C$, e being a boolean-valued expression. The strong similarity between event prefix ($a \longrightarrow C$) and guarded command ($e \longrightarrow C$) is unfortunate, but is part of the official *Circus* syntax [9]. However, it has the same semantics as guarded processes, so, for clarity's sake, we shall use $e \,\&\, C$ in the sequel for both, to avoid confusion with prefixing.

For *Circus* we need a slightly more complicated notion of parallel composition because we have global state. In order to put C_1 and C_2 (say) in parallel, we require that the sets of variables modified by the two commands be disjoint, otherwise the parallel composition is not well-formed. We write $C_1[\![us|cs|vs]\!]C_2$ to indicate that the variables modified by C_1 are contained in us, and those modified by C_2 are in vs, with $us \cap vs = \varnothing$. As with CSP parallel, we also specify the events/channels (cs) on which both sides must synchronise. In addition, when the parallel composition starts, each side gets its own snapshot of the starting variable state, which it then subsequently uses to record its own variable updates. While either of C_1 or C_2 are still running, the state changes made by one are *not* visible to the other. At the end, once both have terminated, then their snapshots are merged to give the overall final state. This means that if C_1 wants to communicate a state change to C_2, while both are still running, then it must use input/output events to achieve this.

One key advantage that *Circus* has over CSP is its ease of handling a large collection of named state components, when most situations only require the update of a few of those components. For example, imaging a process $FIRST$ that waits for an event e and then increments a state component called s, which is one among a large number of such components, and then behaves like $NEXT$. The CSP definition of $FIRST$ and $NEXT$ would be something like the following:

$$FIRST(\dots, c, \dots) \mathrel{\widehat{=}} e \longrightarrow NEXT(\dots, c+1, \dots)$$
$$NEXT(\dots) \mathrel{\widehat{=}} \dots$$

The *Circus* equivalent would be

$$FIRST \mathrel{\widehat{=}} e \longrightarrow c := c+1 \; ; NEXT$$
$$NEXT \mathrel{\widehat{=}} \dots$$

3   Approach

In order to formalise the HD machine [4], a few decisions were made regarding the system environment, sensors and the kind of responses that are required. In this section, we present our decisions about how we deal with timing, the overall structure of the HD machine, as well as how we capture the sensing functionality, and how the system should respond to events according to the requirements.

3.1   Timing Properties

A first consideration we need to take into account is how we handle time, as there are several safety requirements for the machine that deal with time. For example, the software requirement **R-2** deals with the absence of blood flow in the machine for a period of 120 s. After that period is over, the machine should respond right away by stopping the blood flow and raising an alarm.

We did not feel that the timing issues in the specification warranted the complexities of using a timed variant of *Circus*. Also, the precise times are not that important for our model —rather than waiting for a model time to elapse corresponding to 120 s, we would simply treat It more symbolically. All we really need to be able to distinguish is between a time when haven't reached such a limit, and that limit time.

We defined a *Circus* process called *SysClock* that starts with the *ResetClock* process that initialises the *time* variable and then calls *Clock*. On its turn, *Clock* repeatedly issues *tick* events, and increments a state component called *time*, storing the current time. The current *SysClock* time is made available through the channel *getCurrentTime*.

$$Clock \ \widehat{=} \ \mu \ X \ \bullet \ \left(\begin{array}{l} tick \longrightarrow time := time + 1 \\ [\![getCurrentTime!time \longrightarrow \textbf{Skip} \end{array} \right) \ ; X$$

$$ResetClock \ \widehat{=} \ time := 0 \ ; Clock$$

Another feature we need to have in our model is a wait period in order to comply with requirements such as **R-16** that specifies a period of time between two phases of the therapy. We therefore define a *Circus* process *Wait* that counts *n tick* cycles. The variable *n* is decremented after each *tick* and the entire process ends when the value of *n* reaches zero, here defined as a **Skip** action.

$$Wait \ \widehat{=} \ \textbf{var} \, n : \mathbb{Z} \ \bullet$$

$$\left(\textbf{if} \, n > 0 \longrightarrow (tick \longrightarrow Wait(n - 1)) [\![\, n = 0 \longrightarrow \textbf{Skip} \, \textbf{fi} \right)$$

3.2   State Components

The notion of machine state is essential in order to record key values of the various components of the system. Reading the software requirements section led us to identify over 20 state components used during the execution of the system. These are related to sensor measurements, sensor limits and switches that allows the physician to adjust the parameters for the therapy.

We also take into account some components that we decided to include in our specification as state components. These are used, for example, to register the activity and therapy phases of the HD machine. We also create records of the time in the system.

In our model, we define the Z schema *HDGenComp*, composed of the state components that we identified whilst reading the system requirements. For instance, we identify the *airVol* and *airVolLimit* parameters from **R-28–32**, as detailed in Sect. 4.

```
┌─ HDGenComp ─────────────────────────────────
│   airVolLimit : ℤ; airVol : ℤ; alarm : SWITCH; ...
│
```

describing with the description of many components of the system. For instance, [HMCS, Table 2] describes the rinsing parameters that are defined and entered into the machine during therapy. These parameters have specific ranges of values. We model the content of Table 2 as the *RinsingParameters* schema with its components. Moreover, we define the value ranges of the components as state invariants. For example, the range for the *Filling BP rate* is 0–6000 mL and is modelled in Z as the state variable *fillingBPRate*, with an invariant of *fillingBPRate* $\in \{ x : \mathbb{Z} \bullet 0 \leq x \leq 6000\}$. The overall state component *RinsingParameters* is illustrated below.

$\boxed{\begin{array}{l} \underline{RinsingParameters} \\ \quad fillingBPRate \; : \; \mathbb{Z}; \; rinsingBPRate \; : \; \mathbb{Z}; \; \dots \\ \hline \\ \quad fillingBPRate \in \{\, x \, : \, \mathbb{Z} \mid 0 \leq x \leq 6000 \,\} \\ \quad rinsingBPRate \in \{\, x \, : \, \mathbb{Z} \mid 50 \leq x \leq 300 \,\} \\ \quad \dots \end{array}}$

The entire state of the HD machine is modelled as the schema $HDState$, which is composed by the above described $HDGenComp$ schema, along with the $RinsingParameters$ and all the other schemas, $DFParameters$, such as $UFParameters$, $PressureParameters$, and $HeparinParameters$, modelling the variables detailed in the [HMCS, Tables 3–6].

$$HDState \; \hat{=} \; HDGenComp \wedge \; RinsingParameters \wedge \; DFParameters$$
$$\wedge \; UFParameters \wedge \; PressureParameters \wedge \; HeparinParameters$$

We also initialise the $HDGenComp$ components modelled as the $HDGenCompInit$ schema that basically sets the numerical values to zero, along with switching off the alarm and closing sensors and tubes.

$$HDGenCompInit \; \hat{=} \; [\Delta HDState \mid airVolLimit' = 0 \; \wedge \; airVol' = 0 \wedge \dots]$$

As part of our HD machine model, we want to detect the values from the various sensors in the system and update these into the state components described above. In order to achieve that task, we define the *Circus* process *SensorReadings* that watches a number of *Circus* channels each of them responsible for sensing a specific value arising from the machine sensors. These values are then stored in the state components, as described by $HDGenComp$.

$$SensorReadings \; \hat{=}$$
$$senApTransdPress?apTransdPress \longrightarrow SensorReadings$$
$$\square \; senInfVol?infVol \longrightarrow SensorReadings$$
$$\square \; \dots$$

In addition, this process also makes the current recorded readings available

$$SensorReadings \; \hat{=}$$
$$\dots$$
$$\square \; repApTransdPress!apTransdPress \longrightarrow SensorReadings$$
$$\square \; repInfVol!infVol \longrightarrow SensorReadings$$
$$\square \; \dots$$

The process *SensorReadings* is basically a large external choice over all sensor readings and reading reports, that repeats endlessly.

This approach is fine for sensor readings for which the time at which they occur is not important and so we are happy for an interested process to poll the relevant state component at regular intervals. Some sensor readings require some form of timestamping with possible timeouts, and some of these are handled separately, as explained later.

In order to capture the transition between the therapy phases of the HD machine, we provide a *Circus* process called *StatePhase*, that changes the value of the state variable *hdMachineState* depending on signals received during the therapy. Basically, when the process for a specific therapy phase begins, it immediately uses a special event to announce its commencement. Process *StatePhase* monitors these and updates state variables accordingly. For example after a signal *preparationPhase*, produced by the *Circus* process *TherapyPreparation*, the state variable *hdMachineState* is changed to *prepPhase* and will be used by the software requirements as described in the Sect. 4.

$$StatePhase \;\widehat{=}$$

$$\mu\,X \bullet \left(\begin{array}{l} preparationPhase \longrightarrow hdMachineState := prepPhase \\ \square\; connectingToPatient \longrightarrow \\ \qquad hdMachineState := connectThePatient \\ \square\; therapyInitiation \longrightarrow hdMachineState := initPhase \\ \square\; therapyEnding \longrightarrow hdMachineState := endPhase \end{array}\right) \;;\; X$$

3.3 Response to the Requirements

Whilst modelling the software requirements, we were able to identify twelve different kinds of behaviours that are expected among the 36 listed requirements. For each of them, we needed to provide an action that is equivalent to the intended behaviour of the requirement. For example, the expected behaviour for the requirement **R-1** to be satisfied is to stop the blood flow and raise an alarm. We formalise those two responses as Z schemas *StopBloodFlow* and *RaiseAlarm*: the former produces a signal *stopBloodFlow* stopping the current flow and the latter sets the *alarm* to *ENABLED* and then triggers the buzzer of the system.

$$StopBloodFlow \;\widehat{=}\; stopBloodFlow \longrightarrow \textbf{Skip}$$
$$RaiseAlarm \;\widehat{=}\; [\,\Delta\,HDState \mid alarm' \;=\; ENABLED\,]\;;$$
$$produceAlarmSound \longrightarrow \textbf{Skip}$$

This two process capture a common behaviour when a requirement error condition arises.

4   Model Development

In this section we give details of the modelling that resulted from our chosen approach: namely to model each of the safety requirements **R-1** to **R-36** as a *Circus* Action that "enforces" that requirement. It allows us to show how the sensors are integrated into our model and how the system should behave accordingly. We then present an overview of the therapy phases of the machine, describing how we structure our model with respect to the activities performed.

4.1 Software Requirements

A first example of how we model the requirement is illustrated by the requirement **R-1**. According to the description, during the application of arterial bolus, the system monitors the volume of saline infusion and if the volume exceeds 400 ml, the system should stop the blood flow and raise an alarm signal.

$$
\begin{array}{l}
\underline{\quad PreR1 \quad\rule{6cm}{0pt}} \\
\quad \Delta\, HDState \\
\hline
\quad hdActivity \in \{applicationArterialBolus\} \wedge infSalineVol > 400
\end{array}
$$

We define a schema $PreR1$ for the alarm-state precondition, and if it is satisfied, the system will perform $StopBloodFlow$ and $RaiseAlarm$. If the precondition is not satisfied ($\neg\, PreR1$), it waits for a defined period of time (parameter $CheckInterval$) and checks again. This illustrates the general approach here for many of these monitoring requirements. They check state variables at regular intervals and raise alarms if required. This decouples them from the process of doing sensor readings and recording the results.

$$
\begin{aligned}
R1 \,\widehat{=}\, &(PreR1 \;;\; (StopBloodFlow \parallel RaiseAlarm)) \\
&\vee \neg\, PreR1 \;;\; Wait(CheckInterval) \;;\; R1
\end{aligned}
$$

The second software requirement, **R-2**, monitors the blood flow and in the event that no flow is detected for a period longer than 120 s, the system should raise an alarm and stop. We formalise the requirement with help of two interleaved processes, $NoFlowWatchDog$ and $BloodFlowSample$.

$$
R2 \,\widehat{=}\, NoFlowWatchDog \parallel BloodFlowSample
$$

The former process monitors the time interval during which no blood flow occurs. if this exceeds the timeout, then the system will stop the blood pump $StopBP$ and $RaiseAlarm$.

$$
\begin{aligned}
NoFlowWatchDog \,\widehat{=}\,& getCurrentTime?time \longrightarrow \\
& time - lastNonZeroBF > 120000 \;\&\; tick \longrightarrow StopBP \;;\; RaiseAlarm \\
\square\;& time - lastNonZeroBF \leq 120000 \;\&\; tick \longrightarrow NoFlowWatchDog
\end{aligned}
$$

The second helper process is $BloodFlowSample$, which monitors the blood flow arising from the channel $senBloodFlowInEBC$. Whenever the value of $bloodFlowInEBC$ is different, the state component $lastNonZeroBF$ is updated with the current time in the system.

$$
\begin{aligned}
BloodFlowSample \,\widehat{=}\,& getCurrentTime?time \longrightarrow \\
& senBloodFlowInEBC?bloodFlowInEBC \longrightarrow \\
& \left(
\begin{array}{l}
\left(\begin{array}{l}
\mathbf{if}\; bloodFlowInEBC \neq 0 \;\&\; lastNonZeroBF := time \\
[\!]\, bloodFlowInEBC = 0 \;\&\; \mathbf{Skip}\; \mathbf{fi}
\end{array}\right) \;; \\
BloodFlowSample
\end{array}
\right)
\end{aligned}
$$

This is an example of a sensor reading that is not handled by *SensorReadings*, because it needs to record a timestamp for the most recent non-zero reading.

Requirement **R-9** is an example of how we create helper processes in order to capture the intended behaviour of the system. During the phase *connecting the patient*, the machine should monitor the pressure at the VP transducer and if the value measured exceeds 450mmHg for more than 3 s, the machine should respond by stopping the blood pump and raising an alarm signal.

In our model, we first create a helper process $TrackTimervpTransdPressR9$ that monitors such pressure values through the channel $senvpTransdPress$, after one *tick* event and updates the timer interval with the following condition: when the sensed VP transducer pressure is higher than 450, the timer interval is incremented; otherwise, the timer is reset until the condition is satisfied again.

$$TrackTimervpTransdPressR9 \;\widehat{=}$$
$$\left(\left(\begin{matrix} tick \longrightarrow senvpTransdPress?x \longrightarrow \\ \left(\begin{matrix} \textbf{if } x > 450 \;\&\; timerIntervalR9 := timerIntervalR9 + 1 \\ []\, x \leq 450 \;\&\; timerIntervalR9 := 0 \;\textbf{fi} \end{matrix}\right) \\ TrackTimervpTransdPressR9 \end{matrix}\right) ; \right)$$

The next step is to define a Z schema *PreR9*, in which we define the alarm precondition for the requirement itself. The requirement is specified for use during the initiation phase and is satisfied if the value of $vpTransdPress$ is higher than 450 for a period of 3 s, captured by the $timerIntervalR9$ state variable with a value higher than 3000 ms.

PreR9

$\Delta HDState$

$hdMachineState \in \{\,connectThePatient\,\}$
$vpTransdPress > 450 \wedge timerIntervalR9 > 3000$

Should the precondition of the **R-9** requirement be satisfied, the system does *StopBP* and *RaiseAlarm*. Otherwise, it waits for a predefined time interval before checking again.

$$R9 \;\widehat{=}\; \left(\mu X \bullet \left(\begin{matrix} (PreR9 \;;\; (StopBP \parallel RaiseAlarm)) \\ \vee \neg\, PreR9 \;;\; Wait(CheckInterval) \;;\; X \end{matrix}\right)\right)$$
$$\parallel TrackTimervpTransdPressR9$$

During the *connecting the patient* phase, **R-16** specifies a time interval of 310 s for that phase. When the specified time ends, the machine should change to the *initiation phase*. We formalise that requirement through a signal $conToPatient$, followed by a wait period of 310000 ticks, and ended with a signal $therapyInit$ that is triggered at the beginning of that phase.

$$R16 \;\widehat{=}\; conToPatient \longrightarrow Wait(310000) \;;\; therapyInit \longrightarrow \textbf{Skip}$$

4.2   Therapy Processes

We now describe how we model the therapy phases of the HD machine—the top level "workflow", so to speak. According to the requirements, the system starts with the preparation phase, followed by the initiation phase, and an ending phase:

$$MainTherapy \ \widehat{=} \ TherapyPreparation \ ; \ TherapyInitiation \ ; \ TherapyEnding$$

The initiation phase also contains the "perform therapy" phase for some reason that is unclear to us[2]. Each of these phases is further broken down.

For the therapy preparation phase, we define a *Circus* process that starts with a signal *preparationPhase*, followed by a sequence of activities according to [HMCS, Sect. 3.2]. A key idea here is each phase signals that it has started, on a channel, so that requirements and activities that are phase-dependent can ascertain when they should be active.

We capture the steps that compose the therapy preparation phase, each one, with a *Circus* process that behaves accordingly.

$$TherapyPreparation \ \widehat{=}$$
$$preparationPhase \longrightarrow AutomatedSelfTest \ ;$$
$$ConnectingTheConcentrate \ ; \ SetRinsingParameters \ ;$$
$$InsertingRinsingTestingTubSystem \ ; \ PrepHeparinPump \ ;$$
$$SetTreatParameters \ ; \ RinsingDialyzer$$

A similar pattern is used for the other phases.

As an example of a phase activity, we show the specification *SetRinsingParameters*, which collects parameter settings from the clinician, here modelled as a Z schema with inputs:

__ *SetRinsingParameters* _____
$setFBPRate? \ : \ \mathbb{Z}; \ setRBPRate? \ : \ \mathbb{Z}$
$setRTime? \ : \ \mathbb{Z}; \ setUFRFRinsing? \ : \ \mathbb{Z}$
$setUFVFRinsing? \ : \ \mathbb{Z}; \ setBFFCPatient? \ : \ \mathbb{Z}$
$\Delta HDState$

$fillingBPRate' = setFBPRate? \ \wedge \ rinsingBPRate' = setRBPRate?$
$rinsingTime' = setRTime? \ \wedge \ ufVolForRinsing' = setUFVFRinsing?$
$ufRateForRinsing' = setUFRFRinsing?$
$bloodFlowForConnectingPatient' = setBFFCPatient?$
_____

4.3   Putting it all Together

We conclude this section by detailing how we put all the pieces together. We start the main process of the HD machine, the *HDMachine* process, with the schema

---
[2] Particularly, because it should be the phase that lasts longest!.

HDGenCompInit that initialises the state variables of the system. Then, the system is modelled as a parallelism between the *MainTherapy* process and the *SoftwareRequirements* process.

A second parallelism is required between the above and the *StatePhase Circus* process. The latter is a process that watches the changes of state, through signals like *preparationPhase* and *therapyInitiation*, and after these, the state variable *hdMachineState* are set accordingly, in order to be used by the requirements defined for the *SoftwareRequirements* process. Then the components are put in parallel with the *SensorReadings* process, used for example, to update the values of the state components.

$$HDMachine \mathrel{\widehat{=}} HDGenCompInit\ ;$$
$$\left(\left(\left(\begin{array}{l} MainTherapy \\ [\![\,HDGenCompStChanSet\,]\!]\ SoftwareRequirements \\ [\![\,TherapyPhaseChanSet\,]\!]\ StatePhase \\ [\![\,SensorReadingsComm\,]\!]\ SensorReadings \end{array}\right)\right)\right)$$

Finally, the entire system is put in parallel with the *SysClock* process, synchronising on the channels *tick* and *getCurrentTime*, denoting the time elapsed, and the output of the time value for the rest of the therapy, respectively.

$$HDMachine\ [\![\ \{\!|\ tock, getCurrentTime\ |\!\}\ ]\!]\ SysClock$$

5 Checking the Model

Currently, there is limited tool support for *Circus*, and nothing that can be used for direct model-checking of machine-readable *Circus*. Limited support can found as part of the Community Z Tools (CZT) project [1], as extensions to the Z support there. This is facilitated by the way that the machine-readable syntax of *Circus* takes the form of LaTeX documents in that same way as that of Z.

We were able to use the *Circus* extension to CZT, which include a parser and type-checker to assess our model. After minor revisions correcting typos and small type errors, we obtained a model that satisfied both the parser and type-checker.

The current approach to model-checking *Circus* is to translate it into machine-readable CSP (CSP_M), and use FDR3 [3] to do the model checking. Unfortunately, there is no automated way to do this, so such translations have to be done by hand. Fully-, or even semi-, automatic translation from *Circus* to CPSm is difficult, but is an active research topic. FDR3 itself is described as a refinement checker, which basically means is that it is a model checker, where the models are labelled transition systems derived from the operational semantics of CSP, and the properties to be checked are assertions about the existence of a refinement relation between two distinct models, one for the specification, the other for the implementation.

Manual Translation. We manually translated our *Circus* specification into CSP_M, in order to do some basic checks, particularly regarding deadlock freedom. Here we give a brief description of the translation and the challenges we encountered, most notably that of avoiding state-space explosion.

For most of the *Circus* constructs, we have a pretty straightforward translation into CSP_M, as we know that *Circus* is derived from CSP. For example, a lot of the type and channel declarations are very simple, so the following *Circus* fragment:

$$STATEPHASE ::= connectThePatient \mid initPhase \mid prepPhase \mid endPhase$$

channel $preparationPhase, therapyInitiation$

would become the following CSPM fragment:

```
datatype STATEPHASE
  = connectThePatient | initPhase | prepPhase | endPhase
channel preparationPhase, therapyInitiation
```

We then translate Z schemas into CSP_M as tuples. Each component of a Z schema is translated as a nametype represented by a tuple. For example, the following state schema fragment

```
┌─ RinsingParameters ────────────────────────────────
│  fillingBPRate : ℤ; rinsingBPRate : ℤ
├────────────────────────────────────────────────────
│  fillingBPRate ∈ {0 .. 6000} ∧ rinsingBPRate ∈ {50 .. 300}
└────────────────────────────────────────────────────
```

is translated into a tuple where, for example, the first component of the tuple is the range of values for the $fillingBPRate$, of type $\mathbb{N}$, restricted to the values 0 up to 6000.

```
nametype RinsingParameters = ({0..6000},{50..300})
```

5.1 Translation of Circus Processes Containing State

In *Circus*, when we want to manipulate the values of a component of a state component, we can freely access it, even as an assignment, as it is within the context of the *Circus* process. However that is not possible in CSP_M.

We need to adopt a different approach for the translation. Basically we concert each state schema into a CSP process that has *get* and *set* events for each state component, so that, for example, the *Circus* action

$$x := y + 1$$

would become something like

```
getY?myY -> setX!(myY+1) -> Skip
```

All the actions accessing global state would run in parallel synchronising on the relevant *get* and *set* events.

5.2   Checking of the Model Using FDR

We checked our CSP_M model of the HD Machine using FDR3, with assertions regarding deadlock and livelock freedom. These helped us re-factor our model, mainly to avoid deadlocks that occurred because of errors in specifying synchronisation events. As a result of this we are very confident that our *Circus* model is deadlock-free.

A big issue we had to deal with was the fact that our original *Circus* model had a huge state-space: we had a clock that ticked every millisecond, together with a timeout in one of the requirements of 310 s. We also had checks for values in large numeric ranges, typically for fluid volumes. We had to carefully decide how to shrink the state-space by reducing the range of values as low as possible without having an impact on the integrity of the model.

Fortunately, most of the uses for numbers are to specify limits outside of which special action needs to be taken. Also, in many cases the number values are specific to just one requirement or a small coherent group. This makes is easy to shrink the range of values for one such requirement without worrying about its effect on another.

For example, **R-20–21** talks about measures that span the range of normal human body temperature, but are only explicit about two boundaries, one at 33 °C, the other at 41 °C. Error conditions arise if the temperature outside those bounds. This defines three regions of interest, but we do not need to model temperature values in the range 32 . . . 42 (say), but simply have three values that denote: "too cold", "too hot" and "just right".

We were able to check small clusters of requirement processes against the full machine model, and show their interaction was deadlock free. Even running the tests on a virtual machine cluster with 16 cores and 32 GB of RAM we found that we handle at most about 3 requirements at a time. However as they are all independent, it didn't prevent us form checking them all.

Further ways to attempt overcome the state-space problem are described by Roscoe *et al.* [11], suggesting the use of compression techniques in order to model-check larger CSP specifications in FDR, allowing the reduction of both the number of states and the transitions to be visited. We will explore this as part of future work.

5.3   Back-Annotation

Where analysis with FDR3 exposed any structural issues, we modified the *Circus* version, as would be expected. However we have not, at this stage, made the changes to number ranges needed to make model-checking feasible. Simply put, since there are no model-checkers for *Circus*, or automated translation to any other modelling notation such as CSP_M, we felt that we would keep the "full story" in the model. Clearly this would need to be addressed should automatic checking become possible.

6   Observations

One of the often-touted advantages of building formal models is that the rigour, level of detail and completeness that they require, results in the exposure of a lot of ambiguities and incompletenesses in the informal requirements and specifications. Here we collect a number of such issues that arose as part of our rigorous, detailed analysis of the case study.

We did look at the safety requirements and future work will look at formalising those with a view of being able to check the requirements model against the safety one. We shall start with a few observations regarding safety:

- **S-4** talks about draining saline solution to a bag/bucket attached to the venous connector. Presumably this means the patient isn't connected here. But this is when the patient is connected to EBC, and **S-1** requires both arterial and venous connectors are connected simultaneously.
- **S-5** makes a very ambiguous use of the phrase "can be connected". This can refer to a state, of being connected (up) to something or to a process, that involves making the connection with something (which will of course result in being in the state of being connected). We believe that what **S-5** intended to state was that the process (connecting to) could only occur during initiation, but that the state (now connected to) would continue to hold during the main therapy portion.
- **S-7** talks about power-loss. We are given no information about how the machine might cope, or the hardware's power-down state. How might an alarm be raised without power? Also, surely **S-7** should be about blood flow stopping for any reason, not just power failure?
- **S-9** talks about the difference between actual and measured blood flow, mentioning low or negative AP as an issue. Is there a well-defined relationship linking actual flow to measured flow and AP?

Next, issues that arose while looking at requirements and activities:

- **State Components:** We identified a bunch of parameters and sensors that are described in the *software requirements* section. Some of these are not mentioned elsewhere in the entire text and therefore we do not know what the restrictions are regarding expected values for them. We modelled these items based on our limited understanding, defining types for each of them. These are modelled as components of the $HDGenComp$ schema, as part of the $HDState$. For instance, the pressure at the VP transducer and AP transducer are modelled as $vpTransdPress$ and $apTransdPress$ respectively.
- **Alarm:** In several requirements, the expected behaviour of the system is to raise an alarm. However, we don't know precisely what is the overall behaviour of the system for most of these requirements. Once the alarm is raised, what is the expected behaviour of the system? Does the system stop entirely until the alarm is acknowledged, or are some functions not available?
- **R-35** is not formalised in our current version of the HD machine model. There seems to be a safety issue in this requirement. The system *"shall monitor the*

net fluid removal volume and if the net fluid removal volume exceeds (UF set volume + 200 mL)", the machine goes into bypass and the alarm is raised. However, once the alarm is acknowledged by the user, the requirement says that *"the software shall increase the UF set volume by 200 mL"*. Is it really the intention to set a limit monitored by an alarm where the response mandate for the user is to raise the limit to the level were the alarm signal is (just) disabled?

7   Conclusions

We started this *Circus* case-study as a response to the HD machine case study, proposed for the ABZ 2016 conference [7]. We summarise our approach thus: we capture the communication between the sensors and the system and also define a structure for the data used around the system; we make use of Z schemas in order to define the model state and also operations that change that state, all related to the various sensor readings and parameters defined for the therapy.

We have around a thousand lines of *Circus* specification for the model, with about sixty state variables and forty events, and we have around ninety processes/actions used for modelling the requirements and the overall therapy procedures. We found that the ability to read and write small state components by name (using small Z schemas and assignment) coupled with the usual ability to structure CSP as small parallel processes made it easy to restrict any formal text to only parts of the system that were immediately relevant.

In our investigation, we were able to model almost all the software requirements with exception of **R-35** whose description leads us to see a contradiction that may be a safety issue, as discussed in Sect. 6.

Due to the current lack of tool support for direct checking of *Circus* specifications, we needed to translate our model, by hand, into CSP in order to be able to perform model checking using FDR3 [3]. In our translation, we had to adapt the *Circus* model for CSP_M because *Circus* programs has explicit state-based features (such as assignment), which are not present in CSP_M, which instead relies on process parameter-lists to handle state. The equivalent specification written in CSP_M has around 24 hundred lines, more than twice the size of the *Circus* version, due to the inclusion of auxiliary functions and new channels for communicating with the new state-modelling processes. We were able to perform model checking using FD3R and perform checks that the requirements could run in parallel with the therapy model, synchronising on common events, without any deadlocks.

For future work, we intend to build the corresponding model of the safety requirements, and link it to the requirements model in order to look for inconsistencies. Other possible avenues of investigation would include deriving a formal software specification structured around the architecture and functional decomposition of a realistic implementation design. We would expect this to be structured differently to the model we have just derived from the requirements, and it would raise interesting issues regarding their refinement relation and verification.

Another interesting piece of future work would be to derive software prototypes from our *Circus* model to allow us to simulate the system execution. In the long term, we have interest in working with theorem proving for the refinement of *Circus* programs and possible proofs of test-cases. It is also in our plans to explore the development of tools that allows us to perform model-checking directly with *Circus* programs.

Acknowledgments. We would like to thank Thomas Gibson-Robinson for his help in assisting us in achieving the state-space reduction we needed, and the anonymous reviewers for their perceptive comments and pointed questions, which have help to improve this paper. Finally we re-iterate our thanks to our sponsors, CNPq of Brazil, and Science Foundation Ireland.

References

1. Community Z Tools Project: CZT: Community Z Tools, September 2015. http://czt.sourceforge.net/manual.html, checked 14 Mar 2016
2. Dijkstra, E.W.: Guarded commands, nondeterminacy and formal derivation of programs. Commun. ACM **18**, 453–457 (1975)
3. Gibson-Robinson, T., Armstrong, P., Boulgakov, A., Roscoe, A.W.: FDR3 — a modern refinement checker for CSP. In: Ábrahám, E., Havelund, K. (eds.) TACAS 2014. LNCS, vol. 8413, pp. 187–201. Springer, Heidelberg (2014)
4. Gomes, A.O., Butterfield, A.: HD-Machine Case Study Repository (2016). https://bitbucket.org/artur1109/hdmachine/
5. He, J., Hoare, C.A.R.: Unifying theories of programming. In: Orlowska, E., Szalas, A. (eds.) RelMiCS, pp. 97–99 (1998)
6. Hoare, C.A.R.: Communicating Sequential Processes. Computer Science. Prentice-Hall International, Englewood Cliffs (1985)
7. Mashkoor, A.: The Haemodialysis Machine Case Study. Software Competence Center Hagenberg GmbH (SCCH) (2015). http://www.cdcc.faw.jku.at/ABZ2016/HD-CaseStudy.pdf
8. Morgan, C.C.: Programming From Specifications. Prentice Hall International Series in Computer Science, 2nd edn. Prentice Hall, Upper Saddle River (1994)
9. Oliveira, M.V.M.: Formal Derivation of State-Rich Reactive Programs using Circus. Ph.D. thesis, Department of Computer Science - University of York, UK (2005)
10. Oliveira, M., Cavalcanti, A., Woodcock, J.: A UTP semantics for *Circus*. Formal Asp. Comput. **21**(1–2), 3–32 (2009)
11. Roscoe, A.W., Gardiner, P.H.B., Goldsmith, M.H., Hulance, J.R., Jackson, D.M., Scattergood, J.B.: Hierarchical compression for model-checking CSP or how to check 1020 dining philosophers for deadlock. In: Brinksma, E., Steffen, B., Cleaveland, W.R., Larsen, K.G., Margaria, T. (eds.) TACAS 1995. LNCS, vol. 1019, pp. 133–152. Springer, Heidelberg (1995)
12. Woodcock, J., Cavalcanti, A.: The semantics of Circus. In: Bert, D., Bowen, J.P., C. Henson, M., Robinson, K. (eds.) ZB 2002. LNCS, vol. 2272, pp. 184–203. Springer, Heidelberg (2002)
13. Woodcock, J., Davies, J.: Using Z, Specification, Refinement, and Proof. Prentice Hall International Series in Computer Science. Prentice Hall, Upper Saddle River (1996)

Author Index

Printed in the United States
By Bookmasters